HIPPOCRATES

I

LCL 147

HIPPOCRATES

VOLUME I

EDITED AND TRANSLATED BY

PAUL POTTER

HARVARD UNIVERSITY PRESS
CAMBRIDGE, MASSACHUSETTS
LONDON, ENGLAND
2022

LOEB CLASSICAL LIBRARY® is a registered trademark
of the President and Fellows of Harvard College

Library of Congress Control Number 2021946098
CIP data available from the Library of Congress

ISBN 978-0-674-99747-9

*Composed in ZephGreek and ZephText by
Technologies 'N Typography, Merrimac, Massachusetts.
Printed on acid-free paper and bound by
Maple Press, York, Pennsylvania*

CONTENTS

LIST OF HIPPOCRATIC WORKS SHOWING THEIR DIVISION INTO VOLUMES IN THIS EDITION

SERIES INTRODUCTION

1. THE HIPPOCRATIC COLLECTION[1]

The Hippocratic Collection (*Corpus Hippocraticum*) consists of between sixty and seventy medical writings in the Ionic Greek dialect[2] dating for the most part from a hundred year period around 400 BC. The earliest extant copies of these works are preserved in five vellum manuscripts written between the early tenth and the late twelfth centuries AD, and in a few cases in paper manuscripts written toward the end of this period or a little later (cf. §4 below). Since the first century BC, the writings have in whole or in part been labeled "of Hippocrates," and today the ad-

---

[1] Sections 1 and 2 of this Series Introduction contain material reprinted from my *Short Handbook of Hippocratic Medicine* (Québec, 1988) with permission of Les Éditions du Sphinx. Translations are my own unless otherwise noted.

For fuller accounts of the collection, see Jouanna, *Hippocrates*; the revised French edition of the same work, idem. *Hippocrate* (Paris, 2017); idem. *Hippocrate. Introduction générale.* Budé I (1) (Paris, 2020).

[2] On Hippocratic language in general and the Ionic dialect in particular, see E. Schwyzer, "Zur Sprache," in *Hippokrates. Über Entstehung und Aufbau des menschlichen Körpers* (ΠΕΡΙ ΣΑΡΚΩΝ), edited by K. Deichgräber (Leipzig and Berlin, 1935), 62–97.

jective "Hippocratic" is applied widely in the scholarly literature to the writings and the medicine they represent.

The following brief characterization of the individual Hippocratic writings is meant to give a preliminary impression of the breadth, depth, and variety of this literature. The order of presentation is that of A. Foes in his Greek collected edition (1595), which he adapted from a census of Hippocratic writings that Erotian includes in the introduction to his *Collection of Hippocratic Words* (I AD).[3] The word counts are those of the TLG texts; for the arrangement of the treatises within the edition's eleven volumes, see the List of Hippocratic Works, above. The abbreviations in square brackets are from LSJ.

Practice of Medicine

Oath [*Jusj.* = *Jusjurandum*]. This document appears to be an authentic historical formula once used to induct new members into a physicians' guild. These new members, in taking upon themselves financial and teaching obligations toward other guild members, and by committing themselves to certain principles of benevolent practice and to the protection of their patients' interests, enter an exclusive healing fraternity. The role of warrantors of the oath is filled by the patron deities of medicine, Apollo, Asclepius, and the latter's daughters "Health" and "Panacea." (262 words)

Law [*Lex*]. This tract begins by ascribing the current

---

3 See §3 below, pp. l–lii.

low level of medicine's esteem to its practitioners' lack of education and then proceeds to outline the requirements of a good medical preparation. (335 words)

The Art [*de Arte*]. Scholars are divided on whether this highly rhetorical piece, whose object is to prove that the art of medicine exists at all, is the work of a sophist (Th. Gomperz says Protagoras, and Jones suggests Hippias)[4] or a physician. (2,801 words)

Ancient Medicine [*VM* = *de Vetere (or Prisca) Medicina*]. The author advocates traditional dietetic medicine and denounces the recent tendency to explain diseases in terms of such hypothetical factors as heat, cold, dryness, and moisture. An account of how humans first discovered the fundamental dietetic principle—that health is best gained and kept by adapting food intake to the needs and digestive capacities of the individual—is followed by a brief discussion of the role humors (bodily fluids) and bodily structures play in disease. (5,705 words)

Physician [*Medic.* = *de Medico*]. A brief guide for the practitioner that outlines professional behavior and gives instructions on setting up an office and on performing various manual procedures: bandaging, surgical incisions, cupping. (1,552 words)

Decorum [*Decent.* = *de Habitu Decenti*]. Perhaps a lecture to students delivered by a medical teacher, this text depicts how a wise physician conducts his practice, with particular attention to the visiting of patients. Much

---

[4] See Gomperz, pp. 21–31, and Jones *Loeb*, vol. 2, pp. 187f.; Jouanna *Art*, pp. 179–83, argues for authorship by a medical writer.

quoted in antiquity was the sentence (ch. 5) "For a physician who is a philosopher (i.e., a lover of wisdom) is the equal of a god." (1,558 words)

Precepts [*Praec.* = *Praecepta*]. A somewhat disconnected collection of thoughts on the empirical basis of medicine, fees, consultation with colleagues, modest behavior, etc. Often quoted is (ch. 4): "For where there is love of man, there is also love of the art." (1,384 words)

Semiotics (Clinical Signs)

Prognostic [*Prog.* = *Prognosticon*]. "I hold that it is an excellent thing for a physician to practice forecasting. For if he discovers and declares unaided by the side of his patients the present, the past and the future, and fills in the gaps in the account given by the sick, he will be more believed to understand the cases, so that patients will confidently entrust themselves to him for treatment. Furthermore, he will carry out the treatment best if he knows beforehand from the present symptoms what will take place later" (ch. 1). True to this purpose, *Prognostic* elucidates in detail the predictive significance of physical signs (e.g., posture, respiration, fevers, sweating, pains, sleep patterns) and the state of the excreta (e.g., stools, urine, sputum). Implicit in the discussion and explicitly stated in the final chapter is the principle that prognostic signs are independent of variations in weather, geographic location, and—even more surprisingly—diagnosis. (5,363 words)

Humors [*Hum.* = *de Humoribus*]. A collection of notes with little rhyme or reason; subjects include humors, symptoms, excreta, mental states, remedies, causes of disease, weather. (2,330 words)

Crises [*Judic.* = *de Judicationibus*]. A collection of material on disease crises drawn from other Hippocratic writings: e.g., *Prognostic*, *Epidemics II* and *VI*, *Regimen in Acute Diseases* (*Appendix*), *Aphorisms*; varying degrees of textual reworking. (2,025 words)

Critical Days [*Dieb.Judic.* = *de Diebus Judicatoriis*]. A compilation of passages on critical days from various Hippocratic writings: *Epidemics III*, *Internal Affections*, *Diseases III*, *Sevens*;[5] virtually no original text. (1,316 words)

Prorrhetic I [*Prorrh.I* = *Prorrheticon I*]. In this work are collected 170 chapters, which elaborate on the symptoms and prognoses of various diseases; although statements tend to be expressed in the form of generalizations (e.g., ch. 23: "Loss of speech in conjunction with hiccups is a very bad sign"), specific cases are sometimes referred to. For an indication of the range of conditions handled, see *Coan Prenotions* below.

Prorrhetic II [*Prorrh.II* = *Prorrheticon II*]. After casting doubt on the miraculous kind of prognostications that some physicians claim to make, the writer sets out what in his opinion *is* possible, and, after indicating how to discover when patients are deviating from their prescribed regimen, gives an account of conclusions that can be drawn from various signs and symptoms in specific conditions: e.g., dropsy, epilepsy, ulcerations, wounds, ophthal-

---

[5] *Sevens* has survived in its entirety only in Latin translations transmitted by the manuscripts Ambrosianus Lat. G 108 (IX c.) and Parisinus Lat. 7027 (X c.). Cf. Roscher, pp. 45 and 68f.

mias, dysentery, diarrheas, infertility. All statements are expressed as generalizations, and no individual cases are mentioned. (*I* and *II* together total 10,563 words)

Coan Prenotions [*Coac.* = *Coacae Praenotiones*]. A compilation from many Hippocratic writings (e.g., 153 chapters from *Prorrhetic I*) of 640 independent prognostic generalizations arranged by subject matter (13,170 words):

1–30	Chills
31–136	Symptoms in fevers
137–55	Crises
156–84	Headache
185–207	Signs of the ears
208–39	Signs of the face and mouth
240–55	Language and respiration
256–72	Signs in the throat and neck
273–97	Signs in the hypochondrium
298–319	Signs in the loins
320–40	Hemorrhages
341–56	Spasms and convulsions
357–72	Sore throat (angina)
373–436	Diseases of the lungs
437–65	Diseases in the abdomen
466–76	Nervous disorders
477–87	General disease signs
488–501	Wounds
502	Importance of age
503–44	Diseases of women
545–60	Vomiting
561–63	Sweating
564–88	Urines
589–640	Excreta

Etiology (Disease Causation)

Nature of Man [*Nat.Hom.* = *de Natura Hominis*]. The first part of this treatise (chs. 1–8) contains the classic Hippocratic exposition of the theory of the four humors. After disposing of one-element theories of the body (e.g., air, fire, or blood) as irreconcilable with observed facts, the writer sets out his four-element theory: the human body consists of blood, phlegm, yellow bile, and dark bile. As long as these preserve a balance of quantity and an even distribution, health prevails, but when they become disturbed in one of these ways, a disease arises. Furthermore, relationships exist between the four humors, their degrees of heat and moisture, the seasons of the year, and the prevalence of specific diseases. Chapters 9–15 consider a number of questions of etiology, diagnosis, and therapy. (4,017 words)

Generation [*Genit.* = *de Genitura* (or *Semine*)]. A theoretical account of the origin of the generative fluid in the male and the female, and of how their mixture in the uterus determines the constitution of the individual offspring. This treatise and the following two are linked among themselves and to *Diseases of Women I* by several explicit cross-references (e.g., *Generation*, chs. 3 and 4; *Nature of the Child*, ch. 4 (15 L.); *Diseases IV*, ch. 26 (57 L.); *Diseases of Women I*, chs. 1, 44, and 73. (This work is combined with the following two works by Littré, for a total of 19,474 words)

Nature of the Child [*Nat.Puer.* = *de Natura Pueri*]. A detailed speculative system of embryology emphasizing the roles that breath, blood, and heat play in the growth and differentiation of the fetus from conception to birth.

Many special points are explained by a generous application of analogical reasoning and some experimental observation.

Diseases IV [*Morb. IV* = *de Morbis IV*]. From the food and drink ingested, four humors—phlegm, blood, bile, and water—separate off and are drawn to their reservoirs in the body: the head, heart, gallbladder, and spleen, respectively. As long as the cycle of ingestion, digestion, and excretion remains in balance, there is health; when, however, some excess develops, there is either immediate illness or a propensity to illness, which unfavorable weather or an injury of some kind brings to pass. The last part of the treatise (chs. 23–26) discusses in a disconnected way intestinal worms, bladder stones, the fact that drinks do not pass to the lungs, and dropsy.

Fleshes [*Carn.* = *de Carnibus*]. After a brief depiction of the origin of the cosmos, this essay explains how heat acted on the moist, cold, fatty, and gluey components of the earth to produce the different parts of the human body, by drying, melting, burning, etc. The argument is supported by many persuasive observations and analogies. The final chapter establishes the importance of the seven-day cycle in health and disease. (3,467 words)

Eight Months' Child [*Oct.* = *de Octimestri Partu*]. The author contends that live birth during a forty-day period around the eighth month of pregnancy is impossible. Central to the treatise is an unsuccessful attempt to integrate several *a priori* numerological theories and many correct clinical observations into a coherent whole. (2,456 words)

Superfetation [*Superf.* = *de Superfetatione*]. A compilation of obstetrical knowledge loosely organized by topic (e.g., superfetation, premature birth, stillbirth, conception, abortion, treatment), this text is clearly directed toward practice: therapy and prognosis predominate to the virtual exclusion of theoretical considerations. *Superfetation* has verbatim correspondences to texts in *Nature of Women*, *Barrenness*, and *Diseases of Women I*. (3,485 words)

Dentition [*Dent.* = *de Dentitione*]. A collection of thirty-two short chapters on the teething, nursing, and weaning of infants, and on ulcerations of the tonsils and throat. Prognosis predominates. (403 words)

Heart [*Cord.* = *de Corde*]. A detailed presentation of the anatomy and physiology of the human heart and great vessels, apparently based on dissection, this treatise is probably of a somewhat later date than the main body of the collection. The valves of the heart are mentioned, and arteries are differentiated from veins. (1,062 words)

Glands [*Gland.* = *de Glandulis*]. Structure, distribution, function, and pathology of the tonsils, lymph nodes, adrenals, brain, and breasts. The principal function attributed to the glands—apparently on account of their spongy texture—is the absorption and redirection of excessive moisture. (1,810 words)

Nature of Bones [*Oss.* = *de Natura Ossium*]. A cento comprising five fragments drawn from at least four different sources, unified by an interest in the descriptive and functional anatomy of the heart and great vessels: chs. 1–7, source unknown, perhaps the same as chs. 11–19; ch. 8, from the writings of Syennesis of Cyprus, quoted in Aristotle, *Historia Animalium* 511b 24–30; ch. 9, text also present in the Hippocratic work *Nature of Man* 11 and in Aristotle, *Historia Animalium* 512b 13–513a 7, where it is attributed to Polybius or Polybus; ch. 10, from the Hippocratic treatise *Epidemics II* 4.1; chs. 11–19, source unknown, but mentioned in Galen's *Glossary* (Perilli) s.v. κοτυληδόνα and παραστάτας as appended to *Instruments of Reduction*. (3,411 words)

Airs Waters Places [*Aër.* = *de Aëre, Aquis, Locis*]. Chapters 1–11 instruct the itinerant physician on how to draw conclusions concerning the environmental factors that affect health and disease in the places he visits. Chapters 12–24 compare the physique, customs, and mentalities of various European and Asian peoples, all explained on the basis of the differences in their geography and climate. (7,685 words)

Breaths [*Flat.* = *de Flatibus*]. A public lecture that sets out to prove that the source of all human diseases is air that enters the body in breathing, eating, and drinking, or through wounds. (2,923 words)

Sacred Disease [*Morb.Sacr.* = *de Morbo Sacro*]. It is argued that epilepsy was named "sacred" by magicians and charlatans in order to conceal their own ignorance. In fact, this disease is no more sacred than any other, and it is amenable to medical treatment. The site of human thought is the brain; when phlegm blocks the breath-carrying vessels leading to the brain, convulsions result. (4,876 words)

Dietetics

Regimen in Health [*Salubr.* = *de Salubri Diaeta*]. A carefully organized summary of how to maintain health. It instructs the layman on foods and drinks, walks, and emetics, as well as giving special directions for infants, women, and athletes in training. (1,509 words)

Regimen I–III [*Vict.I–III* = *de Victu I–III* (or *Diaeta*)].

Book 1 is devoted to the proposition that the human body is composed of the elemental pair fire and water and that a person's birth, sex, intelligence, etc. are all to be understood in terms of these. Book 2 catalogs the dietetic significance of factors of geography and climate, and of foods and drinks, baths, emetics, sleep, and exercises. Book 3 presents the author's discovery ($\dot{\epsilon}\mu o\grave{\iota}\ \tau\hat{\omega}\ \epsilon\dot{\upsilon}\rho\acute{o}\nu\tau\iota$) that food and exercise are an antithetical pair upon whose balance health depends, followed by two sets of practical instructions—one for the ordinary and one for the well-situated person—on how to establish and maintain this balance through the year. (Combined with the following by Littré for a total of 20,472 words)

Regimen IV = Dreams [Insomn. = de Insomniis]. Many dreams have their origin from the internal functions of the body; these reveal the dreamer's state of health and give indications for an appropriate regimen. *Dreams* contains the only approbative references to prayer in the Hippocratic Collection (chs. 87, 89, 90).

Nutriment [Alim. = de Alimento]. This collection of fifty-five short, probably intentionally enigmatic, pronouncements on various aspects of nutriment and growth belongs more to philosophy than to medicine: e.g., ch. 9, "The beginning of all things is one, and the end of all things is one, and the same thing is end and beginning." (1,339 words)

Regimen in Acute Diseases [Acut. = de Diaeta Acutorum]. This treatise is unified more by its diffusely polemical tone than by any line of argument. At the base of the author's remarks are several widely held dietetic precepts: e.g., the unique value of barley gruel in acute diseases, the necessity of correctly timing therapeutic administrations, the great damage done in both health and disease by rapid changes of diet, the importance of taking each particular patient's habits into account when prescribing. (6,381 words, including the *Appendix*)

Regimen in Acute Diseases (Appendix) [*Acut.(Sp.)* = *de Diaeta Acutorum (Spurium)*]. Not a treatise, but an unconnected series of chapters discussing various diseases, prognostic signs, and treatments. This text gives the impression of being a collection of rough notes that was never worked into a finished form.

Places in Man [*Loc.Hom.* = *de Locis in Homine*]. Pathology—both general and specific—is the subject of this less than systematic conglomeration of utterances on normal structure, the processes through which normal structure becomes abnormal, and methods of therapy. Conditions handled include diseases of the eyes, pleurisy, dropsy, jaundice, wounds in the head, and diseases of women. (8,723 words)

Use of Liquids [*Liqu.* = *de Liquidorum (or Humidorum) Usu*]. The reader is instructed on the external application of liquids—fresh water, salt water, vinegar, wines—to warm, cool, clean, soften, moisten, dry, and soothe. (1,532 words)

Therapeutics

Diseases I [*Morb.I* = *de Morbis I*]. Chapters 1–10 analyze a number of common medical phenomena and concepts: e.g., "whence all diseases arise . . . from which diseases there are changes into other diseases. What physicians treating patients achieve by luck . . . what is said and done with precision in medicine, which things are correct in it, and which are not correct . . . what is dexterity, and what is awkwardness." Chapters 11–34 portray the courses of various acute diseases, especially of the chest, and explain the processes through which bile and phlegm—the ultimate causes of all diseases—produce the observed symptoms. (*I* through *III* are combined by Littré for a total of 26,143 words)

Diseases II [*Morb.II = de Morbis II*]. Each of the chapters 12–75 deals with one specific disease under the following heads: name or defining event, symptoms and course, prognosis (optional), treatment. The order of diseases is more or less anatomical beginning from the head. Varieties of the same disease are sometimes handled as independent entities, e.g., chs. 26–28,: angina, another angina, another angina. Each of chapters 1–11 explains the origin and course of one of the same diseases handled in chapters 12–31.

Diseases III [*Morb.III = de Morbis III*]. Chapters 1–16 deal with sixteen specific diseases under the following headings: name or identifying feature, symptoms and course, prognosis, treatment. The diseases are swelling of the brain, intense headache, stroke, necrosis of the brain, lethargy, ardent fever, swelling of the lung, loss of speech, phrenitis, angina, jaundice, tetanus, opisthotonus, ileus, pneumonia, pleurisy. Chapter 17 lists twenty-five cooling agents and applications.

Affections [*Aff. = de Affectionibus*]. The first part (chs. 1–38) presents—interspersed with various more general pathological considerations—roughly a score of specific diseases, giving their symptoms and course, treatment, and etiology (e.g., ch. 1, "All human diseases arise from bile and phlegm"). The second part (chs. 39–61) is a mixture of instructions for the use of particular dietetic agents (e.g., gruels, wines, cereals, honey, baths, fomentations) and theoretical arguments regarding their properties. (7,640 words)

Internal Affections [*Int.* = *de Affectionibus Interioribus*]. The fifty-four chapters of this textbook of pathology, each devoted to a different disease, are structured: name/identifying feature, etiology (optional), symptoms and course, treatment, prognosis. The order of presentation is in principle according to anatomy: lung and side, abdomen, entire body. Frequently, several varieties of the same disease are handled in separate chapters, often distinguished by their respective cause (e.g., chs. 47–50, "thick" disease: phlegm and bile, bile, phlegm, white phlegm), pathological process (e.g., chs. 14–17, kidney disease: lithiasis, rupture of the vessels, ulceration, suppuration), symptoms (e.g., chs. 27–29, liver disease: livid complexion, color of pomegranate peel, dark complexion), or season of occurrence (e.g., chs. 44–46, ileus: winter, summer, late autumn). (16,553 words)

Girls [*Virg.* = *de Virginum Morbis*]. With the arrival of puberty, girls become subject to various physical and mental disturbances provoked by an overabundance of blood; recommended cure: pregnancy. (472 words)

Nature of Women [*Nat.Mul.* = *de Natura Muliebri*]. This textbook, which shares a number of verbatim parallel texts with other gynecological treatises, such as *Superfetation* and *Diseases of Women I* and *II*, has a double structure: enclosed between a leading chapter on edema of the uterus (chs. 2 and 35) and a final collection of prescriptions (chs. 32–34 and 95–109), the two main sections of *Nature of Women* (chs. 3–31 and 36–94) consist of unordered series of chapters devoted to displacements of the uterus, amenorrhea, infertility, vaginal discharges, disorders associated with childbirth, etc. Treatment predominates. (12,199 words)

Diseases of Women I [*Mul. I = de Mulierum Affectibus I*]. A loosely organized compendium dealing at considerable length with conditions particular to the female sex: chs. 1–9, physiology and pathology of menstruation; 10–24, infertility; 25–34, disorders associated with pregnancy; 35–54, disorders of childbirth; 55–67, pathology of the uterus; 68–73, miscellaneous gynecology; 74–91, prescriptions relating to chs. 1–73; 92–109, miscellaneous prescriptions. (*I* and *II* are combined with *Barrenness* by Littré for a total of 50,007 words)

Diseases of Women II [*Mul. II = de Mulierum Affectibus II*]. A textbook of female reproductive pathology: chs. 1–13, vaginal discharges; 14–44, displacements of the uterus (e.g., hysterical suffocation, prolapse); 45–75, pathology of the uterus; 76–82, conditions not in the uterus; 83–103, prescriptions to chs. 1–82.

Barrenness [*Steril. = de Sterilibus*]. After a well-organized first chapter handling eleven disorders of the uterus and menstruation that result in infertility, this treatise dissolves into a more or less random collection of texts for determining pregnancy and the sex of the fetus (chs. 2–4), prescriptions for promoting conception (chs. 5–20), and treatments for habitual abortion, prolapse of the uterus, etc. (chs. 21–36).

Sight (*Vid. Ac. = de Videndi Acie*]. These fragmentary remains of a handbook of ophthalmology describe several important eye diseases: e.g., cataracts, trachoma, conjunctivitis, amaurosis. Cautery is the most frequently prescribed treatment, generally applied to sites of the body other than the eyes. (806 words)

Surgery

In the Surgery [Off. = de Officina Medici]. Chapters 3–6 give instructions for the positioning of the patient, the surgeon, and the instruments; 7–25 elaborate in some detail on the principles and practice of bandaging and splinting. Chapters 1–2 are a brief introduction. (2,221 words)

Fractures [Fract. = de Fracturis]. This is a highly technical but incomplete surgeon's manual devoted to the setting and bandaging of various broken and dislocated bones: chs. 1–7, fractures of the forearm; 8, fractures of the upper arm; 9–23, dislocations of the foot and ankle, and fractures of the lower limb; 24–37, compound fractures; 38–48, dislocations of the elbow. (11,593 words)

Joints [Art. = de Articulis]. This treatise, which has suffered notable dislocations, losses, and restorations, complements *Fractures* by dealing with: chs. 1–12, dislocations of the shoulder joint; 13–16, fractures and dislocations of the clavicle; 17–29 (taken verbatim from *Instruments of Reduction*, chs. 7–19), dislocations of the elbow, wrist, and finger joints; 30–40, injuries of the jaw, nose, and ears; 41–50, disorders of the spine and ribs; 51–61, dislocations of the hip; 62, club foot; 63–69, compound dislocations, amputation, and necrosis; 70–78, dislocations of the hip; 79–87 (82–87 taken verbatim from *Instruments of Reduction*, chs. 27–31), miscellaneous notes. The author of *Fractures* and *Dislocations* assumes a thorough knowledge of the anatomy of bones on the part of his readers. (21,905 words)

Instruments of Reduction [*Mochl.* = *Mochlicus* (or *Vectiarius*)]. The main body of this epitome of an earlier and more complete version of *Fractures* and *Joints* (cf. Withington, pp. 84–89) presents an orderly account of fractures and dislocations arranged by anatomy: chs. 2–4, nose, ear, jaw; 5–6, shoulder, scapula; 7–15, elbow; 16–18, wrist; 19, finger joints; 20–25, thigh joint; 26, knee; 27–31, ankle and foot; 32, club foot; 33–35, compound dislocations, amputation, and necrosis; 36, spine and ribs. Chapter 1 gives a concise summary of the bones of the body; chapters 37–43 are a collection of miscellaneous notes. (5,091 words)

Ulcers [*Ulc.* = *de Ulceribus*]. Based on his belief that the fundamental characteristic of surface lesions is an abnormal wetness, the author offers a plan of treatment that promotes dryness. Chapters 11–24 contain remedies applicable to specific lesions; chapters 25–26, dealing with bloodletting, are extraneous. (3,434 words)

Hemorrhoids [*Haem.* = *de Haemorrhoidibus*]. Instructions are given for various surgical interventions (including postoperative care) effective in removing hemorrhoids and condylomas: cautery, excision, scraping, putrefactant medications, fomentations. (928 words)

Fistulas [*Fist.* = *de Fistulis*]. This is a practical manual of therapies—mainly operative—for anal fistula and its complications (e.g., strangury, prolapse of the rectum). (1,603 words)

Wounds in the Head [*VC* = *de Vulneribus Capitis*]. This book is highly regarded on account of its clinical insight and clarity of exposition: chs. 1–2, anatomy of the skull (including an account of the sutures); 3–9, wounds of the head; 10–12, their examination and diagnosis; 13–21, prognosis and treatment. (5,130 words)

Excision of the Fetus [*Foet.Exsect.* = *de Exsectione Fetus*]. The first part of this short collection of notes gives instructions for the dismemberment and removal of a dead fetus; miscellaneous data regarding childbirth follow. (495 words)

Anatomy [*Anat.* = *de Anatomia*]. A short sketch of the gross features of the internal parts of the body. (260 words)

Mixed Books

Epidemics I–VII are combined by Littré for a total of 43,404 words; Books *I–VI* are roughly equal in length, Book *VII* about double that.

Epidemics I and *III* [*Epid. I/III* = *Epidemiarum I/III*]. These two titles are in fact a single work consisting of two kinds of clinical records: annual health reports (= constitutions) and individual case histories. The health reports include a survey of one year's weather in a specific location, followed by a detailed discussion of which diseases were observed and how frequently, whom they affected, and what the overall morbidity and mortality amounted to. The case histories follow one patient through an acute disease from the initial onset of fever to definitive remission, or death. Attention is sharply focused on the patient's symptoms and excreta, to the exclusion of diagnosis, prognosis, and treatment.

Epidemics I describes three constitutions on the island of Thasos, and fourteen case histories.

Epidemics III has twelve case histories, one constitution for an unnamed place, and sixteen more case histories.

Epidemics II [*Epid.II* = *Epidemiarum II*]. No clear

structure is discernible in this heterogeneous collection of clinical material. Some sections show a level of exposition that suggests a medical textbook, while others border on the incoherent. Two-thirds of the text consists of medical generalizations illustrated by specific diseases, the rest contains forty-one case histories, instructions for therapy and diet, and part of a constitution. *Epidemics II* shares verbatim passages with *Epidemics IV* and *VI*.

Epidemics IV [Epid.IV = Epidemiarum IV]. The bulk of this book comprises 112 case histories, ranging in length from one line to over a page, and is completed by a little medical theory and a few remarks about the seasons. The whole gives the impression of being a physician's personal notebook meant to help him recall his bedside observations.

Epidemics V [Epid.V = Epidemiarum V]. The hundred-odd case histories that constitute this haphazard collection vary greatly in arrangement, content, and length and include a wide variety of conditions: e.g., pleurisy (ch. 3), indigestion (6), stiff joints (23), loss of appetite (51), toothache (67), injury of the right forefinger followed by spasms (74), phobias (81–82), cancer of the breast (101). *Epidemics V* shares verbatim passages with *Epidemics VII*.

Epidemics VI [Epid.VI = Epidemiarum VI]. Four-fifths of this collection is occupied by medical theories and generalizations that culminate in the much-quoted and enigmatic nexus of Hippocratic wisdom (section 5, ch. 1): "Of diseases: natures: healers" (νούσων φύσιες ἰητροί). Thirty-four case histories, part of a constitution, and various therapeutic and dietetic indications complete the book.

Epidemics VII [*Epid.VII* = *Epidemiarum VII*]. The nearly two-hundred case histories recorded in this work generally receive a more minute attention than is common in other *Epidemics* volumes. At their fullest, the descriptions include the patient's name, place of origin or occupation, the time of year, diagnosis, symptoms and course of the disease, treatment and its effect, final outcome. The collection is unordered, and the diseases included vary greatly in their character, length, and severity.

Aphorisms [*Aph.* = *Aphorismi*]. This collection of 422 pithy statements on the medical art occupied, as long as Hippocratic medicine retained its esteem, the first place in popularity, being printed in innumerable editions, translated into many languages, and commented on without end. Section 1 concentrates on therapy by diet; section 2 on prognosis; 3 on the influence of the seasons and the predispositions of the different ages of life; 4 on purges and the diagnosis of febrile diseases; 5 on convulsions, heat and cold, and conditions of women; 6 on the symptoms of various conditions; and 7 on complications, sequences of diseases, and prognosis. (7,374 words)

Aphorism I, 1 acquired a fame extending beyond medicine: "Life is short, the art is long, opportunity is fleeting, experience treacherous, and judgment difficult."

2. HIPPOCRATIC MEDICINE

As will be evident from this survey, the Hippocratic writings represent a rich storehouse of historical sources for the study of classical medicine. The medicine that emerges is unified by a basic stock of shared conceptions regarding patient and physician, body and disease, care and cure. At the same time, numerous and often contradictory specific views are elaborated from this common tradition: for example, it is generally accepted that the body is composed of some kind of primary constituents, but about what these may be, many different opinions are expressed.

The summary that follows is an attempt to establish the fundamental principles that can be assumed to apply for the collection as a whole, even if some of them are only rarely or never stated explicitly . But beyond what is universally accepted, I record differing ideas on specific points, in order both to exemplify the vitality of the medical thought and also to acquaint the reader with certain particular ideas that play roles beyond medicine, for example, in contemporary philosophic thought, or that became important in later medicine. It must be emphasized that in the Hippocratic writings there is no generally valid, worked-out system of medical belief, but rather varying amalgams of independent and divergent trains of thought.

This sketch is organized under the headings of theory, practice, and profession: these three categories are less the parts than the dimensions of a medicine conceived of as a unity. Influences among the various aspects are continually at work: theory both grows out of and reflects back on to practice; a practitioner's activities form the basis of his social recognition, while at the same time his freedom of practice depends on his place in society; the profession emphasizes its roots in theory in order to assure its status, but the reasoning mentality that then comes to dominate its members cannot but influence its relationship to other branches of learning.

Theory

Hippocratic writers explain the workings of the body in terms of structures and fluids, the interactions of which are regarded as an anatomico-physiological unity: nature (*physis*).

The structures are the "parts" of the body. Many of these are recognized and named in the eighth-century BC Homeric epics: e.g., brain (*enkephalos*), vessel (*phleps*), heart (*kardia*), lungs (*pneumones*), diaphragm (*phrenes*), liver (*hēpar*), intestines (*entera*), kidney (*epinephridios*), bladder (*kystis*), bones (*ostea*), marrow (*myelos*) collarbone (*klēis*), ribs (*pleura*), vertebrae (*sphondylia*), hip joint (*ischion*). In the writings of the second half of the fifth century, both lay and medical, this traditional store of knowledge appears in augmented and refined form. The historian Herodotus names the spleen (*splēn*), uterus (*hysterē*), omentum (*epiploön*), and cavity (*koiliē*) = intestinal tract, clearly differentiating the last from the other abdominal viscera (*splanchna*); the tragedians Sophocles (*Trach.* 1054) and Euripides (*Ion* 1011) name, respectively, the bronchial tubes of the lung (*artēriai pleumonos*) and the "hollow vessel" (*phlebs koilē*) = vena cava. In the Hippocratic Collection such terms and many others are common coin: e.g., diploe (*diploē*), palate (*hyperōiē*), uvula (*staphylē*), tonsils (*antiades*), epiglottis (*klēithron*) = closer, trachea (*bronchos*), esophagus (*stomachos*), shoulder blade (*ōmoplatē*), hypochondrium (*hypochondrion*), perineum (*perineos*), urethra (*ourēthra* or *ourētēr*).

Besides this general anatomical knowledge, several treatises contain detailed anatomical expositions intended as preparation for a surgical text to follow (*Joints* 45, *In-*

struments of Reduction 1, *Wounds in the Head* 1–2) or to provide the necessary background for a theoretical argument the writer is developing (*Places in Man* 2–6, *Sacred Disease* 6–7, *Diseases IV* 7–9, *Nature of Man* 11). Four works are devoted mainly to anatomy: *Fleshes, Heart, Nature of Bones, Anatomy*. In the first three of these, attention is concentrated on the heart and vascular system; this predilection, which is present through much of the collection, is further evidenced by a wide range of new and not yet settled terminology: e.g., wide vessel, narrow vessel, large vessel, small vessel, blood vessel, hollow vessel, splenic vessel, hepatic vessel; *artéria* is used variously to denote artery, trachea, bronchus, ureter, and aorta.

Although the impulse to perform human dissection probably had its origin in the fifth and fourth centuries (cf. *Joints* 1 and 46), none seems to have been actually performed by Hippocratic writers (*Heart* is an exception), as is shown both by their exaggerated belief in the anatomical variation among individual members of the same species (see, e.g., *Wounds in the Head* 1, on the cranial sutures; *Places in Man* 6, on the number of vertebrae) and by the reporting of several imaginary structures (e.g., the course of the "hollow vessel" in *Internal Affections* 18).

The internal parts of the body are regarded as the sites of its physiological functions. Food enters the upper cavity, moves down to the lower cavity passing through the intestines, and is finally excreted (*Regimen in Acute Diseases* [*Appendix*] 11, *Diseases III* 14); urine flows out through the kidneys and bladder (*Aphorisms* 4, 75; *Diseases I* 8); blood and/or breath flow through the vessels (*Diseases I* 24, *Sacred Disease* 7); sound enters the ear and traverses an internal air-space to arrive at the brain, where it is perceived (*Diseases II 4*, *Places in Man* 2). Active roles are often assigned to parts such as the bladder, head, uterus, spleen, lungs, and breasts, on the basis of their form and texture (cf. *Ancient Medicine* 22–23, *Diseases I* 15, *Glands* 3, and *Diseases IV* 2).

The bodily processes taking place in the parts might be summarized as follows: from the three forms of nourishment entering the body—food, drink, air (*pneuma*)—its strength and growth accrue. Health and natural function require a suitable and sufficient intake of these nutriments into the lungs (*Diseases I* 11) and upper cavity (*Diseases I* 34), their proper adaptation to the body's texture and needs through a process of digestion (*Affections* 47), and their free distribution through the vessels as breath and blood (*Diseases IV* 8–14).

Fluids, beginning with breath and blood, play important roles in the body's actions in both health and disease. Other fluids are milk sent from the uterus to the breasts in order to nourish the neonate (*Glands* 16, *Nature of the Child* 10), synovial fluid in the articular spaces lubricating the joints (*Places in Man* 7), and semen in both sexes separated out from all the liquids in the body (*Generation* 1). Opinions on the role of the important pair "phlegm" and "bile" differ widely among various treatises. Often phlegm and bile appear only during illnesses, in the role of "disease substances" with the respective qualities of cold (*Sacred Disease* 10) and hot (*Diseases I* 29), presumably being always present in the body in some form, but under normal circumstances imperceptible (cf. *Diseases I* 2). *Affections* 1 (cf. also 37) states categorically that "all human diseases arise from bile and phlegm," while *Generation* 3 (cf. *Diseases IV* 1–10) posits a quaternary system of humoral pathology based on the four innate substances of blood, bile, water, and phlegm. In *Nature of Man* 4 the four (later canonical) humors (blood, phlegm, yellow bile, and dark bile) are virtual elements, whose due proportion and mixing constitute perfect health.

At the center of Hippocratic pathology is, at least ideally, the concept of the specific disease: illness does not occur in indefinite random configurations, but in a limited number of definite patterns, each defined by a particular anatomico-physiological disturbance at its origin. In the collection more than two hundred such specific diseases and conditions are handled. Many of these have names derived either from a prominent symptom (e.g., consumption, tertian fever, prolapse of the uterus), from the location of the disorder (e.g., peripneumonia, ophthalmia, hysteria), or from some observed or hypothesized mechanism (e.g., dislocation of the shoulder, hydropsy = accumulation of water, ileus = twisting, lientery = slipperiness of the intestine). Diseases without actual names are often referred to by a phrase indicating one of the same three phenomena: e.g., "when a sharp pain beginning from the head suddenly makes a person speechless" (*Diseases III* 8), "disease of the kidney" (*Internal Affections* 15), "if fluid forms on the brain" (*Diseases II* 15). This diagnostic system is applied most rigorously to acute internal diseases and surgical disorders that present clearly identifiable symptoms and courses; in conditions with vaguer and more easily confused presentations, such as infertility and various vaginal discharges, its application is less feasible, with treatment then frequently being deduced directly from signs without any connecting diagnostic (i.e., identification of the specific disease) step (*Diseases of Women I* 77, *Barrenness* 30). In its most complete form, the concept of a specific disease includes: a precise delimitation of the disease's symptoms, course, prognosis and varieties; a clear expression of the fundamental anatomico-physiological disturbance at its origin; hypotheses explaining how this fundamental disturbance produces each observed symptom. Actual nosological accounts in the various Hippocratic treatises, however, tend to concentrate on particular points, and rarely or never are all aspects of a nosological entity fully expounded in a single passage (*Diseases III* 3, *Diseases II* 8, *Affections* 7).

These examples represent the kinds of observations, assumptions, and reasonings that predominate in Hippocratic pathology, mirroring the current theories of normal function presented above: fundamental pathological disorders fall into those of structures (e.g., *Internal Affections* 15, *Wounds in the Head* 4–9), of fluids (e.g., *Sacred Disease* 10, *Affections* 19), or of both (*Diseases II* 1, *Affections* 21). Often a distinction is made between an external cause and the actual pathological process (*Affections* 1); external causes include not only effects experienced personally by an individual but also broader environmental factors such as geographical location (*Airs Waters Places* 1–2), season (*Wounds in the Head* 2, *Nature of Man* 7–8), and specific noxious influences (*Breaths* 6). The pathological process itself may be influenced by further factors such as the age and the sex of the patient (*Nature of Man* 9, *Diseases of Women I* 62), and several times the inheritance of specific conditions is attributed to a direct transmission of hereditary material (*Sacred Disease* 5; *Generation* 8, 11; *Diseases IV* 1).

The reestablishment of health is regarded partly as a natural achievement of the body and partly as the result of appropriate medical intervention; and on several occasions the healing process is depicted explicitly as a combat (*Epidemics I* 11). The specific processes of cure—whether spontaneous or at the hand of the healer—follow logically from Hippocratic theories of disease: humoral or qualitative (e.g., hot, cold, moist, dry) imbalances are equalized (*Affections* 36, *Breaths* 1); disease substances are evacuated by the appropriate route, a process often called "cleaning" (*Internal Affections* 3), or inactivated by dilution (*Diseases I* 20); physical damage due to injury or a disease is repaired (*Diseases I* 21), or tissue overgrowth is removed (*Diseases II* 33–37).

There is a fundamental belief that diseases have definite chronological patterns and tend to develop toward a decisive point ("crisis"), at which a patient either dies or escapes from danger (*Diseases I* 6, *Internal Affections* 41). Where this crisis is associated with the removal from the body of a disease product ("apostasis" or "abscession"), the preparation of the product inside the body is likened to cooking ("pepsis" or "[con]coction"), and the substance is said to have passed from a state of "rawness" to a state of "maturity" (cf. *Regimen in Acute Diseases* [*Appendix*] 32), generally from colorless and odorless to purulent and putrid. Careful observation of a disease's course enables a physician to coordinate treatment with the efforts of the body (*Diseases I* 5), important because wrongly timed treatment is held to be not merely useless but harmful (*Regimen in Acute Diseases* [*Appendix*] 5 and 8).

Ultimately, Hippocratic disease management has three basic components: when the situation is critical, to leave the patient at rest with an innocuous diet (*Regimen in Acute Diseases* 38, [*Appendix*] 54); when the situation is not critical, to strengthen the patient with a mild but nutritious diet like barley gruel (*Aphorisms* 1, 10; *Fractures* 7; *Affections* 44); to promote specific healing processes as the clinical situation or the disease diagnosis requires (*Nature of Man* 2, 9, 13; *Places in Man* 42; *Regimen II* 68).

Practice

Although no systematic exposition of physical examination and history taking is to be found in the collection (cf. *Regimen in Acute Diseases* [*Appendix*] 22), a good idea of the procedures employed may be gained from chance remarks and from the signs and symptoms that appear regularly in case histories and in the textbooks of pathology: the art of examination thus revealed is comprehensive, complex, and refined. Many physical signs are observed directly: e.g., posture, respiration, sleep, drowsiness, listlessness, excitement, delirium, tremors, convulsions; weeping, sneezing, coughing, hiccups, eructation, flatulence, vomiting, bleeding; pain, sensitivity, weakness, paralysis, paresthesias, speechlessness, blindness, deafness; fever, chills, sweating, cold extremities, darkening of the tongue; color, temperature, and texture of the skin, swelling, scabs, scales, scars, papules, pustules; lesions, tumors, wounds, deformities, dislocations, fractures. So too are the characteristics of the excreta: e.g., (stools) color, texture, presence of blood, odor; (urines) color, clarity, density, precipitates, scum on the surface; (sputum) color, consistency, odor, amount, streaked with blood; (vomitus) amount, color, contents, acidity.

Internal phenomena inaccessible to direct investigation are detected by special procedures, such as shaking a patient's shoulders with the physician's ear applied to his chest wall (Hippocratic succussion) in order to hear the fluctuation of fluids inside the chest (e.g., *Diseases III* 16), the application of fine potter's clay to locate a region of raised temperature over a collection of pus (ibid.), or pinching a patient in order to determine whether a convulsion is hysterical or epileptic (*Regimen in Acute Diseases* [*Appendix*] 68). A medical history obtained from the patient or other witnesses adds subjective complaints (e.g., *Diseases II* 51) and supplies information about the time when the physician has been absent. The case histories recorded in *Epidemics I* and *III* illustrate how the results of physical examination and history taking can be integrated into a concise narrative unity. In a number of treatises, a knowledge of seasonal and geographic factors also plays an essential role in evaluating the individual patient (e.g., *Epidemics I* 23; *Airs Waters Places* 1–10).

On the basis of the data collected in these ways, practitioners seek to recognize which specific disease a given patient is suffering from. In the generalized disease descriptions, reference is rarely made to the actual process of diagnosis: the signs, symptoms, and course are recorded, recognition is assumed but not expressed, and prognosis and/or treatment follow directly, often introduced by the expression "when the case is such" (e.g., *Diseases II* 58; *Internal Affections* 48). Only in cases where some particular difficulty exists in recognizing the specific disease is diagnosis brought to the reader's attention (e.g., *Diseases III* 15, "if the patient has only a few of the signs, do not be deceived into thinking that it is not really pneumonia, for it is, only a mild case"; *Internal Affections* 14, "many physicians who do not understand this disease, when they see the sand [sc. in the urine] think that the patient is suffering from stones of the bladder, which he is not, but rather from stones of the kidney").

Prognosis may be based on diagnosis alone, being founded on the empirical evidence of many past cases of the particular specific disease (e.g., *Internal Affections* 41; *Diseases III* 1), or be computed by weighing a wide range of prognostic signs that possess their validity independent of which specific disease a patient happens to be suffering from (e.g., *Prognostic* 25; cf. *Prorrhetic I* and *Coan Prenotions*). Diagnosis is also valuable for prognosis by providing a "typical" disease pattern, against which the signs and symptoms of a particular case can be evaluated to judge its relative severity and likely development (e.g., *Diseases III* 6 and 15).

For treatment, diagnosis is significant in two ways. First, it often identifies the fundamental anatomico-physiological disturbance from which the disease has arisen, and against which specific therapeutic measures must be directed. Second, it furnishes a generalized disease course that may help to determine the critical periods of the present case (i.e., the times when strengthening remedies must be discontinued).

Additional forms of therapy are often applied in response to particular clinical situations independent of diagnosis: e.g., local pain is countered with bloodletting (*Regimen in Acute Diseases* [*Appendix*] 31), cautery (*Diseases II* 12), warm sponges (*Diseases II* 14), fomentations (*Diseases II* 54), or moxibustion (*Affections* 31); inflamed joints are kept mobile by passive movement (*Internal Af-*

fections 51); growth of tissue is promoted by the application of plasters (*Affections* 38); dangerously high fevers are blunted by bloodletting (*Diseases III* 15); blockage of the airway is prevented by the insertion of tubes into the throat (*Diseases III* 10); recalcitrant diarrhea is combated by the stimulation of vomiting (*Affections* 27); the spread of disease fluids through the body is prevented by the application of cautery across the vessels (*Internal Affections* 18).

Conduct of life ("diet" or "regimen") is considered to be of capital importance for both the maintenance and the reestablishment of health. This includes the timing and the content of meals (*Regimen in Acute Diseases* [*Appendix*] 42 and 44; *Affections* 61); the consumption of or abstinence from wine (e.g., *Diseases III* 1, 3, 4; *Affections* 10; *Internal Affections* 18); activity and rest (e.g., *Internal Affections* 17, 30); baths, anointing, and massage (e.g., *Affections* 42; *Internal Affections* 48, 52; *Joints* 9; *Regimen in Acute Diseases* 65–68); sleep and waking (e.g., *Affections* 43; *Internal Affections* 30); and sexual activity (*Diseases II* 51).

The number of medications—mostly of plant origin—prescribed in the collection is almost three hundred. These can be classified according to their actions as emetics (e.g., white hellebore, thapsia juice), laxatives (e.g., whey, cabbage juice), purges (e.g., black hellebore, spurge), narcotics (e.g., poppy juice, mandrake, night-shade), astringents (e.g., oak bark, willow), dermatological agents (e.g., sulfur, seawater), putrefactants (e.g., flower of copper, white lead), and expectorants (e.g., mustard, hyssop). Forms of application include potions, powders, pills, lozenges, enemas, pessaries, salves, plasters, baths, and fumigations (e.g., *Diseases II* 26). Sometimes prescriptions specify the exact amounts of their ingredients (e.g., *Diseases III* 11).

Surgical treatment is limited, in general, to the cleaning and setting of broken and dislocated bones, the removal and repair of accessible lesions (e.g., tumors of the skin, subcutaneous tissue, breast, or cervix uteri; nasal polyps; hemorrhoids), and the cutting or burning of pathways to drain disease substances through a pipe from internal spaces (e.g., thoracentesis, *Diseases II* 47; paracentesis, *Affections* 22), and trephining is employed in some skull wounds (e.g., *Wounds in the Head* 14, 21). There is every indication that these operations were carried out with great care and skill.

Profession

Hippocratic medicine was the activity of a group of practitioners sharing a common theoretical basis, a set of traditional practices, and some sort of professional identity. Professional regulation lacked the finite, formal character known in medieval and modern times: legally anyone could practice medicine. Midwives are occasionally referred to for their expertise in conditions of women (*Fleshes* 19, *Diseases of Women I* 68), but it is difficult to determine how extensive a role female practitioners actually played in gynecological practice. Furthermore, Hippocratic "scientific" medicine was only one of an indistinct plurality of alternative healing arts. These included the temple medicine of the cult of Asclepius (cf. Aristophanes *Plutus* 653–747; Herodas *Mime* 4), the medical divination of inspired "wisemen" (cf. the *iatromantis* in Aeschylus,

Supplices 252 ff.) such as Empedocles (cf. frag. B111 and B112 in DK), and all manner of charlatan (cf. the polemics in *Sacred Disease* 2–4 and *Joints* 42, 78).

In the inevitable competition, the Hippocratic physician was routinely compelled to convince prospective patients of his competence and of the advantages of his kind of medical thought and practice. This task of winning and maintaining patients' trust, the importance of which to clinical success was clearly appreciated (cf. *Precepts* 6), was addressed in several ways: by a seemly appearance and conduct (e.g., *Physician* 1), by being able to state and defend his medical position convincingly (e.g., *Diseases I* 1; *Ancient Medicine* 13, 17; *Fleshes* 6; *Prorrhetic II* 1), by a demonstrated competence in manual procedures (e.g., *Diseases I* 6), by being able to predict accurately a disease's course and outcome (cf. *Fractures* 36 and *Diseases of Women I* 71).

With no licensing of physicians by the state, it is not surprising that attempts at self-regulation were made. This appears to be the purpose of the first section of the *Oath*, which lays down the rules of guild membership in clear terms. Notable in the *Oath* are the essentially familial model, the exclusion of women, and the religious overtones (cf. *Law* 5). The question of how prevalent this type of guild organization was at various times in the various polities of ancient Greece has been widely discussed, but without any clear resolution; the same can be said for the validity of the prohibitions against euthanasia and induced abortion set out in the second section of the document.

Medical education seems often to have assumed the teacher-pupil form common both in crafts perpetuated by an apprenticeship system and in other branches of learning at the time (cf. *Law* 2–3). The value of literacy for medicine was recognized early in the history of Greek prose, as the great variety of uses to which it is put in the writings of the collection attests. Moreover, the importance of medical literature in education was well established (e.g., *Epidemics III* 16). Bedside experience was gained by students when they were assigned to watch over patients between their teacher's visits (*Decorum* 17), and the necessity of firsthand experience for the aspiring surgeon was well appreciated (e.g., *Physician* 14). Itinerancy of physicians, too, is mentioned or implied in several treatises (e.g., *Airs Waters Places* 1; *Epidemics*; *Law* 4), but was probably not the rule: *Physician* 2, in any case, gives instructions for setting up an office (*iatreion*) in the physician's house, and *Joints* 72 refers to a surgeon who "practices in a populous city." Finally, professional etiquette regarding consultation and fees emphasized the good of the patient, and the physician's reputation, over immediate gain (cf. *Precepts* 4, 6, 8; *Joints* 78).

3. FORMATION OF THE COLLECTION[6]

How these disparate writings came to form a collection, and to be attributed to the historical figure Hippocrates of Cos, is not known, although a scene in Xenophon's *So-*

---

[6] Cf. A. Roselli, "Un corpo che prende forma: l'ordine di successione dei trattati ippocratici dall'età ellenistica fino all'età bizantina," *Annali Istituto Orientale Napoli (AION)* 22 (2000): 167–95.

cratic Memoirs (4, 2, 10) confirms that medical books were plentifully available in fifth-century Athens (πολλὰ γὰρ ἰατρῶν ἐστι συγγράμματα). Early evidence concerning Hippocrates the physician is meager, but unanimous in recognizing his professional preeminence. He is referred to in two of Plato's dialogues; the first passage is in the *Protagoras* (311b) (trans. W. K. C. Guthrie):

> Socrates: Suppose for instance you had it in mind to go to your namesake Hippocrates of Cos, the Asclepiad,[7] and pay him a fee for teaching you, and someone asked you in what capacity you thought of Hippocrates with the intention of paying him, what would you answer?
>
> Hippocrates: I should say, in his capacity as a physician.
>
> Socrates: And what will he make you into?
>
> Hippocrates: A physician.

This passage testifies to Hippocrates' fame in medicine, which Plato compares to the fame of Polyclitus of Argos and Phidias of Athens in sculpture, the two most renowned Greek sculptors of the second half of the fifth century.

A second Platonic passage at *Phaedrus* (270c–e) presents Hippocrates as employing a method in medicine—investigation of the nature (φύσις) of the body—that is

---

[7] Whether the term "Asclepiad" here means simply "physician" or also implies adherence to a medical guild tracing its descent from the Homeric hero Asclepius, whose sons Machaon and Podaleirius fought at Troy and were healers (ἰατροί), is not agreed upon by scholars. The Asclepieion of Cos was founded after Hippocrates lived; cf. Herzog, pp. ix–xii and 75.

parallel to how a student of rhetoric investigates the nature of the mind (ψυχή). However, when Socrates actually explains how this "investigation of nature" operates, it becomes clear that Plato is projecting his own dialectical method of reasoning on to medicine in a form unknown in the Hippocratic writings.[8] Plato is playing with an apparent parallel of method that does not in fact exist, which makes this passage of no value in identifying any method of the historical Hippocrates, which could be used to connect him with a particular Hippocratic text (trans. R. Hackworth):

> Socrates: Rhetoric is in the same case as medicine, don't you think?
>
> Phaedrus: How so?
>
> Socrates: In both cases there is a nature we have to determine, the nature of the body in the one, and of the mind in the other, if we mean to be scientific (τέχνη) and not content with mere empirical routine (τριβῇ μόνον καὶ ἐμπειρίᾳ) when we administer medicine and diet to secure health and strength.
>
> Phaedrus: You are probably right, Socrates.
>
> Socrates: Then do you think it is possible to understand the nature of the mind satisfactorily apart from the nature of the whole (ἄνευ τῆς τοῦ ὅλου φύσεως)?

---

[8] Cf. W. K. C. Guthrie, *A History of Greek Philosophy* (Cambridge, 1962–1981), vol. 4 (1975), p. 432: "What 'Hippocrates and the true account' say about nature turns out to be something that surely Hippocrates did not say, namely that to be experts (*technikoi*) ourselves and make others the same we must apply the dialectical method of division."

Phaedrus: If we are to believe Hippocrates of the Ascle-
piads, we cannot even gain understanding about the
body without this method.

Socrates: For, my friend, he is telling the truth. And
indeed, besides what Hippocrates says, we must
examine the argument to see if it agrees.

Phaedrus: Yes.

Socrates: Then look what Hippocrates and the true argument have to say about nature. Is it not that one
must inquire in the following way, concerning the
nature of anything? First, whether the thing, concerning which we ourselves wish to be craftsmen
and capable of making someone else one too, is
simple or complex. Then, if it is simple, to examine
its property of being able to have an effect on something else, or its ability to suffer the effect of some
other thing. If, on the other hand, it has many
forms, to count these, and to see for each of them,
in the way we did for the simple thing, what natural
property it has to act on what, and what property to
be acted upon by what.

The writings of Aristotle contain only one reference to
Hippocrates, alluding at *Politics* (1326a 12–17) to his
greatness in medicine (trans. B. Jowett):

A city too, like an individual, has a work to do; and
that city which is best adapted to the fulfillment of
its work is to be deemed greatest, in the same sense
of the word "great" in which Hippocrates might be
called greater, not as a man, but as a physician, than
someone else who was taller.

In the third book of the *Historia Animalium*, where he is about to present his own opinion on the vessels of the human body, Aristotle quotes texts of three of his scientific predecessors describing their vascular systems: the second of these is from Diogenes of Apollonia, an early Greek philosopher, and need concern us no further here, while the first and the third texts are both to be found virtually verbatim in the writings of the Hippocratic Collection:

H.A. 511b 24–30[9] ~ *Nature of Bones* 8, 1–7
H.A. 512b 3–513a 7[10] ~ *Nature of Bones* 9, 1–34 ~
   Nature of Man 11, 1–28.

Furthermore, Aristotle attributes the two texts explicitly to Syennesis, a physician of Cyprus (Συέννεσις μὲν ὁ Κύπριος ἰατρὸς τόνδε τὸν τρόπον λέγει· αἱ φλέβες αἱ παχεῖαι ὧδε πεφύκασιν . . .) and a Polybus or Polybius —the name differs among the independent Aristotle manuscripts—(Πόλυβος/Πολύβιος δ᾽ ὧδε· τὰ δὲ τῶν φλεβῶν τέτταρα ζεύγη ἐστίν . . .), respectively.

As mentioned above, by the first century BC it was commonly believed that the writings—many, most, or all— were "of Hippocrates." Between the time of Aristotle and the first century, there is little contemporary evidence shedding light on how this change came about, the exception being reports about a Greek scholarly movement centered in Alexandria at the court of the Ptolemies, which included an intensive occupation with an-

9 Aristotle, p. 76.
10 Aristotle, pp. 78–80.

xlvi

cient texts of all kinds; among these, early medical texts
too received the attention of scholars, collecting and explaining obsolete and otherwise notable words in them.[11]
Our main evidence for this medical scholarship is contained in a writing of Erotian, who lived at the time of
the emperor Nero and dedicated his glossary *Collection of
Hippocratic Words* (Τῶν παρ' Ἱπποκράτει λέξεων συνα-
γωγή) to Nero's archiater Andromachus.[12] He gives a
detailed account of the movement in the introduction to
the *Collection*:

> . . . many highly regarded men, not only physicians
> but also grammarians, went seriously to work to ex-
> plain the man (i.e. Hippocrates) and to bring his
> words closer to standard usage. Xenocritus of Cos,
> a grammarian, as Heraclides of Tarentum says, was
> the first to work on explaining such terms; as Apol-
> lonius of Citium reports, Callimachus of the house
> of Herophilus did the same. After him, they say,
> Bacchius of Tanagra (ca 275–200 BC) contributed
> to the field, and in three *syntaxes* (διὰ τριῶν συντά-
> ξεων) completed the project by adding many pieces
> of evidence from the poets. This work, however, the
> contemporary Empiricist Philinus (ca 270–200) at-
> tacked in a writing of six books. Epicles of Crete (in
> the first century BC) also abridged Bacchius' *Words*
> (Λέξεις).[13]

[11] Cf. H. von Staden, "Lexicography in the Third Century
B.C.: Bacchius of Tanagra, Erotian, and Hippocrates," in López-
Férez *Trat.Hip.*, pp. 549–69.

[12] Erotian, pp. 3 and 116.

[13] Erotian, pp. 4f.

But not only does Erotian furnish this historical overview he also includes about sixty definitions from Bacchius' *Words* among his own entries in the *Collection*: e.g.,

α 1. ἀλυσμόν· . . . Βακχεῖος μέντοι ἐν τῷ α´ τῶν Λέξεών φησι· "τὸ ἀλύειν ἀδυνατεῖν, πλανᾶσθαι, ἄχθεσθαι."[14]

α 1. Distressed: . . . Bacchius in the first (sc. *syntaxis*) of the *Words* says: "to be distressed (sc. means) to be disabled, unsettled, grieved."

By examining Bacchius' entries cited by Erotian, it is sometimes possible to identify the exact Hippocratic treatise and passage from which he has taken the word he is defining, and also to know in which of his *syntaxes* the definition was recorded. From these identifications, some of the Hippocratic sources of Bacchius' three *syntaxes* can be determined:

Syntaxis 1: *Prognostic, Sacred Disease, Joints, Instruments of Reduction, Epidemics I, Epidemics VI*

Syntaxis 2: *Prognostic, Prorrhetic I, Joints, In the Surgery, Instruments of Reduction, Regimen in Acute Diseases, Epidemics II*

Syntaxis 3: *Nature of Bones, Fractures, Joints, In the Surgery, Places in Man, Epidemics V*

And even where Erotian does not record from which *syntaxis* of Bacchius' *Words* he is taking a particular entry, it is often possible to determine its Hippocratic source (cf. *Herophilus*, p. 487). In this way, four more Hippocratic

14 Erotian, p. 10.

works Bacchius studied are revealed: *Wounds in the Head, Diseases I, Use of Liquids, Epidemics III.*

From this knowledge of Bacchius' *Words*, then, it can be concluded that by the end of the third century BC at least eighteen of the writings we know as part of the Hippocratic Collection belonged to some kind of a group. It must be emphasized, however, that, as quoted by Erotian, no entry of Bacchius (Ba 13–36, 38–54, 56–75)[15] ever mentions the name of Hippocrates, nor does the title of his *Words* as reported by Erotian ever contain any reference to Hippocrates. Bacchius is certainly studying words derived from treatises that *in Erotian's time* were attributed to Hippocrates of Cos. We also know that in the first century BC Epicles of Crete alphabetized and shortened Bacchius' entries, apparently removing references to their sources:[16] but we have no way of knowing whether Bacchius had attributed the texts to Hippocrates, or as in Aristotle's writings they were assigned to other authors. Furthermore, as cited by Erotian, the title of Bacchius' glossary is always simply *Words* (Λέξεις), the same as that of a similar work by the contemporary lexicographer Aristophanes of Byzantium. How the Hippocratic treatises, or at least some of them, lost their true authors' names and received in return Hippocrates' is a tantalizing mystery that has stimulated an immense amount of speculation and ingenuity—but no satisfactory solution.[17]

[15] Herophilus, pp. 495–500.   [16] Erotian, pp. 7f.

[17] Cf. L. Edelstein, "The Genuine Works of Hippocrates," *Bulletin of the History of Medicine* 7 (1939): 236–48; G. E. R. Lloyd, "The Hippocratic Question," reprinted with a new introduction in Lloyd, pp. 194–223; H. von Staden, "Interpreting 'Hippocrates' in the 3rd and 2nd centuries BC," in Müller, pp. 15–47.

Once Hippocrates reappears in history—the name Hippocrates of Cos is never mentioned in a written source between Aristotle and Apollonius of Citium in the middle of the first century BC—the iconic physician and medical teacher of Greece and his formidable collection of medical texts are universally acclaimed. This new commonplace is evidenced in Apollonius' *On Joints According to Hippocrates*, the first extant medical commentary directed to any Hippocratic work: for Apollonius, Bacchius' work entitled *Words* (Λέξεις) by Erotian has become *About Hippocratic Words* (Περὶ τῶν Ἱπποκρατείων λέξεων).[18] Another contemporary Empiricist writer, Heraclides of Tarentum, calls the same work *About the Words of Hippocrates* (Περὶ τῶν Ἱπποκράτους λέξεων).[19]

From Apollonius' time forward, there are few extant medical writers who fail to quote, refer to, comment on, or react against texts of the Hippocratic Collection. Examples are found in the first-century AD Greek writers Aretaeus of Cappadocia, Soranus of Ephesus, and Rufus of Ephesus, as well as in the anonymous Greek works in the papyri Anonymus Parisinus and Anonymus Londinensis, and in the Latin writers Celsus, Scribonius Largus, and Pliny the Elder.[20]

Erotian, who with his own *Collection* continues the long tradition of medical lexicography examined above, provides in its introduction the sole surviving census of Hip-

---

[18] Apollon. Cit., p. 28.
[19] Heracl. Tar., fr. 352.
[20] See the "Quellenindex" in *Testimonien* vol. I, pp. 534–70.

pocratic works from antiquity, arranged according to six categories:

> These are the semiotic works (σημειωτικά): *Prognostic, Prorrhetic I* and *II*—that the latter is not by Hippocrates, I will show on another occasion—, *Humors.*
>
> Etiological and scientific works (αἰτιολογικὰ καὶ φυσικά): *Breaths, Nature of Man, Sacred Disease, Nature of the Child, Places and Seasons* (=*Airs Waters Places*).
>
> Therapeutic works (θεραπευτικά), pertaining first to surgery: *Fractures, Joints, Ulcers, Wounds and Missiles, Wounds in the Head, In the Surgery, Instruments of Reduction, Hemorrhoids, Fistulas*; then to diet: *Diseases I II* (=*Diseases III*), *Barley-gruel* (=*Regimen in Acute Diseases*), *Places in Man, Diseases of Women I II, Nutriment, Barrenness, Use of Liquids.*
>
> Mixed works (ἐπίμικτα): *Aphorisms, Epidemics 7.*
>
> Works belonging to the discussion of medicine (τῶν δ' εἰς τὸν περὶ τέχνης τεινόντων λόγον): *Oath, Law, Art, Ancient Medicine.*
>
> *Embassy* and *Speech from the Altar* show the man more as a patriot than as a physician (φιλόπα-τριν μᾶλλον ἢ ἰατρὸν).
>
> Thus, since we assert that these writings are definitely (βεβαίως) by Hippocrates, instruction should start, for the sake of clarity, from the semiotic ones, because an understanding of signs must precede all

understanding of etiology and therapy. Well then, let us start with *Prognostic*.[21]

Noteworthy in this document are:

1. its increase from Bacchius' eighteen treatises to more than twice that number;
2. its practical organization by medical content rather than according to some lexicographical motivation;
3. its definite attribution of the treatises to Hippocrates, except for
4. its promise to enter into the question of Hippocratic genuineness in one case (*Prorrhetic II*)—the first example of such a consideration in the Hippocratic literature—although Erotian does not in fact return to the subject in his surviving text;
5. inclusion of the two pseudepigraphic titles *Embassy* and *Speech from the Altar.* Other early references to this body of mainly or wholly fictive literature are found, e.g., in Varro's *De re rustica* 1, 4, 5 (Hippocrates saves cities from a pestilence) and Strabo's *Geography* 14, 2, 19 (Hippocrates derived most of his dietetic practices from cures recorded on votive tablets in the Asclepieion of Cos).[22]

In the writings of Galen of Pergamum (129–ca. 210), the figure "Hippocrates" and his works—Galen knows about fifty titles, but considers some of these to be "less than genuine"—are often employed as instruments to support and validate Galen's own medical and historical

[21] Erotian, p. 9.

[22] These works are edited and translated into English in Smith *Pseud.*; for an account of their history and contents, see Pinault, pp. 1–93.

views. Besides innumerable references scattered through all his works,[23] Galen composed a glossary *Explanation of Difficult Words in Hippocrates* (Τῶν παρ᾽ Ἱπποκράτει γλωττῶν ἐξήγησις),[24] as well as lengthy commentaries on many Hippocratic treatises, including those on *Fractures/Joints, Aphorisms, Prognostic, Humors, Epidemics I, Regimen in Acute Diseases, In the Surgery, Epidemics II, Prorrhetic I, Epidemics III, Epidemics VI, Nature of Man,* and *Airs Waters Places,* which have survived in the Greek original, and on *Humors* and *Epidemics II,* which are extant only in Arabic translation.[25]

While Galen is generally recognized as an exceptionally competent medical scientist and scholar, his reliability as a historical source—in particular concerning the earliest centuries of the Hippocratic Collection's existence—is assessed very differently by different modern scholars, some assuming that he possessed genuine early evidence on which he based his interpretations, but others doubting this and attributing many of the opinions Galen expresses not to any unique evidence he had at his disposal but rather to his preconceived notions about what Hippocrates and his writings "must" have been.[26]

In the late ancient period, when much of Greek medical literature took the form of compilations made from

---

[23] See *Testimonien* vols. II,1 and II,2.

[24] See Perilli.

[25] See Boudon, pp. 159–62; Manetti/Ros.

[26] Cf. Smith, pp. 175–246; H. von Staden, "Galen as Historian: His Use of Sources on the Herophileans," In López-Férez *Galeno,* pp. 205–22; G. E. R. Lloyd, "*Galen on Hellenistics and Hippocrateans: Contemporary Battles and Past Authorities,*" In Lloyd, pp. 398–416; Jouanna, *Hippocrates,* in general, but esp. pp. 353–57.

earlier writers, Galen's medical authority and its ostensible foundation on Hippocratic medicine is illustrated by Oribasius in the introduction to his *Medical Collections* requisitioned by the Byzantine emperor Julian:

> . . . according to your desire, when we were stationed in western Gaul, I completed the abstract you asked me to make from the writings of Galen alone. After you had praised that work, you requested a second labor, that I should search out and collect all the writings useful for the purpose of medicine, and I am happy to do this because I think that such a collection will be most useful, making readers able in each instance easily to find what is necessary in diseases . . . I will collect only from the best writers, while omitting nothing of what was provided by Galen before, since he excels all other writers on these subjects, because he employs the most precise methods and definitions in following the principles and opinions of Hippocrates.[27]

In Byzantine medical education centered in Alexandria, a more general engagement with Hippocratic texts persisted, resulting ca. 600 AD in the writing of many commentaries, including the still extant works by Palladius on *Fractures*[28] and *Epidemics VI*,[29] by John of Alexandria on *Epidemics VI* (surviving only in Latin translation[30] ex-

[27] Oribasius VI 1,1, p. 4.
[28] Cf. Irmer.
[29] Cf. Dietz, vol. 2, pp. 1–204.
[30] Cf. Pritchet.

cept for fragments in Greek) and *Nature of the Child*,[31] and by Stephanus of Athens on *Aphorisms*[32] and *Prognostic*.[33]

The Hippocratic Collection continued to exert an important influence on medicine through the medieval period in the Byzantine,[34] Islamic[35] and Latin[36] cultures, as is shown by the continued copying of texts; by translations into Syriac, Arabic, Hebrew, and Latin; and by the role the figure Hippocrates played in the medical discussions of the time.

With the Renaissance, a process of constant scientific and technological advance was set in motion, which over the succeeding centuries led to the development of modern scientific medicine, definitively superseding the medicine of the ancients. When W. Osler in his textbook *Principles and Practice of Medicine* (New York, 1892) quotes fourteen Hippocratic passages, this is clearly not for their special scientific importance, but rather to evoke a sense

[31] Cf. Duffy *Epid.*, pp. 9–117; 127–75.

[32] Cf. Westerink.

[33] Cf. Duffy *Prog.*

[34] See J. M. Duffy, "Byzantine Medicine in the Sixth and Seventh Centuries: Aspects of Teaching and Practice," In Scarborough, pp. 21–27.

[35] See Ullmann, pp. 25–35; Sezgin, pp. 23–47; U. Weisser, "Das Corpus Hippocraticum in der arabischen Medizin," in Baader, pp. 377–408; P. Pormann, ed., *Epidemics in Context. Greek Commentaries on Hippocrates in the Arabic Tradition* (Berlin, 2012).

[36] See G. Baader, "Die Tradition des Corpus Hippocraticum im europäischen Mittelalter," in Baader, pp. 409–19; and see also Kibre.

of professional continuity from classical times. Similarly, the recital of the Hippocratic *Oath* at medical ceremonies today serves mainly to emphasize the timeless ethical framework within which our medical art strives still to practice. In fact, it is predominantly in the popular mind and in "alternative" medical discourse that specific Hippocratic ideas such as "the healing power of nature," the special significance of the environment for health, and the importance of the biological "idiosyncrasy" of the individual retain their influence in the twenty-first century.

4. TRANSMISSION OF THE COLLECTION

Today, most Hippocratic texts are preserved in approximately thirty Greek manuscripts, which date from the early tenth century to the middle of the sixteenth century. About twenty of these have been demonstrated by scholars to be derived solely from other manuscripts of the group, and can thus be eliminated as possible sources of independent authentic readings.

Five of the remaining dozen are proven by their age, contents, or texts to be independent from the rest, and therefore the primary witnesses to the texts they each contain:[37]

[37] For comprehensive overviews of these manuscripts and their relationships, see J. Irigoin, "Tradition manuscrite et histoire du texte," in Bourgey, pp. 3–18; J. Irigoin, "1987/8. Hippocrate et la Collection hippocratique," in idem. *Tradition et critique des textes grecs* (Paris, 1997), pp. 193–210.

B   Laurentianus Graecus 74, 7. Vellum (370 x 270
    mm). early X c. 406 folios[38]

The verses on folios 8f. (cf. Cocchi, pp. 34–42) celebrate
an unidentifiable Byzantine physician named Nicetas for

---

[38] These 406 folios contain two slightly different numbering
systems in their lower-outside corner: one, written by hand, from
folio 1 to folio 405, which skips over the leaf between its numbers
83 and 84; the other, stamped, from folio 2 to folio 407: I follow
the latter system. On the manuscript in general, see Schöne
Apoll., pp. v–xxiv.

[39] See Cocchi, pp. 2–33.

copying the texts from worn-out manuscripts with a view to restoring interest in ancient surgery.

B was purchased by Janus Lascaris in Crete in 1492, on behalf of Lorenzo de' Medici, and brought to Florence.

M Marcianus Venetus Graecus 269. Vellum (360 x 235 mm). X c. 463 folios (including a first leaf numbered independently, supernumerary leaves 57B and 293B, and missing a leaf numbered 345)[40]

[40] See Irigoin *Bess.*

[41] This index—two or three centuries later than the manuscript itself—has no intrinsic connection with the rest of the volume. It provides a relatively accurate record of the manuscript's contents. See Irigoin *Bess.*, pp. 164–67.

Between folios 408^v and 409^r the texts *Excision of the Fetus* (repeated: see fol. 297^r–98^r above), *Prorrhetic I* and *II*, *Fistulas*, *Hemorrhoids*, *Coan Prenotions*, and *Epidemics I–V* (part) are lost.

M was donated to St. Mark's Library in Venice by Cardinal Bessarion on his death in 1472, not being a part of his original legacy of Greek manuscripts to the library recorded in an inventory of 1468.

Θ   Vindobonensis Medicus Graecus 4. Vellum (340 x 280 mm). X or XI c. 418 folios

Θ was among 240 manuscripts, Greek and oriental, acquired by Augier Ghislain de Busbecq, while he was ambassador in Istanbul between 1558 and 1562, for the Imperial Library in Vienna.

A   Parisinus Graecus 2253. Vellum (235 x 180 mm).
     XI c. 192 folios

1^r–33^v	*Coan Prenotions*
34^r–36^v	*Nutriment*
37^r–64^v	*Regimen in Acute Diseases*
65^r–70^v	*Humors*
70^v–74^v	*Use of Liquids*
74^v–75^r	*Speech from the Altar*
75^r–81^r	*The Art*
81^r–93^v	*Nature of Man*
93^v–100^r	*Breaths*
100^r–117^v	*Places in Man*
117^v–30^r	*Ancient Medicine*
130^v–44^v	*Epidemics I*
145^r–92^v	Galen, *On the Usefulness of the Parts*, Books 10 and 14

The manuscript A was in the collection of Cardinal Domenico Grimani in the convent S. Antonio di Castello in Venice at the time of his death in 1523 and served as printer's copy for two treatises in the Aldine *Hippocrates*, which appeared in 1526; it entered the French Royal Library around the middle of that century.[42]

[42] See V. Nutton, *John Caius and the Manuscripts of Galen* (Cambridge, 1987), p. 52.

V   Vaticanus Graecus 276. Vellum (395 x 277 mm). XI
     or XII c. 207 folios[43]

[43] See Irigoin *Vatic.*

[44] This index, which is part of the original manuscript, begins
with a numbered list of the first twenty-three treatises in it, while
omitting any reference to the text of *Regimen in Acute Diseases*,
which appears between treatises 4 (*Prognostic*) and 5 (*In the
Surgery*). In the manuscript, treatise 23 is followed by a second
version of *Superfetation*, which is not mentioned in the index, and
then the texts numbered 56, 49, 46, 48, 42, and 59–62 in the index
in this sequence. The texts numbered 24–41, 43–45, 47, 50–
55, and 57–58 in the index are absent from the manuscript; of
these most are extant in other Hippocratic manuscripts, but *Sevens* (περὶ ἑβδομάδων), *Deadly Wounds* (περὶ τρωμάτων ὀλε-
θρίων), *Extraction of Projectiles* (περὶ βελῶν ἐξαιρέσιος), *Hellebore* (περὶ ἑλλεβόρου), *Enemas* (περὶ κλυσμῶν), and *Coition*
(περὶ ἀφροδισίων) are today lost in Greek. Cf. Irigoin *Vatic.*,
pp. 272–76.

V seems to be written by two (or possibly three) different hands—Va and Vb—whose succession occurs within the treatise *Diseases of Women I* around folio 149 between quires 19 and 20. This change of scribes may coincide with a change of textual sources, since the two versions of *Superfetation* in Va and Vb derive from different manuscript traditions.

A Latin translation of *Nature of the Child* was made from this manuscript by Bartholomaeus of Messina in Palermo for Manfred, king of Sicily (1258–1266).

A remaining handful of Greek manuscripts provide in

specific instances authentic textual evidence not contained in the five manuscripts just described.

C' Parisinus Graecus Supplement 446 X c.
    45ᵛ–59ᵛ   *Aphorisms*[45]
    60ʳ–69ᵛ   *Prognostic*[46]

Urb Urbinas Graecus 64 XII c.
    102ʳ–2ᵛˢ   *Breaths* 1[47]
    102ᵛ   *Medications*[48]
    103ʳ–3ᵛ   *The Art* 5[49]

P   Parisinus Graecus 2047 A XIII c.
    85ʳ–85ᵛ, 120ᵛ *Airs Waters Places* 24[50]

Hb Parisinus Graecus 2142 (newer part) XIV c.
    547ᵛ   *Sevens* 1–5[51]

Ambᵃ Ambrosianus Graecus 134 (B 113 sup.) XIII/XIV c.
    1ʳ–2ᵛ   *Oath, Law*[52]

Vind Vindobonensis Medicus Graecus XXXVII XIV c.
    66ʳ–67ʳ   *Letters 3–5, Oath*[53]

---

[45] Cf. Magdelaine, vol. 1, pp. 91–93.
[46] Cf. Alexanderson, pp. 84f. and 92f.
[47] Cf. Heiberg, pp. vi f.
[48] Cf. Schöne.
[49] Cf. Heiberg, pp. vi f.
[50] Cf. Jouanna *Airs*, pp. 108f.
[51] See note 5 above; Roscher, pp. 2–10.
[52] Cf. Jouanna *Serm.*, pp. lxxxii–lxxxvii, 205–8.
[53] Cf. Jouanna *Serm.*, pp. lii f.

Furthermore, Parisinus Graecus 2140 (= I) XIV c.[54] supplies on folios 323ᵛ–83ʳ copies of the texts lost from M between folios 408ᵛ and 409ʳ, while Hb provides on folios 540ᵛ–43ᵛ a piece of text from *Embassy* (ending in ch. 9 at μεγάλοι προεδεήθησαν [καὶ οἱ καρτεροὶ]) lost from M after folio 461ᵛ.

On August 14, 1515, Marcus Fabius Calvus of Ravenna completed a Latin translation of all the Hippocratic texts available to him in the original Greek. This translation was based, for the M manuscript tradition, on a manuscript (Vaticanus Graecus 278 = W) that Calvus had transcribed for himself—dating the completion of the task to July 24, 1512—out of a number of manuscripts, including apparently Vaticanus Graecus 277 = R, which he at one time owned, and for the V manuscript tradition, on V itself. This translation, which was published in Rome in 1525 under the patronage of Pope Clement VII, lacks the second part of *Coan Prenotions* and all of *Use of Liquids*, which are transmitted only in the manuscript A. Calvus' acquaintance with V's index, which he transcribed at the head of W after R's index, may explain his introduction in the printed book of the titles of seven treatises listed in the V index but no longer available in Greek.[55]

Franciscus Asulanus' publication in 1526 of the *Opera*

---

[54] See Mondrain for a detailed investigation of fourteenth century Hippocratic manuscripts, and in particular Parisinus Graecus 2140.

[55] See *Hippocratis . . . opera . . . per M. Fabium . . . Latinitate donata* (Basel, 1526): e.g., 323 (*Sevens*); 328 (*Potions*); 329 (*Hellebore*); 448 (*Dangerous Wounds*); 465 (*Extraction of Projectiles*).

omnia Hippocratis in Greek by the Aldine press, which his father Andreas Asulanus had managed since Aldus' death in 1515, made the collection widely available in a single volume. The editor's preparatory notes in the manuscripts from which the various texts were printed give valuable insight into his methods and choices: these printer's copies are, with a few exceptions, Parisinus Graecus 2141 (= G) for the M tradition, Holkhamensis Graecus 92 (= Holk) for the V tradition, and A for the two treatises extant only there. From various individual textual readings adopted by Asulanus, it is clear that he had direct access to and exploited not only A but three more of the five primary witnesses presented above—B, M, and V; only the Viennese manuscript Θ, which at the time had not yet arrived in Western Europe, remained untapped, being used for the first time in the edition of Stephan Mack, *Hippocratis opera omnia cum variis lectionibus* (Vienna, 1743–1749).

E. Littré, *Oeuvres complètes d'Hippocrate* (Paris 1839–1861), although based directly on only one of the five primary Greek manuscripts, A, and betraying no evidence that the editor had any awareness of the stemmatic theory of textual criticism being developed over the period of the edition's appearance, represents the first scholarly edition in the modern sense, and set the pattern for all future research. H. Kühlewein's Teubner edition (1894–1902) marks the beginning of serious study of the relationships among the individual Greek manuscripts, while W. H. S. Jones' Loeb edition (1923–1931) evolves in its methodology, moving from simply taking over Kühlewein's and Littré's texts in the first volume—except for the *Oath*—to conducting independent manuscript investigations in volumes 2 and 4;[56] for volume 3, E. T. Withington based his text on the well-documented edition of J. E. Petrequin in his *Chirurgie d' Hippocrate* (Paris, 1877–1878).

In the intervening century, a great deal of work by many scholars in many countries has resulted in a more-informed historical understanding of the transmission of the texts, including those in papyri,[57] in medieval translations, and in the Greek manuscripts, and a surer and more comprehensive awareness of the collection's contents, significance, and development. The editing and translating of seven further Loeb *Hippocrates* volumes (5–11) between 1988 and 2018 benefited significantly from this progress, which suggested that revising the first four volumes in the same light could furnish for the first time an up-to-date, complete edition and English translation of the corpus, in line with the Loeb Classical Library's mandate.

[56] See vol. 2, pp. xlviii–lxvi.

[57] See Marganne; Marganne/Mertens; Andorlini; *Index Hipp.*, pp. xi–xii; *Index Hipp. Supp.*, pp. viii–ix.

GENERAL INTRODUCTION

W. H. S. Jones' first Loeb *Hippocrates* volume (1923), which the present volume replaces, was derived almost entirely from text editions dating from the nineteenth century: *Ancient Medicine*, *Airs Waters Places*, and *Epidemics I and III* are based on volume 1 of H. Kühlewein's *Hippocratis Opera* (Leipzig, 1894), while *Precepts* and *Nutriment* go back to Littré (1861) and Ermerins (1864). Only for the *Oath* did Jones consult Greek manuscripts, a study which he presented in fuller form in his monograph *The Doctor's Oath* (Cambridge, 1924). In the intervening century, all these texts have been the object of much further study, including the appearance of new critical editions and commentaries.

The works in volume 1 vary greatly in their form and content. *Ancient Medicine* and *Airs Waters Places* are essays formulated in elegant expository prose, in which their authors treat central questions of medical method, the former investigating the possible sources of medical knowledge and its relationship to contemporary scientific thought, the latter expounding in detail the importance of geographical factors on human health and disease. *Epidemics I and III* are primary witnesses to Hippocratic observation and experience, giving every appearance of being—even if with some degree of literary polishing—actual studies of disease occurrence in specific populations, both collectively (annual public health reports) and individually (forty-two patient histories). These texts are expressed in concise medical terminology, apply a rigid chronology as their framework, and hold their narrator's role to a minimum. The *Epidemics* have exerted a profound influence on medical narrative down to the present day. The Hippocratic *Oath* is, or pretends to be, a historical document, although its lack of historical context often leaves its interpreter at a loss. Unquestionably, however, its identification of many of the perennial ethical problems attached to medical practice and its high moral ideal justify the reverence it has received for two millennia. *Precepts* is interesting for its author's thoughts on medical ethics, etiquette, and professionalism, in spite of the fact that the text often lacks logical coherence, and its language on several occasions defies comprehension. *Nutriment*, which is also often difficult to understand—although in this case the mystery may be partly the result of an intentional imitation of the language of the early Greek philosopher Heraclitus of Ephesus—contains many speculative explanations of how the three basic forms of nutriment (food, drink, and breath) interact with the body.

MANUSCRIPT TRADITION

M	Marcianus Venetus Graecus 269 (X c.)
A	Parisinus Graecus 2253 (XI c.)
V	Vaticanus Graecus 276 (XI/XII c.)

I	Parisinus Graecus 2140 (XIII c.)
H	Parisinus Graecus 2142
Ha	older part[58] (XII/XIII c.)
Hb	newer part (XIV c.)
R	Vaticanus Graecus 277 (XIV c.)
Recentiores	approximately twenty manuscripts (XV/ XVI c.)

The *stemma codicum* appearing as Figure 1 provides an overview of the interdependencies among the manuscripts containing the seven treatises in this volume. The particular treatises are transmitted in the following independent witnesses:

Ancient Medicine	M A
Airs Waters Places	V[59]
Epidemics I	A V I
Epidemics III	V I
Oath	M V[60]
Precepts	M
Nutriment	M A

In *Epidemics I* and *III* the text of M is lost and must be reconstructed from M's descendants. In both cases, the text is contained in the newer part (Hb) of the manuscript H (fols. 466^r–74^v and 482^r–90^v) dependent on I, and thus has no independent textual authority.

[58] Folios 46, 49, 55–78, and 80–308.
[59] For a more detailed account of this complex textual transmission, see below, pp. 64f.
[60] For additional witnesses, see the account below, pp. 290f.

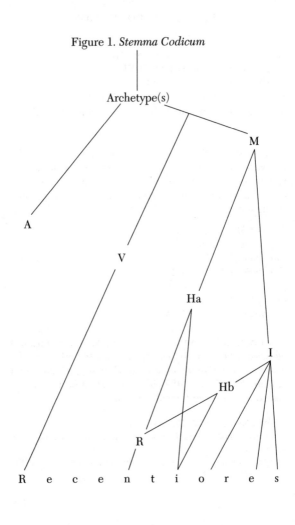

Figure 1. *Stemma Codicum*

NOTE ON TECHNICAL TERMS[61]

The following expressions require clarification as they are employed in this volume with particular significance.

ἀριστᾶν/lunch: this early meal was apparently not universally taken; cf., however, *OCD* s.v. "meals": "In Classical Athens two meals—a light lunch (ἄριστον) and dinner (δεῖπνον) in the evening—appear to have been the usual."

ἀτέραμνος, ἀτεραμνίη/refractory, refractoriness: these words refer several times in *Airs Waters Places* to a characteristic of hard, salty waters that combines a physical property—being difficult to bring to a boil—and a biological effect on the intestines, uterus, and breast—drying. The author associates the two features as follows (ch. 7): "Waters that tend most to boil and melt are likely to relax the cavity and melt its contents away, whereas those that are refractory, hard and least suitable for boiling generally contract the cavities and dry them."

δύναμις/potency: often the strength or potency inherent in a nutriment, which through the process of digestion is taken over by the body. In a more technical sense either a specific quality of a foodstuff—e.g., astringent, insipid, irritating, bitter, acid, salty, sweet—or a more elemental property—e.g., hot, cold, moist, dry.

ἐπικρατεῖν/assimilate: literally, "to rule over, command, or master," as applied in this volume to the process of diges-

---

[61] For the explanation of other technical terms occurring in the present volume, see also Loeb *Hippocrates* vol. 8, p. 10f.; vol. 10, pp. x–xiii; and vol. 11, p. xvi f.

tion, which is imagined as a contest between the intestinal tract (κοιλίη) and the ingested nutriment (τροφή), the verb signifying the victory of the former in assimilating the latter, which provides the body with strength and growth.

ἱερὴ νοῦσος/sacred disease: convulsive disorders, including epilepsy, have several designations in the collection. Among these the most common is "sacred disease," but others include "epilepsy" (e.g., *Nutriment* 26), "childhood condition" (e.g., *Airs Waters Places* 3), and "Heracles' disease" (e.g., *Diseases of Women I* 7).

μετεωρολόγα, ἀστρονομίη/meteorology, astronomy: the two Greek terms, which seem in *Airs Waters Places* 2 to be used synonymously, refer to both the celestial bodies and the terrestrial atmosphere, whose direct interrelationship is assumed: cf., ibid., chapters 5 and 8. Besides the solstices and equinoxes, three sidereal events are used in *Airs Waters Places* and *Epidemics I/III* to delineate points in the year: the heliacal rising of Sirius (July 19) and Arcturus (September 17), and the cosmic setting of the Pleiades (November 6). How directly these references are related to actual observations, and whether they are understood as specific dates or periods of time related to those dates, is unclear.[62]

πέσσειν, πέψις/to concoct, (con)coction: these terms are applied to both physiological and pathological processes within the body. In the healthy state, coction is the application of heat by the intestinal tract (κοιλίη) to the in-

---

[62] Cf. Wenskus, pp. 93–102 and 111–14.

gesta in order to render them milder and more assimilable. In a disease, coction is the process by which the body applies heat to some collected "disease substance" in order to make it harmless, and to prepare it for expulsion from the body through abscession, apostasis.

REFERENCES

EDITIONS, TRANSLATIONS, AND COMMENTARIES

Hippocratic Collection

Early Works

Calvus	*Hippocratis Coi . . . octoginta volumina . . . per M. Fabium Calvum, Rhavennatem . . . latinitate donata . . .* Rome, 1525.
Aldina	*Omnia opera Hippocratis . . . in aedibus Aldi & Andreae Asulani soceri.* Venice, 1526.
Froben	*Hippocratis Coi . . . libri omnes . . . [per Ianum Cornarium].* Basel, 1538.
Cornarius	*Hippocratis Coi . . . opera . . . per Ianum Cornarium . . . Latina lingua conscripta.* Venice, 1546.
Cornarius in marg.	Marginal notes by Janus Cornarius in a copy of the Aldine edition. Presently kept in the Göttingen University Library.
Zwinger	*Hippocratis Coi . . . viginti duo commentarii . . . Theod. Zvingeri studio & conatu.* Basel, 1579.
Foes	*Magni Hippocratis . . . opera quae extant . . . Latina interpretatione & annotationi-*

bus illustrata, Anutio Foesio . . . authore.
Geneva, 1657–1662.

Linden   *Magni Hippocratis Coi opera omnia graece
& latine edita . . . industria & diligentia
Joan. A. Vander Linden.* Leiden, 1665.

Mack   *Hippocratis opera omnia . . . studio et opera Stephani Mackii.* Vienna, 1743–1749.

Cocchi   *Graecorum chirurgici libri . . . e collec-
tione Nicetae . . . descripti conversi atque
editi ab Antonio Cocchio.* Florence, 1754.

Grimm   Grimm, J. F. K. *Hippokrates Werke aus
dem griechischen. . . .* Altenburg, 1781–
1792.

Coray   Coray, A. *Traité d'Hippocrate. Des airs, des
eaux et des lieux.* Paris, 1800.

Post-Eighteenth Century

Adams   Adams, F. *The Genuine Works of Hippocrates.* London, 1849.

Alexanderson   Alexanderson, B. *Die hippokratische Schrift
Prognostikon.* Gothenburg, 1963.

Bourbon   Bourbon, F. *Hippocrate. Nature de la
femme.* Budé XII (1). Paris, 2008.

Chadwick   Chadwick, J., and W. N. Mann. *The Medical Works of Hippocrates.* Oxford, 1950.

Daremberg   Daremberg, Ch. *Oeuvres choisies d' Hippocrate.*[2] Paris, 1855.

Deichgräber   Deichgräber, K. *Pseudhippokrates Über
die Nahrung.* Wiesbaden, 1973.

Diller   Diller, H. *Hippokrates. Über die Umwelt.*
CMG I 1,2. Berlin, 1970.

Diller *Schr.* ————. *Hippokrates. Schriften . . . über-setzt.* Hamburg, 1962.

Ecca Ecca, G. *Die hippokratische Schrift Prae-cepta.* Wiesbaden, 2016.

Ermerins Ermerins, F. Z. *Hippocratis . . . reliquiae.* Utrecht, 1859–1864.

Festugière Festugière, A.-J. *Hippocrate. L'ancienne médecine.* Paris, 1948.

Fuchs Fuchs, R. *Hippokrates, sämmtliche Werke. Ins Deutsche übersetzt. . . .* Munich, 1895–1900.

Gardeil Gardeil, J. B. *Traduction des Œuvres mé-dicales d' Hippocrate, sur le texte grec, d' après l' édition de Foës.* Toulouse, 1801.

Gomperz Gomperz, Th. *Die Apologie der Heil-kunst.*[2] Leipzig, 1910.

Heiberg Heiberg, J. *Hippocratis Opera.* CMG I 1. Leipzig and Berlin, 1927.

Joly *Aliment* Joly, R. *Hippocrate . . . De l'Aliment.* Budé VI (2). Paris, 1972.

Jones *Loeb* Jones, W. H. S. *Hippocrates with an Eng-lish Translation.* London and New York, 1923–1931.

Jones *Oath* ————. *The Doctor's Oath.* Cambridge, 1924.

Jones *Phil.* ————. *Philosophy and Medicine in An-cient Greece.* Baltimore, 1946.

Jouanna Jouanna, J. *Hippocrates.* Translated by M. B. DeBevoise. Baltimore, 1999.

Jouanna *Airs* ————. *Hippocrate. Airs, Eaux, Lieux.* Budé II (2). Paris, 1996.

Jouanna *Anc.*	———. *Hippocrate. L'ancienne médecine.* Budé II (1). Paris, 1990.
Jouanna *Art*	———. *Hippocrate. Des Vents. De l'Art.* Budé V (1). Paris, 1988.
Jouanna *Epid.*	———. *Hippocrate. Épidémies I et III.* Budé IV (1). Paris, 2016.
Jouanna *Serm.*	———. *Hippocrate. Le Serment . . . La Loi.* Budé I (2). Paris, 2018.
Kapferer	Kapferer, R., and G. Sticker. *Die Werke des Hippokrates.* Stuttgart, 1933–1940.
Kühlewein	Kühlewein, H. *Hippocratis Opera.* Leipzig, 1894–1902.
Langholf	Langholf, V. *Syntaktische Untersuchungen zu Hippokrates-Texten.* Mainz, 1977.
Littré	Littré, E. *Oeuvres complètes d'Hippocrate.* Paris, 1839–1861.
Magdelaine	Magdelaine, C. *Histoire du texte et édition critique, traduite et commentée, des Aphorismes d'Hippocrate.* Diss., Paris-Sorbonne, 1994.
Moisan	Moisan, M. *Edition critique et commentée, avec traduction française, de deux oeuvres de la collection hippocratique: Préceptes et Médecin.* Diss., Paris-Sorbonne, 1993.
Petrequin	Petrequin, J. E. *Chirurgie d'Hippocrate.* Paris, 1877–1878.
Roscher	Roscher, W. H. *Die hippokratische Schrift von der Siebenzahl.* Paderborn, 1913.
Schiefsky	Schiefsky, M. *Hippocrates on Ancient Medicine.* Leiden, 2005.

REFERENCES

Schöne Schöne, H. "Hippokrates ΠΕΡΙ ΦΑΡΜΑ-ΚΩΝ." *Rheinisches Museum* N.F. 73 (1924): 434–48.

Schwyzer Deichgräber, K., and E. Schwyzer. *Hippokrates über Entstehung und Aufbau des menschlichen Körpers (ΠΕΡΙ ΣΑΡΚΩΝ).* Leipzig and Berlin, 1935.

Smith *Pseud.* Smith, W. D. *Hippocrates. Pseudepigraphic Writings: Letters – Embassy – Speech from the Altar – Decree.* Leiden, 1990.

Withington Withington, E. T. *Hippocrates with an English Translation.* Vol. 3. London and New York, 1928.

Other Authors

Aëtius Olivieri, A. *Aetii Amideni Libri medicinales I–VIII.* CMG VIII 1–2. Leipzig, 1935–1950.

Apollon. Cit. Kollesch, J., and F. Kudlien. *Apollonios von Kition. Kommentar zu Hippokrates über das Einrenken der Gelenke.* CMG XI 1,1. Berlin, 1965.

Aristotle Louis, P. *Aristote. Histoire des animaux.* Budé I. Paris, 1964.

Boudon Boudon-Millot, V. *Galien . . . Sur ses propres livres.* Budé I. Paris, 2007.

Dietz Dietz, F. R. *Scholia in Hippocratem et Galenum.* Königsberg, 1824.

Dioscorides	Wellmann, M. *Pedanii Dioscuridis . . . De materia medica*. Berlin, 1906–1914.
DK	Diels, H., and W. Kranz. *Die Fragmente der Vorsokratiker.*[6] Berlin, 1951.
Duffy *Epid.*	Duffy, J. M. et al. *John of Alexandria. Commentary on Hippocrates' Epidemics VI. Fragments . . . Commentary on Hippocrates' On the Nature of the Child.* CMG XI 1,4. Berlin, 1997.
Duffy *Prog.*	———. *Stephanus the Philosopher. A Commentary on the Prognosticon of Hippocrates.* CMG XI 1,2. Berlin, 1983.
Erotian	Nachmanson, E. *Erotiani Vocum Hippocraticarum collectio.* Gothenburg, 1918.
Galen	Kühn, C. G. *Claudii Galeni opera omnia. . . .* Leipzig, 1825–1833.
Heracl. Tar.	Deichgräber, K. *Die griechische Empirikerschule.*[2] Berlin, 1965.
Herophilus	von Staden, H. *Herophilus. The Art of Medicine in Early Alexandria.* Cambridge, 1989.
Hesychius	Latte, K. et al. *Hesychii Alexandrini Lexicon.* Copenhagen, 1953–1966, and Berlin, 2005–2009.
Irmer	Irmer, D. *Palladius. Kommentar zu Hippokrates "De fracturis."* Hamburg, 1977.
Oribasius	Raeder, I. *Oribasii Collectionum Medicarum Reliquiae.* CMG VI 1,1–2,2. Leipzig and Berlin, 1928–1933.
Perilli	Perilli, L. *Galeni Vocum Hippocratis Glossarium.* CMG V 13,1. Berlin, 2017.

REFERENCES

Pritchet   Pritchet, C. D. *Iohannis Alexandrini Commentaria in sextum librum Hippocratis Epidemiarum*. Leiden, 1975.

Schöne *Apoll.*   Schöne, H. *Apollonius von Kitium* . . . ΠΕΡΙ ΑΡΘΡΩΝ. Leipzig, 1896.

Scribonius   Sconocchia, S. *Scribonius Largus. Compositiones*. Leipzig, 1983.

Soranus   Ilberg, I. *Soranus*. CMG IV. Leipzig, 1927.

     Temkin, O. *Soranus' Gynecology*. Baltimore, 1956.

Vagelpohl (I)   Vagelpohl, U. *Galen. Commentary on Hippocrates' Epidemics Book I, Parts I–III*. CMG Suppl. Orient. V 1. Berlin, 2014.

Vagelpohl (II)   Vagelpohl, U., and S. Swain. *Galen. Commentary on Hippocrates' Epidemics Book II, Parts I–VI*. CMG Suppl. Orient. V 2. Berlin, 2016.

Wenkebach (I)   Wenkebach, E. *Galeni In Hippocratis Epidemiarum libr. I comm. III*. Pfaff, F. *Galeni In Hippocratis Epidemiarum libr. II comm. V in Germ. linguam trans*. CMG V 10,1. Leipzig and Berlin, 1934.

Wenkebach (II)   ———. *Galeni In Hippocratis Epidemiarum libr. III comm. III*. CMG V 10, 2.1. Leipzig and Berlin, 1936.

Westerink   Westerink, L. G. *Stephanus of Athens. Commentary on Hippocrates' Aphorisms*. CMG XI 1,3,1–3. Berlin, 1985–1995.

REFERENCES

GENERAL WORKS

Andorlini   Andorlini, I., ed. *Greek Medical Papyri I/ II*. Florence, 2001/2009.

Baader   Baader, G., and R. Winau, eds. *Die hippokratischen Epidemien. Verhandlungen des Ve Colloque International Hippocratique (Berlin 1984)*. Stuttgart, 1989.

Bourgey   Bourgey, L., and J. Jouanna, eds. *La Collection Hippocratique et son Rôle dans l'Histoire de la Médecine. Colloque de Strasbourg (1972)*. Leiden, 1975.

Budé   Collection des universités de France publiée sous le patronage de l'Association Guillaume Budé.

CMG   Corpus Medicorum Graecorum

Eijk   van der Eijk, Ph., ed. *Hippocrates in Context. Papers Read at the XIth International Hippocratic Colloquium (Newcastle-upon-Tyne 2002)*. Leiden, 2005.

Fleischer   Fleischer, U. *Untersuchungen zu den pseudohippokratischen Schriften Παραγγελίαι, Περὶ ἰητροῦ, und Περὶ εὐσχημοσύνης*. Berlin, 1939.

Guthrie   Guthrie, W. K. C. *A History of Greek Philosophy*. Cambridge, 1962–1981.

Heringa   Heringa, A. *Observationum criticarum liber*. Leeuwarden, 1749.

Herzog   Herzog, R., ed. *Kos. Band I: Asklepieion. Baubeschreibung und Baugeschichte* (P. Schazmann). Berlin, 1932.

REFERENCES

Index Hipp. Kühn, J.-H., U. Fleischer et al. *Index Hippocraticus*. Göttingen, 1986–1989; Anastassiou, A., and D. Irmer. *Supplement*. 1999.

Irigoin *Bess.* Irigoin, J. "L' Hippocrate du Cardinal Bessarion." In *Miscellanea Marciana di Studi Bessarionei* (Medioevo e Umanesimo 24), pp. 161–74. Padua, 1976.

Irigoin *Vatic.* ———. "Le manuscrit V d' Hippocrate (Vaticanus Graecus 276). Étude codicologique et philologique." In *I testi medici greci: tradizione e ecdotica*, edited by A. Garzya and J. Jouanna, pp. 269–83. Naples, 1999.

Joly Joly, R., ed. *Corpus Hippocraticum. Actes du Colloque Hippocratique de Mons (1975)*. Mons, 1977.

Jouanna *Hippocrates*. Translated by M. B. Bevoise. Baltimore, 1999.

Jouanna/Zink Jouanna, J., and M. Zink, eds. *Hippocrate et les hippocratismes: médecine, religion, société. XIVᵉ Colloque International Hippocratique*. Paris, 2014.

Kibre Kibre, P. *Hippocrates Latinus*. Rev. ed. New York, 1985.

Lloyd Lloyd, G. E. R. *Methods and Problems in Greek Science*. Cambridge, 1991.

López-Férez López-Férez, J. A., ed. *Galeno: Obra, Pensamiento e Influencia*. Madrid, 1991.
Galeno

López-Férez López-Férez, J. A., ed. *Tratados Hipocráticos. Actas del VIIᵉ Colloque International Hippocratique (1990)*. Madrid, 1992.
Trat.Hip.

REFERENCES

LSJ	Liddell, H. G., R. Scott, and H. S. Jones, eds. *A Greek-English Lexicon.*[9] Oxford, 1940.
Manetti/Ros.	Manetti, D., and A. Roselli. "Galeno commentatore di Ippocrate." In *Aufstieg und Untergang der römischen Welt (ANRW)* II 37,2, edited by W. Haase and H. Temporini, pp. 1529–635. Berlin, 1994.
Marganne	Marganne, M.-H. *Inventaire analytique des papyrus grecs de médecine.* Geneva, 1981.
Marganne/ Mertens	Marganne, M.-H., and P. Mertens. "Medici et Medica, 2e edition." In "Specimina" *per il* Corpus *dei Papiri Greci di Medicina*, edited by I. Andorlini, pp. 3–71. Florence, 1997.
Mondrain	Mondrain, B. "Lire et copier Hippocrate . . . au XIVe siècle." In Boudon et al., *Ecdotica e ricezione dei testi medici greci*, pp. 359–410. Naples, 2006.
Müller	Müller, C. W. et al., eds. *Ärzte und ihre Interpreten.* München, 2006.
Nachmanson	Nachmanson, E. *Erotianstudien.* Uppsala, 1917.
OCD	Hornblower, S., and A. Spawforth, eds. *The Oxford Classical Dictionary.*[3] Oxford, 1996.
OED	Murray, J. A. H. et al., eds. *The Oxford English Dictionary.* Oxford, 1933; *Supplement*, 1972–1986.
Pinault	Pinault, J. R. *Hippocratic Lives and Legends.* Leiden, 1992.

REFERENCES

Scarborough   Scarborough, J., ed. *Symposium on Byzantine Medicine*. Washington, DC, 1985.

Sezgin   Sezgin, F. *Geschichte des arabischen Schriftums. Band III. Medizin – Pharmazie – Zoologie – Tierheilkunde bis ca 430 H.* Leiden, 1970.

Smith   Smith, W. D. *The Hippocratic Tradition*. Ithaca, NY, 1979.

Testimonien   Anastassiou, A., and D. Irmer. *Testimonien zum Corpus Hippocraticum*. Göttingen, (vol. I) 2006; (vol. II,1) 1997; (vol. II,2) 2001; (vol. III) 2012.

Ullmann   Ullmann, M. *Die Medizin im Islam*. Leiden, 1970.

Wenskus   Wenskus, O. *Astronomische Zeitangaben von Homer bis Theophrast*. Stuttgart, 1990.

GENERAL BIBLIOGRAPHY

EDITIONS, TRANSLATIONS AND COMMENTARIES

Bourbon, F. *Hippocrate. Femmes stériles, Maladies des jeunes filles, Superfétation, Excision du fœtus.* Budé XII (4). Paris, 2017.

Craik, E. M. *Hippocrates.* Places in Man. Oxford, 1998.

———. *Two Hippocratic Treatises:* On Sight *and* On Anatomy. Leiden, 2006.

———. *The Hippocratic Treatise* On Glands. Leiden, 2009.

García Gual, C. et al. *Tratados hipocráticos . . . introducciones, traducciones y notas.* Madrid, 1983–2003.

Giorgianni, F. *Hippokrates. Über die Natur des Kindes* (De genitura *und* De natura pueri). Wiesbaden, 2006.

Hanson, M. *Hippocrates. On Head Wounds.* CMG I 4,1. Berlin, 1999.

Joly, R. *Hippocrate. De la génération, De la nature de l'enfant, Des maladies IV . . .* Budé XI. Paris, 1970.

Jouanna, J. *Hippocrate. Maladies II.* Budé X (2). Paris, 1983.

———. *Hippocrate. De la maladie sacrée.* Budé II (3). Paris, 2003.

———. *Hippocrate. Pronostic.* Budé III (1). Paris, 2013.

Jouanna, J., and M. D. Grmek. *Hippocrate. Epidémies V et VII.* Budé IV (3). Paris, 2000.

Jouanna, J., and C. Magdelaine. *Hippocrate. L' Art de la médecine.* Paris, 1999.

Lienau, C. *Hippokrates. Über Nachempfängnis, Geburtshilfe und Schwangerschaftsleiden.* CMG I 2,2. Berlin, 1973.

Lloyd, G. E. R., ed. *Hippocratic Writings.* Harmondsworth, 1978.

Overwien, O. *Hippokrates. Über die Säfte.* CMG I 3,1. Berlin, 2014.

Roselli, A. *Ippocrate. La malattia sacra.* Venice, 1996.

Sakalis, D. T. Ἱπποκράτους Ἐπιστολαί. Ἔκδοση κριτική καὶ ἑρμηνευτική. Ioannina, 1989.

GENERAL WORKS

Bliquez, L. *The Tools of Asclepius.* Leiden, 2014.

Bruno Celli, B. *Bibliografía Hipocrática.* Caracas, 1984.

Byl, S. "Les dix dernières années (1982–1992) de la recherche hippocratique." *Université Jean Monnet-Saint Etienne. Centre Jean Palerne. Lettre d'Informations* 22 (1993): 1–39.

Craik, E. M. *The 'Hippocratic' Corpus: Content and Context.* London, 2015.

Dean-Jones, L. A., and R. M. Rosen, eds. *Ancient Concepts of the Hippocratic. Papers Presented at the XIIIth International Hippocratic Colloquium, Austin, Texas 2008.* Leiden, 2015.

Eijk, Ph. J. van der. *Medicine and Philosophy in Classical Antiquity.* Cambridge, 2005.

Eijk, Ph. J. van der, H. F. Horstmanshoff, and P. H. Schrijvers, eds. *Ancient Medicine in Its Socio-cultural Context*. Amsterdam, 1995.

Fichtner, G. *Corpus Hippocraticum. Verzeichnis der hippokratischen und pseudohippokratischen Schriften*. Berlin, 2015.

Flashar, H. *Hippokrates. Meister der Heilkunst*. Munich, 2016.

Garofalo, I., A. Lami, D. Manetti, and A. Roselli, eds. *Aspetti della terapia nel Corpus Hippocraticum. Atti del IXe Colloque International Hippocratique (Pisa 1996)*. Florence, 1999.

Grmek, M. D., and F. Robert, eds. *Hippocratica. Actes du Colloque hippocratique de PARIS (1978)*. Paris, 1980.

Harris, W. V., ed. *Mental Disorders in the Classical World*. Leiden, 2013.

Heidel, W. A. *Hippocratic Medicine. Its Spirit and Method*. New York, 1941.

Horstmanshoff, H. F., ed. *Hippocrates and Medical Education: Selected Papers Read at the XII International Hippocrates Colloquium*. Leiden, 2010.

Jouanna, J. *Greek Medicine from Hippocrates to Galen*. Leiden, 2012.

Lasserre, F., and Ph. Mudry, eds. *Formes de pensée dans la Collection hippocratique. Colloque International Hippocratique (Lausanne 1981)*. Geneva, 1983.

Leven, K.-H., ed. *Antike Medizin. Ein Lexikon*. Munich, 2005.

Maloney, G. *Index Inverses du vocabulaire Hippocratique*. Québec, 1987.

Maloney, G., and W. Frohn. *Concordance des oeuvres Hippocratiques*. Québec, 1984 and Hildesheim, 1986.

Maloney, G., and R. Savoie. *Cinq cents ans de bibliographie Hippocratique*. Québec, 1982.

Manetti, D., L. Perilli, and A. Roselli, eds. *Ippocrate e gli altri*. Rome, 2021.

Nutton, V. *Ancient Medicine*. London, 2004.

Oser-Grote, C. M. *Aristoteles und das Corpus Hippocraticum. Die Anatomie und Physiologie des Menschen*. Stuttgart, 2004.

Perilli, L., Ch. Brockmann, K.-D. Fischer, and A. Roselli, eds. *Officina Hippocratica*. Berlin, 2011.

Pormann, P., ed. *Cambridge Companion to Hippocrates*. Cambridge, 2018.

Potter, P., G. Maloney, and J. Desautels, eds. *La Maladie et les maladies dans la Collection Hippocratique. Actes du VIᵉ Colloque International Hippocratique (1987)*. Québec, 1990.

Thivel, A., and A. Zucker, eds. *Le normal et le pathologique dans la Collection hippocratique. Actes du Xᵉᵐᵉ colloque international hippocratique (1999)*. Nice, 2001.

Wittern, R., and P. Pellegrin, eds. *Hippokratische Medizin und antike Philosophie. Verhandlungen des VIII. Internationalen Hippokrates-Kolloquiums (Kloster Banz/Staffelstein 1993)*. Hildesheim, 1996.

ANCIENT MEDICINE

INTRODUCTION

Ancient Medicine is recorded under this title by Erotian in his list of Hippocratic works, among those "pertaining to the discussion of the art" (τῶν εἰς τὸν περὶ τέχνης τεινόντων λόγον), and mentioned by name in three glosses; seven other of his glosses are also attributed by Nachmanson[1] to the treatise. Galen refers to the treatise by name in his *Commentary on Epidemics II* and includes a number of words in his glossary that may have been drawn from it.[2] In the fifth century, the lexicographer Hesychius glosses a half dozen words present in the treatise.[3] The three Arabic medical authors ʿAlī ibn Rabban aṭ Ṭabarī, Rhazes, and ibn al-Baiṭār all cite passages from the text of *Ancient Medicine* in their writings.[4]

The treatise is an important historical document for the light it sheds on the relationship between Greek medicine and Greek philosophy at some indeterminable time in the decades around 400 BC. The author defends a traditional, dietetic medicine based on experience against a new medicine based on hypothetical entities such as cold,

[1] Pp. 339f.
[2] See *Testimonien* vol. II,1, pp. 455–57.
[3] See *Testimonien* vol. III, p. 386f.
[4] See Ullmann, p. 31.

hot, dryness, and moisture. He denies that the postulated entities have any real existence or clear role to play in medical theory or practice, while defending traditional medicine as being grounded on the determinable "perception of the body" in the individual patient.

The polemical framework of *Ancient Medicine* is established in its first chapter, with the remainder of the chapters falling into three main sections:

A. The Origin and Method of Traditional Medicine

itself are caused by humors that are too strong and are only relieved when these strong humors are concocted and compounded so that they become weak.

C. The True Medical Knowledge Necessary for Practice

20–21 Knowledge of natural science in the broad sense does not pertain to medicine, which is based rather on the knowledge practitioners have amassed relating to the effects of foods, drinks, baths, etc. on individual human bodies.

22 Diseases arise either from the qualities of the humors or from the forms of the parts of the body: examples of these processes.

23–24 Details concerning other qualities and structures.

Ancient Medicine is transmitted in the independent Greek manuscripts M and A but has left no extant medieval translation in Arabic or Latin. It is printed in all the collected editions and translations, including Zwinger, Adams, Kühlewein, Heiberg, and Chadwick. The treatise received much scholarly attention in the twentieth century for the light it casts on the relationship between medicine and philosophy, e.g.:

Jones, W. H. S. *Philosophy and Medicine in Ancient Greece.* Baltimore, 1946.

Festugière, A.-J. *Hippocrate. L'ancienne médecine.* Paris, 1948.

Diller, H. "Hippokratische Medizin und attische Philosophie." *Hermes* 80 (1952): 385–409.

Kühn, J.-H. *System- und Methodenprobleme im Corpus Hippocraticum.* Wiesbaden, 1956.

Lloyd, G. E. R. "Who is Attacked in *On Ancient Medicine*?" *Phronesis* 8 (1963): 108–26; repr. with new introduction in Lloyd, pp. 49–69.

The two most important recent editions and commentaries are:

Jouanna, J. *Hippocrate. De l'ancienne médecine.* Budé II(1). Paris, 1990.

Schiefsky, M. *Hippocrates* On Ancient Medicine. Leiden, 2005.

The present edition takes account of the scholarship cited but depends in particular on Jouanna's edition. The manuscripts M and A have been consulted occasionally from microfilm.

ΠΕΡΙ ΑΡΧΑΙΗΣ ΙΗΤΡΙΚΗΣ

I 570
Littré

1. Ὁκόσοι μὲν ἐπεχείρησαν περὶ ἰητρικῆς λέγειν ἢ
γράφειν, ὑπόθεσιν αὐτοὶ ἑωυτοῖσιν ὑποθέμενοι τῷ
λόγῳ, θερμὸν ἢ ψυχρὸν ἢ ὑγρὸν ἢ ξηρὸν ἢ ἄλλο τι
ὃ ἂν θέλωσιν, ἐς βραχὺ ἄγοντες τὴν ἀρχὴν τῆς αἰτίης τοῖσιν ἀνθρώποισι τῶν νούσων τε καὶ τοῦ θανά-
του, καὶ πᾶσι τὴν αὐτήν, ἐν ἢ δύο ὑποθέμενοι,[1]
ἐν πολλοῖσι μὲν καὶ οἷσι λέγουσι καταφανέες εἰσὶν
ἁμαρτάνοντες, μάλιστα δὲ ἄξιον μέμψασθαι ὅτι ἀμφὶ
τέχνης ἐούσης, ᾗ χρέωνταί τε πάντες ἐπὶ τοῖσι μεγίστοισι καὶ τιμῶσι μάλιστα τοὺς ἀγαθοὺς χειροτέχνας
καὶ δημιουργούς. εἰσὶν δὲ δημιουργοὶ οἱ μὲν φλαῦροι,
οἱ δὲ πολλὸν διαφέροντες· ὅπερ, εἰ μὴ ἦν ἰητρικὴ
ὅλως μηδ᾽ ἐν αὐτῇ ἔσκεπτο μηδ᾽ εὕρητο μηδέν, οὐκ
ἂν ἦν, ἀλλὰ πάντες ἂν ὁμοίως αὐτῆς ἄπειροί τε καὶ
ἀνεπιστήμονες ἦσαν, τύχῃ δ᾽ ἂν πάντα τὰ τῶν
καμνόντων διοικεῖτο. νῦν δ᾽ οὐχ οὕτως ἔχει, ἀλλ᾽
ὥσπερ καὶ τῶν ἄλλων τεχνέων | πασέων οἱ δημιουρ-
γοὶ πολλὸν ἀλλήλων διαφέρουσι κατὰ χεῖρα καὶ κατὰ
γνώμην, οὕτω δὲ καὶ ἐπὶ ἰητρικῆς. διὸ οὐκ ἠξίουν

572

---

[1] ὑποθέμενοι A: προθέμενοι M

ANCIENT MEDICINE

1. All those who have attempted to speak or write about medicine by assuming some postulate for themselves in their account—heat, cold, moisture, dryness, or anything else they might wish—thus narrowing down the primary cause of diseases and death in human beings, and reducing it to the same in every case by postulating these one or two things, are clearly mistaken in many points of their arguments, but they are most deserving of censure concerning the actual existing profession of medicine,[1] to which everyone resorts in the most momentous matters, and whose good craftsmen and practitioners everyone rewards with the highest honors. In truth, some practitioners are mediocre while others greatly excel: but this would not be so, if there were no such thing as an art of medicine, and nothing had been investigated or discovered in it: in that case all people would be equally inexperienced and unlearned in it, and the treatment of the sick would be completely haphazard. This however is not the case, for just as in all the other arts practitioners vary greatly in their manual skill and in their understanding, so too in medicine. For this reason I deny that medicine needs some vain

[1] See the Hippocratic *Art* for an extended discussion of the medical art and its existence.

αὐτὴν ἔγωγε κενῆς[2] ὑποθέσιος δεῖσθαι ὥσπερ τὰ
ἀφανέα τε καὶ ἀπορεόμενα, περὶ ὧν ἀνάγκη, ἤν τις
ἐπιχειρῇ τι λέγειν, ὑποθέσει χρῆσθαι, οἷον περὶ τῶν
μετεώρων ἢ τῶν ὑπὸ γῆν· ἃ εἴ τις λέγοι καὶ γινώσκοι
ὡς ἔχει, οὔτ᾽ ἂν αὐτῷ τῷ λέγοντι οὔτε τοῖσιν ἀκούουσι
δῆλα ἂν εἴη, εἴτε ἀληθέα ἐστὶν εἴτε μή. οὐ γὰρ ἔστι
πρὸς ὅ τι χρὴ ἐπανενέγκαντα[3] εἰδέναι τὸ σαφές.

2. Ἰητρικῇ δὲ πάλαι πάντα ὑπάρχει καὶ ἀρχὴ καὶ
ὁδὸς εὑρημένη, καθ᾽ ἣν καὶ τὰ εὑρημένα πολλά τε καὶ
καλῶς ἔχοντα εὕρηται ἐν πολλῷ χρόνῳ· καὶ τὰ λοιπὰ
εὑρεθήσεται, ἤν τις ἱκανός τε ἐὼν καὶ τὰ εὑρημένα
εἰδὼς ἐκ τούτων ὁρμώμενος ζητῇ. ὅστις δὲ ταῦτα ἀπο-
βαλὼν καὶ ἀποδοκιμάσας πάντα, ἑτέρῃ ὁδῷ καὶ ἑτέρῳ
σχήματι ἐπιχειρεῖ ζητέειν, καί φησί τι ἐξευρηκέναι,
ἐξηπάτηται καὶ ἐξαπατᾶται· ἀδύνατον γάρ· δι᾽ ἃς δὲ
ἀνάγκας ἀδύνατον, ἐγὼ πειρήσομαι ἐπιδεῖξαι, λέγων
καὶ ἐπιδεικνύων τὴν τέχνην ὅτι[4] ἐστίν. ἐκ δὲ τούτου
καταφανὲς ἔσται ἀδύνατα ἐόντα ἄλλως πως τούτων
εὑρίσκεσθαι. μάλιστα δέ μοι δοκέει περὶ ταύτης δεῖν
λέγοντα τῆς τέχνης γνωστὰ λέγειν τοῖσι δημότῃσιν.
οὐ γὰρ περὶ ἄλλων τινῶν οὔτε ζητέειν οὔτε λέγειν
προσήκει ἢ περὶ τῶν παθημάτων ὧν αὐτοὶ οὗτοι νο-
σέουσί τε καὶ πονέουσιν. αὐτοὺς μὲν οὖν τὰ σφέων

[2] κενῆς M: καινῆς A
[3] ἐπανενέγ. M: ἀνενέγ. A
[4] ὅτι MA: ὅ τι a later hand in M

postulate the way subjects that cannot be directly seen or tested do, when someone is attempting to give an account of them, as for example the things in the heavens or those under the earth; for it is never clear either to the person himself or to his hearers whether what he is saying and expounding is really true or not, since there is no reality to which his account can be compared in order to know what is fact.

2. Medicine, though, has existed in its entirety for ages, possessing both a beginning and an established method, by means of which many and excellent discoveries have been made over a long period of time; and what remains will also be discovered, if an inquirer is competent and, being cognizant of the discoveries already made, conducts his researches beginning from these. But anyone who, casting aside and rejecting all these, attempts to conduct research by some other method or in some other manner, and claims to have discovered something, has been deceived himself, and is deceiving others,[2] for it is impossible. Why it is impossible, I will attempt to demonstrate by arguing and proving that the art of medicine truly exists. From this it will become clear that it is impossible to discover anything except by the current method of research. Particularly necessary, in my opinion, for anyone discussing this art is to talk about matters familiar to ordinary people, for his research and discussion concern nothing other than the diseases from which these very patients are ill and suffer. Now to learn by themselves how their own

[2] This is the interpretation of Foes, Littré, Fuchs, and Jones *Phil.*; Cornarius, Jones *Loeb*, Festugière, Jouanna, and Schiefsky give: "he has been deceived, and he continues to be deceived."

ΠΕΡΙ ΑΡΧΑΙΗΣ ΙΗΤΡΙΚΗΣ

574 αὐτῶν παθήματα | καταμαθεῖν, ὥς τε γίνεται καὶ παύ-
εται καὶ δι᾿ οἵας προφάσιας αὔξεταί τε καὶ φθίνει,
δημότας ἐόντας οὐ ῥηΐδιον· ὑπ᾿ ἄλλου δὲ εὑρημένα
καὶ λεγόμενα, εὐπετές. οὐδὲν γὰρ ἕτερον ἢ ἀναμιμνή-
σκεται ἕκαστος ἀκούων τῶν ἑωυτῷ συμβαινόντων. εἰ
δέ τις τῆς τῶν ἰδιωτέων γνώμης ἀποτεύξεται καὶ μὴ
διαθήσει τοὺς ἀκούοντας οὕτως, τοῦ ἐόντος ἀποτεύξεται. καὶ διὰ ταῦτα οὖν ταῦτα οὐδὲν δεῖ ὑποθέσιος.

3. Τὴν γὰρ ἀρχὴν οὔτ᾿ ἂν εὑρέθη ἡ τέχνη ἡ ἰητρικὴ
οὔτ᾿ ἂν ἐζητήθη—οὐδὲν γὰρ αὐτῆς ἔδει—εἰ τοῖσι κάμνουσι τῶν ἀνθρώπων τὰ αὐτὰ διαιτωμένοισί τε καὶ
προσφερομένοισιν, ἅπερ οἱ ὑγιαίνοντες ἐσθίουσί τε
καὶ πίνουσι καὶ τἆλλα διαιτέονται, συνέφερεν, καὶ μὴ
ἦν ἕτερα τούτων βελτίω. νῦν δὲ αὐτὴ ἡ ἀνάγκη ἰητρικὴν ἐποίησεν ζητηθῆναί τε καὶ εὑρεθῆναι ἀνθρώποισιν, ὅτι τοῖσι κάμνουσι ταὐτὰ προσφερομένοισιν,
576 ἅπερ οἱ ὑγιαίνοντες, | οὐ συνέφερεν, ὡς οὐδὲ νῦν συμ-
φέρει. ἔτι δὲ ἄνωθεν ἔγωγε ἀξιῶ οὐδ᾿ ἂν τὴν τῶν ὑγιαινόντων δίαιτάν τε καὶ τροφήν, ᾗ νῦν χρέωνται, εὑρεθῆναι, εἰ ἐξήρκει τῷ ἀνθρώπῳ ταὐτὰ ἐσθίοντι καὶ
πίνοντι βοΐ τε καὶ ἵππῳ καὶ πᾶσιν ἐκτὸς ἀνθρώπου,
οἷον τὰ ἐκ τῆς γῆς φυόμενα, καρπούς τε καὶ ὕλην καὶ
χόρτον· ἀπὸ τούτων γὰρ καὶ τρέφονται καὶ αὔξονται
καὶ ἄπονοι διάγουσιν οὐδὲν προσδεόμενοι ἄλλης διαίτης. καί τοι τήν γε ἀρχὴν ἔγωγε δοκέω καὶ τὸν
ἄνθρωπον τοιαύτῃ τροφῇ κεχρῆσθαι. τὰ δὲ νῦν διαιτήματα εὑρημένα καὶ τετεχνημένα ἐν πολλῷ χρόνῳ
γεγενῆσθαί μοι δοκεῖ, ὡς γὰρ ἔπασχον πολλά τε καὶ

diseases come about and then go away, and the causes that make them either increase or decrease, is not an easy task for ordinary people; but when these things have been discovered and are set forth by someone else, it is simple to grasp; for only an effort of memory is required by the patient as he listens to an account of what has happened to him. But if someone does not succeed in being understood by laymen, and fails to inform his hearers, he is missing the point. Therefore, for the same reasons too, medicine is shown to need no postulate.

3. To begin with, the art of medicine would neither have been discovered, nor even looked for—as there would have been no need for medicine—if sick people had benefited by the same mode of living and regimen as the food, drink and mode of living used by people in health, and if there had been no other things better for the sick than these. But in fact it was sheer necessity that caused humans to seek and to discover medicine, because the sick did not, and do not, benefit by the same regimen as the healthy. To trace the matter yet further back, I personally believe that not even the mode of living and nourishment enjoyed at the present time by people in health would have been discovered, had a human being been satisfied with the same food and drink as satisfy an ox, a horse, and every other animal except a human, for example the products of the earth—fruits, wood and grass. For on these the animals are nourished, grow, and live without pain, having no need at all of any other kind of living. Yet I am of the opinion that in the beginning humans too used this sort of nourishment; but our present ways of living have, I think, been discovered and elaborated during a long period of time. For many and terrible were the sufferings of humans

δεινὰ ὑπὸ ἰσχυρῆς τε καὶ θηριώδεος διαίτης ὠμά τε
καὶ ἄκρητα καὶ μεγάλας δυνάμιας ἔχοντα ἐσφερόμε-
νοι—οἷά περ ἂν καὶ νῦν ὑπ' αὐτῶν πάσχοιεν πόνοισί
τε ἰσχυροῖσι καὶ νούσοισι περιπίπτοντες καὶ διὰ τάχεος θανάτοισιν· ἧσσον μὲν οὖν ταῦτα τότε εἰκὸς ἦν
πάσχειν διὰ τὴν συνήθειαν, ἰσχυρῶς δὲ καὶ τότε. καὶ
τοὺς μὲν πλείστους τε καὶ ἀσθενεστέρην φύσιν ἔχοντας ἀπόλλυσθαι εἰκός, τοὺς δὲ τούτων ὑπερέχοντας
πλείω χρόνον ἀντέχειν· ὥσπερ καὶ νῦν ἀπὸ τῶν ἰσχυρῶν βρωμάτων οἱ μὲν ῥηϊδίως ἀπαλλάσσονται, οἱ δὲ
μετὰ πολλῶν πόνων τε καὶ κακῶν.

Διὰ δὴ ταύτην τὴν χρείην[5] καὶ οὗτοί μοι δοκέουσι
ζητῆσαι τροφὴν ἁρμόζουσαν τῇ φύσει καὶ εὑρεῖν
ταύτην, ᾗ νῦν χρεώμεθα. ἐκ μὲν οὖν τῶν πυρῶν βρέ-
ξαντές σφας καὶ πτίσαντες καὶ καταλέσαντές τε καὶ
διασήσαντες καὶ φορύξαντες καὶ ὀπτήσαντες ἀπε-
τέλεσαν ἄρτον, ἐκ δὲ τῶν κριθέων μᾶζαν· ἄλλα | τε
συχνὰ[6] περὶ ταύτην πρηγματευσάμενοι ἥψησάν τε
καὶ ὤπτησαν καὶ ἔμιξαν καὶ ἐκέρασαν τὰ ἰσχυρά
τε καὶ ἄκρητα τοῖσιν ἀσθενεστέροισι, πλάσσοντες
πάντα πρὸς τὴν τοῦ ἀνθρώπου φύσιν τε καὶ δύναμιν,
ἡγεύμενοι, ὅσα μὲν ἂν ἰσχυρότερα ᾖ ἢ[7] δυνήσεται
κρατέειν ἡ φύσις, ἢν ἐμφέρηται, ἀπὸ τούτων δ' αὐτῶν
πόνους τε καὶ νούσους καὶ θανάτους ἔσεσθαι, ὁπόσων
δ' ἂν δύνηται ἐπικρατέειν, ἀπὸ τούτων τροφήν τε καὶ
αὔξησιν καὶ ὑγιείην. τῷ δὲ εὑρήματι τούτῳ καὶ ζη-

---

5 χρείην M: αἰτίην A   6 συχνὰ M: πολλὰ A

578

14

from the strong and brutish living when they partook of raw foods, uncompounded and possessing strong potencies[3]—the same things in fact as humans would suffer if they used them today, falling into violent pains and diseases quickly followed by death. Formerly indeed they probably suffered less, because they were used to the regimen, but they suffered severely even then. The majority of people, having a weaker constitution, naturally perished, while the stronger resisted longer, just as at the present time some people deal easily with strong foods, while others do so only with many severe pains and troubles.

For this reason the ancients, too, seem to me to have looked for nourishment that harmonized with the human constitution, and to have discovered the diet we use now. So from wheat, after steeping, winnowing, grinding and sifting it, kneading and baking, they made bread, and from barley they made cake. Working on food in many other ways, they boiled it or baked it, and mixed it, combining the strong and uncompounded components with the weaker ones, so as to adapt everything to the constitution and strength of the human being, thinking that from foods which the human constitution cannot assimilate[4] when they are ingested, because they are too strong, will come pain, disease, and death, while from such as can be assimilated will come nourishment, growth and health. To this discovery and research what more accurate or appro-

[3] *Potency*: see Note on Technical Terms.
[4] *Assimilate*: see Note on Technical Terms.

---

[7] ἤ Littré in *app. crit.*: οὐ MA

ΠΕΡΙ ΑΡΧΑΙΗΣ ΙΗΤΡΙΚΗΣ

τήματι τί ἄν τις ὄνομα δικαιότερον ἢ προσῆκον μᾶλλον θείη ἢ ἰητρικήν, ὅ τι[8] γε εὕρηται ἐπὶ τῇ τοῦ ἀνθρώπου ὑγιείῃ τε καὶ σωτηρίῃ καὶ τροφῇ, ἄλλαγμα
ἐκείνης τῆς διαίτης, ἐξ ἧς οἱ πόνοι καὶ νοῦσοι καὶ
θάνατοι ἐγίνοντο;

4. Εἰ δὲ μὴ τέχνη αὕτη νομίζεται εἶναι, οὐκ ἀπεοι-
κός· ἧς γὰρ μηδείς ἐστιν ἰδιώτης, ἀλλὰ πάντες ἐπισ-
τήμονες διὰ τὴν χρῆσίν τε καὶ ἀνάγκην, οὐ προσήκει
ταύτης οὐδένα τεχνίτην καλεῖσθαι· ἐπεὶ | τό γε εὕρημα
μέγα τε καὶ πολλῆς σκέψιός τε καὶ τέχνης. ἔτι γοῦν
καὶ νῦν οἱ τῶν γυμνασίων τε καὶ ἀσκησίων ἐπιμελό-
μενοι αἰεί τι προσεξευρίσκουσι κατὰ τὴν αὐτὴν ὁδὸν
ζητέοντες ὅ τι ἔδων τε καὶ πίνων ἐπικρατήσει τε αὐτοῦ
μάλιστα καὶ ἰσχυρότατος[9] αὐτὸς ἑωυτοῦ ἔσται.

5. Σκεψώμεθα δὲ καὶ τὴν ὁμολογουμένως ἰητρικήν,
τὴν ἀμφὶ τοὺς κάμνοντας εὑρημένην, ἣ καὶ ὄνομα καὶ
τεχνίτας ἔχει· ἦρά τι καὶ αὐτὴ τῶν αὐτῶν ἐθέλει, καὶ
πόθεν ποτὲ ἦρκται. ἐμοὶ μὲν γάρ, ὅπερ ἐν ἀρχῇ εἶπον,
οὐδ᾽ ἂν ζητῆσαι ἰητρικὴν δοκέει οὐδείς, εἰ ταὐτὰ διαιτήματα τοῖσί τε κάμνουσι καὶ τοῖσι ὑγιαίνουσιν
ἥρμοζεν. ἔτι γοῦν καὶ νῦν ὅσοι ἰητρικῇ μὴ χρέωνται,
οἵ τε βάρβαροι καὶ τῶν Ἑλλήνων ἔνιοι, τὸν αὐτὸν
τρόπον, ὅνπερ οἱ ὑγιαίνοντες, διαιτέονται πρὸς ἡδονήν, καὶ οὔτ᾽ ἂν ἀπόσχοιντο οὐδενὸς ὧν ἐπιθυμέουσιν
οὔθ᾽ ὑποστείλαιντο ἄν. οἱ δὲ ζητήσαντές τε καὶ εὑρόντες ἰητρικὴν τὴν αὐτὴν κείνοισι διάνοιαν ἔχοντες,
περὶ ὧν μοι ὁ πρότερος λόγος εἴρηται, πρῶτον μέν,
οἶμαι, ὑφεῖλον τοῦ πλήθεος τῶν σιτίων αὐτῶν τούτων,

16

priate name could be given than "medicine," seeing that it was discovered with a view to the health, salvation and nourishment of the human being, in the place of a mode of living from which resulted pains, diseases, and deaths?

4. That this art is commonly considered not to exist is not surprising, for it seems unfitting to call anyone an artist in a craft in which there are no laymen, but all possess some knowledge gained through practice and necessity. Nevertheless the discovery was a great one, involving much investigation and art. Indeed, even at the present day those who study gymnastics and athletic exercises are constantly making some fresh discovery or other, by investigating, according to the same method, which foods and drinks the athlete will best assimilate, and which will make him grow to his maximum strength.

5. But let us also consider the recognized art of medicine that was discovered for the treatment of the sick, and has both a name and practitioners. Does it have any of the same objects as the other (sc. dietetic) art, and what was its origin? In my opinion, as I said at the beginning, no one would even have looked for medicine, if the same regimen had suited both the sick and the healthy. At any rate, even today people who do not make use of medical science—both foreigners and some Greeks—live (sc. even when they are ill) according to the same regimen as do those in health, just as they please, and would neither forgo nor restrict the satisfaction of any of their desires. But those who searched for and discovered medicine, having the same intention as the people I discussed above, first, I think, lessened the bulk of the foods themselves, diminish-

---

⁸ ὅ τι M: ὅτι A   ⁹ -ότατος M: -ότερος A

καὶ ἀντὶ πλεόνων ὀλίγιστα ἐποίησαν. ἐπεὶ δὲ αὐτοῖσι τοῦτο ἔστι μὲν ὅτε πρός τινας τῶν καμνόντων ἤρκεσε
582 | καὶ φανερὸν ἐγένετο ὠφελῆσαι, οὐ μέντοι πᾶσί γε, ἀλλ' ἦσάν τινες οὕτως ἔχοντες, ὡς μηδ'[10] ὀλίγων σι- τίων δύνασθαι ἐπικρατέειν, ἀσθενεστέρου δὲ δή τινος οἱ τοιοίδε ἐδόκεον δεῖσθαι, εὗρον τὰ ῥυφήματα μίξαν- τες ὀλίγα τῶν ἰσχυρῶν πολλῷ τῷ ὕδατι καὶ ἀφαιρεό- μενοι τὸ ἰσχυρὸν τῇ κρήσει τε καὶ ἑψήσει. ὅσοι δὲ μηδὲ τῶν ῥυφημάτων ἐδύναντο ἐπικρατέειν, ἀφεῖλον καὶ ταῦτα, καὶ ἀφίκοντο ἐς πόματα, καὶ ταῦτα τῇσί τε κρήσεσι καὶ τῷ πλήθει διαφυλάσσοντες ὡς με- τρίως ἔχοι, μήτε πλείω τῶν δεόντων μήτε ἀκρητέ- στερα προσφερόμενοι μηδὲ ἐνδεέστερα.

6. Εὖ δὲ χρὴ τοῦτο εἰδέναι, ὅτι τοῖσι τὰ ῥυφήματα ἐν τῇσι νούσοισιν οὐ συμφέρει, ἀλλ' ἄντικρυς, ὅταν ταῦτα προσαίρωνται, παροξύνονταί σφισιν οἵ τε πυ- ρετοὶ καὶ τὰ ἀλγήματα· καὶ δῆλον τὸ προσενεχθὲν τῇ μὲν νούσῳ τροφή τε καὶ αὔξησις γενόμενον, τῷ δὲ σώματι φθίσις τε καὶ ἀρρωστίη. ὅσοι δ' ἂν τῶν ἀν- θρώπων ἐν ταύτῃ τῇ διαθέσει ἐόντες προσενέγκωνται ξηρὸν σιτίον ἢ μᾶζαν ἢ ἄρτον, καὶ ἢν πάνυ σμικρόν, δεκαπλασίως ἂν μᾶλλον καὶ ἐπιφανέστερον κακω-
584 θεῖεν ἢ ῥυφέοντες, δι' οὐδὲν ἄλλο ἢ διὰ | τὴν ἰσχὺν τοῦ βρώματος πρὸς τὴν διάθεσιν· καὶ ὅτῳ ῥυφεῖν μὲν συμφέρει, ἐσθίειν δ' οὔ, εἰ πλείω φάγοι, πολὺ ἂν μᾶλ- λον κακωθείη, ἢ <εἰ> ὀλίγα.[11] καὶ εἰ ὀλίγα δέ, πονή- σειεν ἄν. πάντα δὴ τὰ αἴτια τοῦ πόνου ἐς τὸ αὐτὸ ἀνάγεται, τὰ ἰσχυρότατα μάλιστά τε καὶ ἐπιφανέσ-

ing their amount from abundant to scanty. But although this regimen was adequate for some patients, and seemed clearly to benefit them, it was not so for them all, as some patients were in such a condition that they could not assimilate even small quantities of food; since these people were evidently thought to need weaker nutriment, gruels were invented by mixing small quantities of strong foods with much water, and by reducing their strength by compounding and boiling them. Finally, from patients who were unable to assimilate even gruels, they withdrew these and moved to drinks, regulating them in their composition and quantity to be moderate, by administering nothing that was either more or less in amount, or less compounded, than it should be.

6. It must be clearly understood that patients who are not benefited in diseases by gruels but just the opposite, on taking them suffer an exacerbation of their fever and pain, and what is taken manifestly turns out to be nourishment and increase for the disease, while for the body it leads to attrition and enfeeblement. And any patients who in this condition take dry food like barley cake or bread, even a very little, will be hurt ten times more, and more obviously, than if they take gruels, simply and solely because the food is too strong for their condition. And a man to whom gruels are beneficial, but not to eat solid food, will suffer much more harm if he eats a larger quantity of it than if he eats only a little, although he will feel pain even if he eats a little. Indeed, all the causes of suffering can be reduced to one, namely, it is the strongest foods

---

[10] μηδ' I: μὴ δι' M: μὴ A
[11] ἢ ‹εἰ› ὀλίγα Ermerins: ἢ ὀλίγα A: om. M

19

τατα λυμαίνεσθαι τὸν ἄνθρωπον καὶ τὸν ὑγιέα ἐόντα
καὶ τὸν κάμνοντα.

7. Τί οὖν φαίνεται ἑτεροῖον διανοηθεὶς ὁ καλεύμενος
ἰητρὸς καὶ ὁμολογεομένως χειροτέχνης, ὃς ἐξεῦρε τὴν
ἀμφὶ τοὺς κάμνοντας δίαιτάν τε καὶ τροφήν, ἢ κεῖνος
ὁ ἀπ' ἀρχῆς τοῖσι πᾶσιν ἀνθρώποισι τροφήν, ᾗ νῦν
χρεώμεθα, ἐξ ἐκείνης τῆς ἀγρίης τε καὶ θηριώδεος
διαίτης εὑρών τε καὶ παρασκευασάμενος; ἐμοὶ μὲν
γὰρ φαίνεται ὁ ωὑτὸς τρόπος[12] καὶ ἓν καὶ ὅμοιον τὸ
εὕρημα. ὁ μέν, ὅσων μὴ ἐδύνατο ἡ φύσις ἡ ἀνθρω-
πίνη ὑγιαίνουσα ἐπικρατέειν ἐμπιπτόντων διὰ τὴν
θηριότητά τε καὶ τὴν ἀκρησίην, ὁ δέ, ὅσων ἡ διάθεσις, ἐν οἵῃ ἂν ἑκάστοτε ἕκαστος τύχῃ διακείμενος, μὴ
δύνηται ἐπικρατέειν, ταῦτα ἐζήτησεν ἀφελεῖν. τί δὴ
586   τοῦτο ἐκείνου διαφέρει ἀλλ' ἢ πλέον τό γε | εἶδος, καὶ
ὅτι ποικιλώτερον καὶ πλείονος πρηγματείης; ἀρχὴ δὲ
ἐκείνη ἡ πρότερον γενομένη.

8. Εἰ δέ τις σκέπτοιτο τὴν[13] τῶν καμνόντων δίαιταν[14] πρὸς τὴν τῶν ὑγιαινόντων, εὕροι ἂν οὐ[15] βλαβε-
ρωτέρην ἤπερ ἡ[16] τῶν ὑγιαινόντων πρὸς τὴν τῶν θη-
ρίων τε καὶ πρὸς τὴν τῶν ἄλλων ζῴων. ἀνὴρ γὰρ
κάμνων νοσήματι μήτε τῶν χαλεπῶν τε καὶ ἀφόρων
μήτε αὖ τῶν παντάπασιν εὐηθέων, ἀλλ' ᾖ[17] αὐτῷ ἐξα-

---

[12] τρόπος M: λόγος A   [13] τὴν A: τίς ἡ M
[14] δίαιταν A: δίαιτα M   [15] οὐ om. M
[16] ἤπερ ἡ Littré in app. crit.: ηπερ ην M: ἡ περὶ A
[17] ἀλλ' ᾖ J. F. Dübner in Littré (vol. 2, p. L): ἀλλ' οὔτε M:
ἄλλη A: ἀλλ' ἢ later hand in A

that hurt a person most and most obviously, whether he is well or ill.

7. What difference then can be seen between the motive of what we call a physician, who is an acknowledged craftsman, the discoverer of the mode of life and of the nourishment suitable for the sick, and the person who originally discovered and prepared the nourishment for everyone which we now use, in place of the savage and brutish diet of old? To me, in fact, their way of thinking seems to be identical and their discovery one and the same. The latter sought to do away with those things which, when taken, the constitution of a healthy person cannot assimilate because of their brutish and uncompounded character, the former those things which the temporary condition in which a sick individual happens to be prevents him from assimilating. How do the two pursuits differ, except in that the latter has more scope and is more complex, and requires the greater application, while the former is the starting point and came first in time?

8. If someone were to consider (sc. persons who needed) the regimen of the ill, but instead (sc. received) that of the healthy, he would discover (sc. that they had found) it no more harmful than (sc. had those who needed) the diet of the healthy (sc. received) in its place the diet of wild beasts and other animals.[5] Take a person sick with a disease that is neither severe and unbearable, nor yet altogether mild, but such that it is likely to become pro-

[5] The meaning of this sentence is assured by the later part of the chapter, but its formulation is very elliptical.

μαρτάνοντι μέλλει ἐπίδηλον ἔσεσθαι, εἰ ἐθέλοι κατα-
φαγεῖν ἄρτον καὶ κρέας ἢ ἄλλο τι ὧν οἱ ὑγιαίνοντες
ἐσθίοντες ὠφελέονται, μὴ πολὺ ἀλλὰ πολλῷ ἔλασσον
ἢ ὑγιαίνων ἂν ἐδύνατο, ἄλλος τε τῶν ὑγιαινόντων φύσιν ἔχων μήτ' παντάπασιν ἀσθενέα μήτε αὖ ἰσχυρήν,
φαγών τι ὧν βοῦς ἢ ἵππος φαγὼν ὠφελοῖτό τε καὶ
ἰσχύοι, ὀρόβους ἢ κριθὰς ἢ ἄλλο τι τῶν τοιούτων, μὴ
πολύ, ἀλλὰ πολλῷ μεῖον ἢ δύναιτο, οὐκ ἂν ἧσσον ὁ
588 ὑγιαίνων τοῦτο ποιήσας πονήσειέ τε καὶ | κινδυνεύ-
σειε κείνου τοῦ νοσέοντος, ὃς τὸν ἄρτον ἢ τὴν μᾶζαν
ἀκαίρως προσηνέγκατο. ταῦτα δὴ πάντα τεκμήρια,
ὅτι αὕτη ἡ τέχνη πᾶσα ἡ ἰητρικὴ τῇ αὐτῇ ὁδῷ ζητεο-
μένη εὑρίσκοιτο ἄν.

9. Καὶ εἰ μὲν ἦν ἁπλοῦν, ὥσπερ ὑφήγηται, ὅσα μὲν
ἦν ἰσχυρότερα, ἔβλαπτεν, ὅσα δ' ἦν ἀσθενέστερα,
ὠφέλει τε καὶ ἔτρεφεν καὶ τὸν κάμνοντα καὶ τὸν ὑγιαί-
νοντα, εὐπετὲς ἂν ἦν τὸ πρῆγμα· πολλὸν γὰρ τοῦ
ἀσφαλέος ἂν ἔδει περιλαμβάνοντας ἄγειν ἐπὶ τὸ
ἀσθενέστατον.[18] νῦν δὲ οὐκ ἔλασσον ἁμάρτημα, οὐδὲ
ἧσσον λυμαίνεται τὸν ἄνθρωπον, ἢν ἐλάσσονα καὶ
ἐνδεέστερα τῶν ἱκανῶν προσφέρηται. τὸ γὰρ τοῦ λιμοῦ μένος ἰσχυρῶς ἐνδύνεται[19] ἐν τῇ φύσει τοῦ ἀνθρώπου καὶ γυιῶσαι καὶ ἀσθενέα ποιῆσαι καὶ ἀπο-
κτεῖναι. πολλὰ δὲ καὶ ἄλλα κακὰ ἑτεροῖα μὲν τῶν ἀπὸ
πληρώσιος, οὐχ ἧσσον δὲ ἀμὰ[20] δεινά, καὶ ἀπὸ κενώσιος διότι πολλὸν ποικιλώτερά τε καὶ διὰ πλείονος

18 -έστατον Μ: -έστερον Α

nounced if he goes astray (sc. in his regimen), and suppose that he were to resolve to eat bread and meat, or any other food that is beneficial to people in health, not much of it, but far less than he could take if he was well; then again take a healthy person, with a constitution neither altogether weak, nor again altogether strong, and suppose he were to eat one of the foods that would be beneficial and strength-giving to an ox or a horse, vetches or barley or something similar, not much of it, but far less than he could take (sc. of normal food). If the healthy person did this he would suffer no less pain and danger than the sick person who took bread or barley cake at a time when he should not. All these facts are proofs that the art of medicine itself, if research is continued by the current method, can be completely elucidated.

9. If the situation were as simple as it has been presented, however, so that too strong foods were harmful, whereas weaker ones were beneficial and nourishing for both the ill and the healthy, the matter would be easy, for simple recourse to the weakest foods would necessarily secure a great degree of safety. But in reality, it is no less a mistake, nor does it harm people any less, if they are given less and weaker foods than they should be; for violent hunger then presses forcefully on their human frame, to enervate, weaken and kill them. Many other evils different from those resulting from repletion, but no less terrible, also come from inanition, which are greater in variety and require a more exact method of treatment. For

---

19 ἰσχυρῶς ἐνδύνεται A: δύναται ἰσχυρῶς M
20 ἁμὰ Jouanna: ἄμα M: om. A

ἀκριβείης ἐστί. δεῖ γὰρ μέτρου τινὸς στοχάσασθαι·
μέτρον δὲ οὐδὲ ἀριθμὸν οὔτε σταθμὸν ἄλλον, πρὸς ὃ
590   ἀναφέρων εἴσῃ τὸ ἀκριβές, οὐκ | ἂν εὕροις ἀλλ᾽ ἢ τοῦ
σώματος τὴν αἴσθησιν. διὸ ἔργον οὕτω καταμαθεῖν
ἀκριβῶς, ὥστε σμικρὰ ἁμαρτάνειν ἔνθα ἢ ἔνθα, κἂν
ἐγὼ τοῦτον τὸν ἰητρὸν ἰσχυρῶς ἐπαινέοιμι τὸν σμικρὰ ἁμαρτάνοντα. τὸ δὲ ἀτρεκὲς ὀλιγάκις ἔστι κατ-
ιδεῖν.

Ἐπεὶ οἱ πολλοί γε τῶν ἰητρῶν τὰ αὐτά μοι δοκέουσι τοῖσι κακοῖσι κυβερνήτῃσι πάσχειν. καὶ γὰρ
ἐκεῖνοι ὅταν ἐν γαλήνῃ κυβερνῶντες ἁμαρτάνωσιν, οὐ
καταφανέες εἰσίν· ὅταν δ᾽ αὐτοὺς κατάσχῃ χειμών τε
μέγας καὶ ἄνεμος ἐξώστης, φανερῶς πᾶσιν ἤδη ἀνθρώποισι δι᾽ ἀγνωσίην καὶ ἁμαρτίην δηλοί εἰσιν
ἀπολέσαντες τὴν ναῦν. οὕτω δὴ καὶ οἱ κακοί τε καὶ οἱ
πλεῖστοι ἰητροί, ὅταν μὲν θεραπεύωσιν ἀνθρώπους
μηδὲν δεινὸν ἔχοντας, ἐς οὓς ἄν τις τὰ μέγιστα ἐξ-
αμαρτάνων[21] οὐδὲν δεινὸν ἐργάσαιτο—πολλὰ δὲ τὰ
τοιαῦτα νοσήματα καὶ πολλόν τι πλείω τῶν δεινῶν
ἀνθρώποισι συμβαίνει—ἐν μὲν δὴ τοῖσι τοιούτοισι
ἁμαρτάνοντες οὐ καταφανέες εἰσὶ τοῖσιν ἰδιώτῃσιν·
ὅταν δ᾽ ἐντύχωσι μεγάλῳ τε καὶ ἰσχυρῷ καὶ ἐπισφα-
λεῖ νοσήματι, τότε σφέων τά τε ἁμαρτήματα καὶ ἡ
ἀτεχνίη πᾶσι καταφανής· οὐ γὰρ ἐς μακρὸν αὐτῶν
ἑκατέρου αἱ τιμωρίαι, ἀλλὰ διὰ τάχεος πάρεισιν.

10. Ὅτι δ᾽ οὐδὲν ἐλάσσους ἀπὸ κενώσιος ἀκαίρου
κακοπάθειαι γίνονται τῷ ἀνθρώπῳ ἢ ἀπὸ πληρώσιος,
καταμανθάνειν καλῶς ἔχει ἐπαναφέροντας ἐπὶ τοὺς

you must aim at some measure, but there is no measure—either number or other standard, you could possibly discover, by referring to which you would know what is exactly correct—other than the perception of the body. For this reason, it is laborious to acquire knowledge so exact that only such small mistakes are made here and there, and I would strongly praise any physician who made only small mistakes, since exact knowledge is hardly ever seen.

Most physicians seem to me to be in the same situation as bad pilots; as long as the latter err while they are sailing in calm waters, they remain unexposed, but when a great storm overtakes them with a violent gale, it immediately becomes clear to everyone that it is their ignorance and incompetence that have lost the ship. So also when bad physicians—who comprise the great majority—treating patients who are suffering from no serious complaint, so that even very great faults would not affect them seriously—such illnesses are frequent and occur far more commonly than serious diseases in human beings—err in such cases, they are not exposed to laymen; but when they meet with a severe, violent and dangerous illness, then their errors and want of skill are manifest to all, for the punishments of both (sc. the pilot and the healer) do not wait long, but are soon there.

10. That the distress which arises in a person from untimely inanition is no less than that of unseasonable repletion, may be learned well by reference to the healthy.

---

21 ἐξαμαρ. A: ἅμαρ. M

ὑγιαίνοντας. ἔστι γὰρ οἷσιν αὐτῶν συμφέρει μονοσιτέειν, καὶ τοῦτο διὰ τὸ συμφέρον οὕτως αὐτοὶ συνε-
τάξαντο, ἄλλοισι δὲ ἀριστᾶν διὰ τὴν αὐτὴν ἀνάγκην.
592 οὕτω γὰρ | αὐτοῖσι συμφέρει, καὶ μὴ τούτοισιν οἳ δι᾽
ἡδονὴν ἢ δι᾽ ἄλλην τινὰ συγκυρίην ἐπετήδευσαν ὁπότερον αὐτῶν· τοῖσι μὲν γὰρ πλείστοισι τῶν ἀνθρώπων οὐδὲν διαφέρει, ὁπότερον ἂν ἐπιτηδεύσωσιν, εἴτε
μονοσιτέειν εἴτε ἀριστᾶν, τούτῳ τῷ ἔθει χρῆσθαι. εἰσὶ
δέ τινες οἳ οὐκ ἂν δύναιντο ἔξω τοῦ συμφέροντος
ποιέοντες ῥηϊδίως ἀπαλλάσσειν, ἀλλὰ συμβαίνει αὐτῶν ἑκατέροισι παρ᾽ ἡμέρην μίαν καὶ ταύτην οὐχ
ὅλην μεταβάλλουσιν ὑπερφυὴς κακοπάθεια.

Οἱ μὲν γὰρ ἢν ἀριστήσωσιν μὴ συμφέροντος αὐτοῖσιν, εὐθέως βαρέες καὶ νωθροὶ καὶ τὸ σῶμα καὶ
τὴν γνώμην χάσμης τε καὶ νυσταγμοῦ καὶ δίψης
πλήρεις· ἢν δὲ καὶ ἐπιδειπνήσωσι, καὶ φῦσα καὶ
στρόφος καὶ ἡ κοιλίη καταρρήγνυται. καὶ πολλοῖσιν
ἀρχὴ νούσου αὕτη μεγάλης ἐγένετο, καὶ ἢν τὰ αὐτὰ
σιτία, ἃ μεμαθήκεσαν ἅπαξ ἀναλίσκειν, ταῦτα δὶς
προσενέγκωνται καὶ μηδὲν πλείω. τοῦτο δέ, ἢν ἀριστᾶν μεμαθηκώς τις καὶ οὕτως αὐτῷ ξυμφέρον[22] μὴ
ἀριστήσῃ, ὅταν τάχιστα παρέλθῃ ἡ ὥρη, εὐθὺς ἀδυ-
ναμίη δεινή, τρόμος, ἀψυχίη· ἐπὶ τούτοισιν ὀφθαλμοὶ
κοῖλοι, οὖρον χλωρότερον καὶ θερμότερον, στόμα πικρόν, καὶ τὰ σπλάγχνα οἱ δοκέει κρέμασθαι, σκοτο-

[22] ξυμφέρον M: συμφέρον ἢν A

26

For some people benefit from taking only one meal each day, and because of this benefit they make it a rule to have only one meal; others again—for the same reason, that they are benefited by doing so—take lunch[6] as well. Both of these have a benefit from their practice, but not others who adopt one or other of the two habits merely for the sake of pleasure or for some other chance reason: to most people, in fact, it makes no difference which of the two practices they adopt as their habit, either to eat once a day or also to take lunch. But there certainly are people who, if they do anything outside the range of what is beneficial, do not get off easily, but if they change their respective ways for a single day, and not even a whole day, are all assailed by extraordinary distress.

Some, if they lunch although lunch does not agree with them, immediately feel heavy and sluggish in body and in mind, and are prey to yawning, drowsiness and thirst; while, if they go on to eat dinner as well, flatulence follows with colic and violent diarrhea. Many have found such an action to result in a serious illness, even if the total quantity of the same food they took at the two meals was no greater than what they were accustomed to consume at one meal. On the other hand, if someone who is accustomed to take lunch, and has found that beneficial, misses taking it, he suffers, as soon as the lunch hour has gone by, from prostrating weakness, trembling and faintness; hollowness of the eyes follows; his urine becomes more yellow and hotter, and his mouth bitter; his viscera seem to hang down; then come dizziness, depression and inability

[6] *Lunch*: see Note on Technical Terms.

δινίη, δυσθυμίη, δυσεργίη.[23] ταῦτα δὲ πάντα, καὶ ὅταν
594 δειπνεῖν ἐπιχειρήσῃ, ἀηδέστερος | μὲν ὁ σῖτος, ἀνα-
λίσκειν δὲ οὐ δύναται ὅσα ἀριστιζόμενος πρότερον
ἐδείπνει. ταῦτα δὲ αὐτὰ μετὰ στρόφου καὶ ψόφου
καταβαίνοντα συγκαίει τὴν κοιλίην, δυσκοιτέουσί τε
καὶ ἐνυπνιάζουσι τεταραγμένα τε καὶ θορυβώδεα.
πολλοῖσι δὲ καὶ τούτων αὕτη ἀρχὴ νούσου ἐγένετο.

11. Σκέψασθαι δὲ χρή, διὰ τίνας προφάσιας[24]
αὐτοῖσιν ταῦτα συνέβη. τῷ μέν, οἶμαι, μεμαθηκότι
μονοσιτέειν, ὅτι οὐκ ἀνέμεινεν τὸν χρόνον τὸν ἱκανόν,
μέχρι αὐτοῦ ἡ κοιλίη τῶν τῇ προτεραίῃ προσενηνεγ-
μένων σιτίων ἀπολαύσῃ τελέως καὶ ἐπικρατήσῃ καὶ
λαπαχθῇ τε καὶ ἡσυχάσῃ, ἀλλ᾽ ἐπὶ ζέουσάν τε καὶ
ἐζυμωμένην καινὰ ἐπεσηνέγκατο. αἱ δὲ τοιαῦται κοιλίαι πολλῷ τε βραδύτερον πέσσουσι καὶ πλείονος
δέονται ἀναπαύσιός τε καὶ ἡσυχίης. ὁ δὲ μεμαθηκὼς
ἀριστίζεσθαι, ὅτι οὐκ,[25] ἐπειδὴ τάχιστα ἐδεήθη τὸ
σῶμα τροφῆς καὶ τὰ πρότερα κατανάλωτο καὶ οὐκ
εἶχεν οὐδεμίαν ἀπόλαυσιν, εὐθέως[26] αὐτῷ προσεγένετο
καινὴ τροφή, φθίνει δὲ καὶ συντήκεται ὑπὸ λιμοῦ.
πάντα γάρ, ἃ λέγω πάσχειν τὸν τοιοῦτον ἄνθρωπον,
λιμῷ ἀνατίθημι. φημὶ δὲ καὶ τοὺς ἄλλους ἀνθρώπους
ἅπαντας, οἵτινες ἂν ὑγιαίνοντες ἄσιτοι δύο ἡμέρας
ἢ τρεῖς γένωνται, ταὐτὰ πείσεσθαι, οἷάπερ ἐπὶ τῶν
ἀναρίστων γενομένων εἴρηκα. |

[23] δυσεργίη Jouanna: δυσεργείη A: δυσοργίην M
[24] τίνας προφάσιας M: τίνα αἰτίαν A

28

to work. These are all symptoms he suffers, and when he attempts to have dinner, his food is less pleasant, and he cannot digest what formerly he used to have for dinner when he had had lunch. The food itself, descending into his cavity[7] with colic and noise, sears it, and disturbed sleep follows, accompanied by wild and troubled dreams. In many of these patients, this has also been the beginning of an illness.

11. It is necessary to inquire into the cause why such things happen to these people. The one who had been accustomed to one meal suffered, I think, because he did not wait long enough until his cavity had completely digested and assimilated the food taken the day before, and until it had become empty and quiet, but took new foods while his cavity was still in a state of hot turmoil and ferment. Such cavities concoct food much more slowly than others do, and require a longer period of rest and quiet. As for the person accustomed also to take lunch, it (i.e., his suffering) arose because, at the time when his body needed nourishment and what had been ingested before was all consumed and no supply was left, no new nutriment was given to it: such a person wastes away and pines from hunger, for everything I describe such a person as suffering, I attribute to inanition. And I assert furthermore that anyone else, who when in good health fasts for two or three days, will suffer the same symptoms I have described as occurring in those who do not take their (sc. customary) lunch.

---

[7] *Cavity*: see Note on Technical Terms, vol. 8, p. 10f.

596   12. Τὰς δὲ τοιαύτας φύσιας ἔγωγέ φημι τὰς ταχέως
τε καὶ ἰσχυρῶς τῶν ἁμαρτημάτων ἀπολαυούσας
ἀσθενεστέρας εἶναι τῶν ἑτέρων. ἐγγύτατα δὲ τοῦ
ἀσθενέοντός ἐστιν ὁ ἀσθενής, ἔτι δὲ ἀσθενέστερος ὁ
ἀσθενέων, καὶ μᾶλλον αὐτῷ προσήκει ὅ τι ἂν τοῦ
καιροῦ ἀποτυγχάνῃ πονεῖν. χαλεπὸν δὲ[27] τοιαύτης
ἀκριβείης ἐούσης περὶ τὴν τέχνην τυγχάνειν αἰεὶ τοῦ
ἀτρεκεστάτου. πολλὰ δὲ εἴδεα κατ᾽ ἰητρικὴν ἐς τοσ-
αύτην ἀκρίβειαν ἥκει, περὶ ὧν εἰρήσεται. οὔ φημι δὲ
δεῖν διὰ τοῦτο τὴν τέχνην ὡς οὐκ ἐοῦσαν οὐδὲ καλῶς
ζητεομένην τὴν ἀρχαίην ἀποβαλέσθαι, εἰ μὴ ἔχει
περὶ πάντα ἀκρίβειαν, ἀλλὰ πολὺ μᾶλλον διὰ τὸ ἐγ-
γὺς οἶμαι τοῦ ἀτρεκεστάτου ὁμοῦ[28] δύνασθαι ἥκειν
598   λογισμῷ | ἐκ πολλῆς ἀγνωσίης θαυμάζειν τὰ ἐξευρη-
μένα, ὡς καλῶς καὶ ὀρθῶς ἐξεύρηται καὶ οὐκ ἀπὸ τύχης.
        13. Ἐπὶ δὲ τῶν τὸν καινὸν τρόπον τὴν τέχνην ζη-
τεύντων ἐξ ὑποθέσιος λόγον ἐπανελθεῖν βούλομαι. εἰ
γάρ τί ἐστιν θερμὸν ἢ ψυχρὸν ἢ ξηρὸν ἢ ὑγρὸν τὸ
λυμαινόμενον τὸν ἄνθρωπον, καὶ δεῖ τὸν ὀρθῶς ἰη-
τρεύοντα βοηθεῖν τῷ μὲν θερμῷ ἐπὶ τὸ ψυχρόν, τῷ δὲ
ψυχρῷ ἐπὶ τὸ θερμόν, τῷ δὲ ξηρῷ ἐπὶ τὸ ὑγρόν, τῷ
δὲ ὑγρῷ ἐπὶ τὸ ξηρόν. ἔστω μοι ἄνθρωπος μὴ τῶν
ἰσχυρῶν φύσει, ἀλλὰ τῶν ἀσθενεστέρων· οὗτος δὲ
πυροὺς ἐσθιέτω, οὓς ἂν ἀπὸ τῆς ἅλω ἀνέλῃ, ὠμοὺς
καὶ ἀργούς, καὶ κρέα ὠμὰ καὶ πινέτω ὕδωρ. ταύτῃ
χρεώμενος τῇ διαίτῃ εὖ οἶδ᾽ ὅτι πείσεται πολλὰ καὶ

12. Such constitutions, I contend, that rapidly and severely feel the effects of errors, are weaker than others. A weak person is only one step removed from a sickly person, but a sickly person is weaker still, and is more apt to suffer distress whenever he misses the correct measure. And when medicine involves such nice exactness, it is difficult always to attain perfect accuracy, although many aspects of the art have already reached that high degree of exactness; about these I will speak later. I insist, however, that we should not reject the ancient art as nonexistent, or as not having discovered anything true, just because it has not attained exactness in every detail; but much rather, because it has been able by reasoning to rise from deep ignorance very close to perfect accuracy, I think we should admire its discoveries as the work, not of chance, but of scientific inquiry honestly and correctly conducted.

13. I would like now to return to the argument of those who conduct their investigations into the art according to the new fashion, by building upon a postulate. So let us begin by assuming that there is such a thing as heat, or cold, or dryness, or moistness, which injures a person, and that the healer must counteract cold with hot, hot with cold, moist with dry and dry with moist. Take a person whose constitution is not strong, but rather of the weaker kind; have him eat wheat straight from the threshing floor, untoasted and unworked, and raw meat, and drink water. The employment of this diet will assuredly cause him

---

[27] δὲ A: δὴ M: δὲ μὴ Erotian s.v. ἀτρεκέως (p. 11,8)
[28] ὁμοῦ M: οὐ A

ΠΕΡΙ ΑΡΧΑΙΗΣ ΙΗΤΡΙΚΗΣ

δεινά· καὶ γὰρ πόνους πονήσει καὶ τὸ σῶμα ἀσθενὲς
ἔσται καὶ ἡ κοιλίη φθαρήσεται καὶ ζῆν πολὺν χρόνον
οὐ δυνήσεται.

Τί δὴ χρὴ βοήθημα παρεσκευάσθαι ὧδ᾽ ἔχοντι;
θερμὸν ἢ ψυχρὸν ἢ ξηρὸν ἢ ὑγρόν; δῆλον[29] ὅτι τούτων
τι. εἰ γὰρ τὸ λυμαινόμενόν ἐστιν τούτων τὸ ἕτερον, τῷ
ὑπεναντίῳ προσήκει λῦσαι, ὡς ὁ ἐκείνων λόγος ἔχει.
τὸ μὲν γὰρ βεβαιότατόν τε καὶ προφανέστατον φάρμακον ἀφελόντα τὰ διαιτήματα, οἷσιν ἐχρῆτο, ἀντὶ
μὲν τῶν πυρῶν ἄρτον διδόναι, ἀντὶ δὲ τῶν ὠμῶν
κρεῶν ἑφθά, πιεῖν τε ἐπὶ τούτοισιν οἴνου. ταῦτα μετα-
βάλλοντα οὐχ οἷόν τε μὴ οὐχ ὑγιέα γενέσθαι, ἤν γε
μὴ παντάπασιν ᾖ διεφθαρμένος ὑπὸ χρόνου τε καὶ
τῆς διαίτης. τί δὴ φήσομεν; πότερον αὐτῷ ὑπὸ[30] ψυ-
χροῦ κακοπαθέοντι θερμὰ ταῦτα προσενέγκαντες |
ὠφέλησαν—ἢ τἀναντία; οἶμαι γὰρ ἔγωγε πολλὴν
ἀπορίην ἐρωτηθέντι παρασχεῖν. ὁ γὰρ τὸν ἄρτον
παρασκευάζων τῶν πυρῶν τὸ θερμὸν ἢ τὸ ψυχρὸν ἢ
τὸ ξηρὸν ἢ τὸ ὑγρὸν ἀφείλατο; ὃ γὰρ καὶ πυρὶ καὶ
ὕδατι δέδοται καὶ πολλοῖσιν ἄλλοισιν εἴργασται, ὧν
ἕκαστον ἰδίην δύναμιν καὶ φύσιν ἔχει, τὰ μὲν τῶν
ὑπαρχόντων ἀποβέβληκεν, ἄλλοισι δὲ κέκρηταί τε
καὶ μέμικται.

14. Οἶδα μὲν γὰρ καὶ τάδε δήπου, ὅτι διαφέρει ἐς
τὸ σῶμα τοῦ ἀνθρώπου καθαρὸς ἄρτος ἢ συγκομι-
στός, ἢ ἀπτίστων πυρῶν ἢ ἐπτισμένων, ἢ πολλῷ
ὕδατι πεφυρημένος ἢ ὀλίγῳ, <ἢ>[31] ἰσχυρῶς πεφυρη-

600

much severe suffering; he will experience pains and weakness of his body, his cavity will be disturbed, and most likely he will be unable to survive for very long.

Well, what remedy should be prepared for a man in this condition—heat or cold or dryness or moistness? One of these, plainly, for if the injury was caused by one of the opposites, it would be requisite to treat it with its contrary, as their theory implies. In reality, of course, the most reliable and most obvious medicine would be for the patient to abandon the regimen he was using, and to give him bread instead of wheat, meat boiled instead of raw, and after that some wine to drink. These changes could not but restore him to health, unless indeed his health had been entirely ruined by long continuance of the diet. What then shall we say? That he has been suffering from cold, and that taking these hot things benefited him? Or the opposite? I believe I have put my interlocutor in quite a quandary! For is it the heat of the wheat, or the cold, or the dryness, or the moistness the baker removed from it? For something that has been exposed to fire and water, and that has been compounded from many different things, each of which has its own particular potency and nature, has lost some of its own qualities, and has been mixed and combined with others.

14. Of course I know that it also makes a difference to a person's body whether his bread is made from bolted or unbolted flour, whether it is made from winnowed or unwinnowed wheat, whether it has been kneaded with much water or with little water, whether it has been thoroughly

---

29 δῆλον A: ἁπλὸν γὰρ M   30 ὑπὸ M: ἀπὸ A
31 ἢ add. Ermerins

μένος ἢ ἀφύρητος, ἢ ἔξοπτος ἢ ἔνωμος, ἄλλα τε πρὸς
τούτοισι μυρία. ὡς δ᾽ αὕτως καὶ περὶ μάζης· καὶ αἱ
δυνάμιες δὲ μεγάλαι τε ἑκάστου καὶ οὐδὲν ἡ ἑτέρη
τῇ ἑτέρῃ ἐοικυῖα. ὅστις δὲ ταῦτα οὐκ ἐπέσκεπται ἢ
σκεπτόμενος οὐκ οἶδε, πῶς ἄν τι οὗτος δύναιτο τῶν
κατὰ τὸν ἄνθρωπον παθημάτων εἰδέναι; ὑπὸ γὰρ ἑνὸς
ἑκάστου τούτων πάσχει τε καὶ ἑτεροιοῦται ὤνθρωπος
ἢ τοῖον ἢ τοῖον, καὶ διὰ τούτων πᾶς ὁ βίος καὶ ὑγιαίνοντι καὶ ἐκ νούσου ἀνατρεφομένῳ καὶ κάμνοντι.

Οὐκ ἂν οὖν ἕτερα τούτων χρησιμώτερα οὐδ᾽ ἀναγ-
καιότερα εἴη εἰδέναι δήπου, ὡς δὲ καλῶς καὶ λογισμῷ
προσήκοντι ζητήσαντες πρὸς τὴν τοῦ ἀνθρώπου φύσιν εὗρον αὐτὰ οἱ πρῶτοι εὑρόντες καὶ ᾠήθησαν
602   ἀξίην τὴν τέχνην θεῷ | προσθεῖναι, ὥσπερ καὶ νομί-
ζεται. οὐ γὰρ τὸ ξηρὸν οὐδὲ τὸ ὑγρὸν οὐδὲ τὸ θερμὸν
οὐδὲ τὸ ψυχρὸν οὐδὲ ἄλλο τούτων οὐδὲν ἡγησάμενοι
οὔτε λυμαίνεσθαι οὔτε προσδεῖσθαι οὐδενὸς τούτων
τὸν ἄνθρωπον, ἀλλὰ τὸ ἰσχυρὸν ἑκάστου καὶ τὸ κρέσσον τῆς φύσιος τῆς ἀνθρωπίνης, οὗ μὴ ἠδύνατο κρατέειν, τοῦτο βλάπτειν ἡγήσαντο καὶ τοῦτο ἐζήτησαν
ἀφελέειν.[32] ἰσχυρότατον δ᾽ ἐστὶ τοῦ μὲν γλυκέος τὸ
γλυκύτατον, τοῦ δὲ πικροῦ τὸ πικρότατον, τοῦ δὲ
ὀξέος τὸ ὀξύτατον, ἑκάστου δὲ πάντων τῶν ἐνεόντων
ἡ ἀκμή. ταῦτα γὰρ ἑώρων καὶ ἐν τῷ ἀνθρώπῳ ἐνεόντα
καὶ λυμαινόμενα τὸν ἄνθρωπον. ἔνι γὰρ ἐν ἀνθρώπῳ
καὶ ἁλμυρὸν καὶ πικρὸν καὶ γλυκὺ καὶ ὀξὺ καὶ στρυφ-

---

[32] -ελέειν M: -αιρεῖν A

kneaded or left unkneaded, whether it has been thoroughly baked or underbaked, and that there are countless other differences too; and barley cake varies in just the same way. The potencies of each variety are also powerful, and none is at all like any other. But how could someone who has not examined these matters, or who examines them without instruction, gain any insight about human ailments? For from each of these different factors a person is affected and changed in one way or another, and his whole life (sc. is affected), whether he is healthy, or recovering from a disease, or ill.

Accordingly, there could surely be nothing more useful or more necessary to know than these things, and how their first discoverers, pursuing their inquiries excellently and with suitable application of reason to human nature, made their discoveries, and even considered this art worthy to be ascribed to a god, as in fact is the usual belief. For they did not believe that the dry or the moist or the hot or the cold or anything else of that kind injures a person, or that he has need of any such thing, but they held that it is the strength of each food too powerful for the human constitution to assimilate which causes harm, and this they sought to remove. The strongest part of the sweet is what is sweetest, of the bitter the most bitter, of the acid the most acid, and each of the constituents has its extreme: these extremes the investigators also discovered inside the human being, as well as the fact that they could be injurious to a person; for there are in a person salty and bitter, sweet and acid, astringent and insipid, and a vast number

νὸν καὶ πλαδαρὸν καὶ ἄλλα μυρία παντοίας δυνάμιας
ἔχοντα πλῆθός τε καὶ ἰσχύν. ταῦτα μὲν μεμιγμένα
καὶ κεκρημένα ἀλλήλοισιν οὔτε φανερά ἐστιν οὔτε
λυπέει τὸν ἄνθρωπον. ὅταν δέ τι τούτων ἀποκριθῇ καὶ
αὐτὸ ἐφ᾽ ἑωυτοῦ γένηται, τότε καὶ φανερόν ἐστι καὶ
λυπέει τὸν ἄνθρωπον· τοῦτο δέ, τῶν βρωμάτων ὅσα
ἡμῖν ἀνεπιτήδειά ἐστιν καὶ λυμαίνεται τὸν ἄνθρωπον
ἐσπεσόντα,[33] τούτων ἓν ἕκαστον ἢ πικρόν ἐστιν καὶ
ἄκρητον ἢ ἁλμυρὸν ἢ ὀξὺ ἢ ἄλλο τι ἄκρητόν τε καὶ
ἰσχυρόν, καὶ διὰ τοῦτο ταρασσόμεθα ὑπ᾽ αὐτῶν,
604 ὥσπερ καὶ ὑπὸ τῶν ἐν | τῷ σώματι ἀποκρινομένων.

Πάντα δὲ ὅσα ἄνθρωπος ἐσθίει ἢ πίνει, τὰ τοιαῦτα
βρώματα ἥκιστα τοιούτου χυμοῦ ἀκρήτου τε καὶ δια-
φέροντος δῆλά ἐστιν μετέχοντα, οἷον ἄρτος τε καὶ
μᾶζα καὶ τὰ ἑπόμενα τούτοισιν, οἷσιν εἴθισται ὁ ἄνθρωπος πλείστοισί τε καὶ αἰεὶ χρῆσθαι, ἔξω τῶν πρὸς
ἡδονήν τε καὶ κόρον ἠρτυμένων τε καὶ ἐσκευασμένων.
καὶ ἀπὸ τούτων πλείστων ἐσιόντων ἐς τὸν ἄνθρωπον
τάραχος τε καὶ ἀπόκρισις τῶν ἀμφὶ τὸ σῶμα δυναμίων ἥκιστα γίνεται, ἰσχὺς δὲ καὶ αὔξησις καὶ τροφὴ
μάλιστα δι᾽ οὐδὲν ἕτερον γίνεται[34] ἢ ὅτι εὖ τε συγ-
κέκρηται[35] καὶ οὐδὲν ἔχει οὔτε ἄκρητον οὔτε ἰσχυρόν,
ἀλλ᾽ ὅλον ἕν τε γέγονε καὶ ἁπλοῦν καὶ ἥσυχον.[36]

15. Ἀπορέω δ᾽ ἔγωγε, οἱ τὸν λόγον ἐκεῖνον λέγοντες
καὶ ἄγοντες ἐκ ταύτης τῆς ὁδοῦ ἐπὶ ὑπόθεσιν τὴν

[33] ἔσπες. Α: ἔκπεσ. Μ   [34] γίνεται Α: om. Μ
[35] συγκέκρ. Μ: κέκρ. Α

36

of other things, possessing potencies of all sorts, varying both in quantity and in strength. These, when mixed and compounded with one another are neither apparent nor do they harm a person; but when one of them is separated off, and exists by itself, then it becomes apparent and causes harm. Moreover, of the foods that are unsuitable for us and cause harm to a person when they are ingested, each one is either bitter and uncompounded, or salty, or acid, or something else uncompounded and strong, and for this reason we become disordered by them, just as we are by things that separate out inside the body.

But all things people eat or drink (sc. regularly) are manifestly quite free from any such uncompounded and potent humor, e.g., bread, barley cake, and suchlike, which a person is accustomed constantly to use in great quantity—with the exception of those dishes which are seasoned and prepared with a view to pleasure and sur-feit—and from such foods, when plentifully partaken of by a person, there arises no disorder at all or isolation of the properties present in the body; but strength, growth and nourishment in great measure result, for the simple reason that they are well-compounded, and contain noth-ing undiluted and strong, but form a single, simple whole at rest.[8]

15. I am at a loss to understand how those who main-tain the other view, and abandon the current method to

---

[8] Littré's change of ἰσχυρόν to ἥσυχον saves the sense of the passage; Kühlewein and Jouanna delete καὶ ἰσχυρόν.

---

[36] ἥσυχον Littré in *app. crit.*: ἰσχυρόν MA

τέχνην τίνα ποτὲ τρόπον θεραπεύουσι τοὺς ἀνθρώπους, ὥσπερ ὑποτίθενται. οὐ γάρ ἐστιν αὐτοῖσιν, οἶμαι, ἐξευρημένον αὐτό τι ἐφ' ἑωυτοῦ θερμὸν ἢ ψυχρὸν
ἢ ξηρὸν ἢ ὑγρὸν μηδενὶ ἄλλῳ εἴδει κοινωνέον. ἀλλ'
οἶμαι ἔγωγε ταῦτα βρώματα καὶ πόματα αὐτοῖσιν
ὑπάρχειν, οἷσι πάντες χρεώμεθα· προστιθέασι δὲ τῷ
606   μὲν εἶναι θερμῷ, | τῷ δὲ ψυχρῷ, τῷ δὲ ξηρῷ, τῷ δὲ
ὑγρῷ· ἐπεὶ ἐκεῖνό γε ἄπορον προστάξαι τῷ κάμνοντι
θερμόν τι προσενέγκασθαι. εὐθὺ γὰρ ἐρωτήσει· τί;
ὥστε ληρεῖν ἀνάγκη ἢ ἐς τούτων τι τῶν γινωσκο-
μένων καταφεύγειν. εἰ δὲ δὴ τυγχάνει τι θερμὸν ἐὸν
στρυφνόν, ἄλλο δὲ θερμὸν ἐὸν πλαδαρόν, ἄλλο δὲ
θερμὸν ἄραδον ἔχον—ἔστι γὰρ καὶ ἄλλα πολλὰ
θερμὰ καὶ ἄλλας πολλὰς[37] δυνάμιας ἔχοντα ἑωυτοῖσιν
ὑπεναντίας[38]—ἢ διοίσει τι αὐτῶν προσενεγκεῖν τὸ
θερμὸν καὶ στρυφνὸν ἢ τὸ θερμὸν καὶ πλαδαρὸν ἢ
ἅμα τὸ ψυχρὸν καὶ στρυφνόν—ἔστι γὰρ καὶ τοιοῦτο—ἢ τὸ ψυχρόν τε καὶ πλαδαρόν· ὡς μὲν[39] γὰρ ἐγὼ
οἶδα, πᾶν τοὐναντίον ἀφ' ἑκατέρου αὐτῶν ἀποβαίνει,
οὐ μοῦνον ἐν ἀνθρώπῳ, ἀλλὰ καὶ ἐν σκύτει καὶ ἐν
ξύλῳ καὶ ἐν ἄλλοισι πολλοῖσιν, ἅ ἐστιν ἀνθρώπου
ἀναισθητότερα. οὐ γὰρ τὸ θερμόν ἐστι τὸ τὴν μεγάλην δύναμιν ἔχον, ἀλλὰ τὸ στρυφνὸν καὶ τὸ πλαδαρὸν καὶ τἄλλα ὅσα μοι εἴρηται καὶ ἐν τῷ ἀνθρώπῳ
καὶ ἔξω τοῦ ἀνθρώπου, καὶ ἐσθιόμενα καὶ πινόμενα
καὶ ἔξωθεν ἐπιχριόμενά τε καὶ προσπλασσόμενα.

[37] πολλὰς Α: om. Μ

38

base medicine on a postulate, could treat their patients according to their postulate. For they have not identified, I believe, anything that is hot or cold, dry or moist, *of and by itself*, sharing no part in some other quality, but as far as I can see they have at their disposal the same foods and drinks we all use. To one of these, however, they assign the attribute of being hot, to another, cold, to another, dry, to another, moist, since it would be futile simply to order a patient to take something "hot," as he would immediately ask: "But what?" So that these practitioners would either have to talk nonsense, or take recourse to some known substance. And if one hot thing happens to be astringent, and another hot thing insipid, and a third hot thing irritating (for there are many different kinds of hot things, possessing many other potencies even contrary to each other), surely it will make a difference whether a patient is given the hot astringent thing, or the hot insipid, or what is cold and astringent at the same time (for there is such a thing), or the cold insipid thing. In fact I am sure that each one of these pairs produces an effect exactly the opposite of that produced by the other, not only in a human being, but even in a leathern or wooden vessel, and in many other things less sensitive than a human being. For it is not heat which possesses such a great potency, but the astringent and the insipid and the other qualities I have mentioned, both inside and outside the human being, both when eaten and when drunk, and when applied externally as an ointment or a plaster.

---

[38] ἔχ. ἑωυτοῖσιν ὑπ. Jouanna: ὑπ. ἑωυτῆσιν ἔχ. M: ἔχ. αὐτοῖς ὑπ. A   [39] ὡς μὲν M: ὥσπερ A

16. Ψυχρότητα δ' ἐγὼ καὶ θερμότητα πασέων ἥκιστα τῶν δυναμίων νομίζω δυναστεύειν ἐν τῷ σώματι
διὰ τάσδε τὰς προφάσιας·[40] ὃν μὲν ἂν δήπου χρόνον
μεμιγμένα αὐτὰ ἑωυτοῖς ἅμα τὸ ψυχρόν τε καὶ θερμὸν
608 ἐνῇ, οὐ λυπέει. κρῆσις γὰρ καὶ μετριότης | τῷ μὲν
ψυχρῷ γίνεται ἀπὸ τοῦ θερμοῦ, τῷ δὲ ψυχρῷ ἀπὸ τοῦ
θερμοῦ. ὅταν δ' ἀποκριθῇ χωρὶς ἑκάτερον, τότε λυπέει. ἐν δὲ δὴ τούτῳ τῷ καιρῷ, ὅταν τὸ ψυχρὸν ἐπιγένηται καί τι λυπήσῃ τὸν ἄνθρωπον, διὰ τάχεος
πρῶτον δι' αὐτὸ τοῦτο πάρεστι τὸ θερμὸν αὐτόθεν ἐκ
τοῦ ἀνθρώπου, οὐδεμιῆς βοηθείης οὐδὲ παρασκευῆς
δεόμενον· καὶ ταῦτα καὶ ἐν ὑγιαίνουσι τοῖσιν ἀνθρώποισιν ἀπεργάζεται καὶ κάμνουσιν. τοῦτο μέν, εἴ τις
θέλει ὑγιαίνων χειμῶνος διαψῦξαι τὸ σῶμα ἢ λουσά-
μενος ψυχρῷ ἢ ἄλλῳ τῳ τρόπῳ, ὅσῳ ἂν ἐπὶ πλέον
αὐτὸ ποιήσῃ—καὶ ἤν γε μὴ παντάπασιν παγῇ τὸ
σῶμα—ὅταν εἵματα λάβῃ καὶ ἔλθῃ ἐς τὴν σκέπην, ἔτι
μᾶλλον καὶ ἐπὶ πλέον θερμαίνεται τὸ σῶμα· τοῦτο δέ,
εἰ ἐθέλοι ἐκθερμανθῆναι ἰσχυρῶς ἢ λουτρῷ θερμῷ ἢ
πυρὶ πολλῷ, ἐκ δὲ τούτου τωὐτὸ εἷμα ἔχων ἐν τῷ αὐτῷ
χωρίῳ τὴν διατριβὴν ποιεῖσθαι ὥσπερ διεψυγμένος,
πολὺ φανεῖται καὶ ψυχρότερος καὶ ἄλλως φρικαλε-
ώτερος· ἢ <εἰ>[41] ῥιπιζόμενός τις ὑπὸ πνίγεος καὶ
610 παρασκευαζόμενος | αὐτὸς ἑωυτῷ ψῦχος ἐκ τούτου
τρόπου διαπαύσαιτο τοῦτο ποιέων, δεκαπλάσιον
ἔσται τὸ καῦμα καὶ πνῖγος ἢ τῷ μηδὲν τοιοῦτο ποιέοντι.

Τόδε δὴ καὶ πολὺ μεῖζον· ὅσοι[42] ἂν διὰ χιόνος ἢ

16. In fact, I believe that of all the potencies none hold less sway in the body than cold and heat, for the following reasons. As long as the hot and cold in the body are mixed together, they cause no pain; for the cold is tempered and moderated by the hot, and the hot by the cold. When, however, either of them becomes separated, and resides apart from the other, then it does cause pain. But at that moment, when cold befalls a person and provokes pain, first heat quickly arrives on the scene spontaneously from inside the body, and without the need of any help or preparation; and it does this both in the healthy and in the ill. For instance, if a person in health will cool his body in winter, either by a cold bath or in any other way, the more he does this—provided that his body is not entirely frozen—the more intensely and thoroughly will he be warmed when he puts on his clothes and enters a shelter. Again, if someone will make himself thoroughly hot by means of either a hot bath or a large fire, and afterward wears the same clothes and stays in the same place where he was when he was chilled, he will feel far colder and besides more shivery than he did before. Or if a person who is fanning himself, because of the stifling heat, and providing coolness for himself in this way, ceases to do this, the heat will seem ten times more intense to him than to someone who has done nothing of the sort.

The following is much stronger evidence still. All who,

---

[40] προφάσιας M: αἰτίας M

[41] εἰ add. Ermerins

[42] μεῖζον· ὅσοι Jouanna: μέζω ὅσοι M: μείζονος· οἱ A

ἄλλου ψύχεος βαδίσαντες ῥιγώσωσι διαφερόντως πό-
δας ἢ χεῖρας ἢ κεφαλήν, οἷα πάσχουσιν ἐς τὴν νύκτα,
ὅταν περισταλέωσί τε καὶ ἐν ἀλέῃ γένωνται ὑπὸ
καύματος καὶ κνησμοῦ· καὶ ἔστιν οἷσι φλύκταιναι
ἀνίστανται ὥσπερ τοῖσιν ἀπὸ πυρὸς κατακεκαυμέ-
νοισι· καὶ οὐ πρότερον τοῦτο πάσχουσιν, πρὶν θερ-
μανθῶσιν. οὕτως ἑτοίμως ἑκάτερον αὐτῶν ἐπὶ θάτερον
παραγίνεται. μυρία δ᾽ ἂν καὶ ἄλλα ἔχοιμι εἰπεῖν. τὰ
δὲ κατὰ τοὺς νοσέοντας, οὐχὶ ὅσοισιν ἂν ῥῖγος γένηται, τούτοισιν ὀξύτατος ὁ πυρετὸς ἐκλάμπει—καὶ οὐχὶ
οὕτως ἰσχυρός, ἀλλὰ καὶ παυόμενος δι᾽ ὀλίγου, καὶ
ἄλλως τὰ πολλὰ ἀσινής; καὶ ὅσον ἂν χρόνον παρῇ,
διάθερμος καὶ διεξιὼν διὰ παντὸς τελευτᾷ ἐς τοὺς πό-
612 δας μάλιστα, οὗπερ τὸ ῥῖγος καὶ | ἡ ψῦξις νεηνικω-
τάτη καὶ ἐπὶ πλεῖον ἐνεχρόνισεν. πάλιν τε ὅταν
ἱδρώσῃ τε καὶ ἀπαλλαγῇ ὁ πυρετός, πολὺ μᾶλλον
διέψυξεν ἢ εἰ μὴ ἔλαβε τὴν ἀρχήν. ᾧ οὖν διὰ τάχεος
οὕτω παραγίνεται τὸ ἐναντιώτατόν τε καὶ ἀφαιρεόμε-
νον τὴν δύναμιν ἀπὸ ταὐτομάτου, τί ἂν ἀπὸ τούτου
μέγα ἢ δεινὸν γένοιτο; ἢ τί δεῖ πολλῆς ἐπὶ τοῦτο βοηθείης;

17. Εἴποι ἄν τις· ἀλλ᾽ οἱ πυρεταίνοντες τοῖσι καύ-
σοισί τε καὶ περιπλευμονίῃσι καὶ ἄλλοισιν ἰσχυροῖσι
νοσήμασιν οὐ ταχέως ἐκ τῆς θέρμης ἀπαλλάσσονται,
οὐδὲ πάρεστιν ἐνταῦθα ἐπὶ τὸ θερμὸν τὸ ψυχρόν. ἐγὼ

---

[9] The text "a fever which . . . longer than elsewhere" is difficult

on walking through snow or some other intense cold, become overchilled in their feet, hands or head—what suffering they endure at night, when they come into a warm place and wrap up, from burning and tingling! In some cases blisters even form, like those caused by burning from a fire. But it is not until these people warm up that they suffer the symptoms. So ready is cold to pass into heat and heat into cold. But I could give a multitude of other proofs. In the ill, for example, does fever not blaze out most acutely in those who are experiencing a chill—a fever which however is not very strong, but which gradually declines, and generally does no harm? And then for as long as it (sc. the chill) persists, (sc. the fever) remains hot, and spreading through the whole (sc. body) comes to an end in most cases in the feet, where the chill and the cold are most violent and last longer than elsewhere.[9] Again, when fever disappears with the breaking out of perspiration, it cools a patient so that he is far colder than if fever had never attacked him in the first place. Now in this case where the most contrary potency arrives spontaneously so quickly to nullify the first one, what great or terrible effect could result? And what great intervention would be called for in such a case?

17. Someone might retort: "But febrile patients in ardent fevers, pneumonias, or other virulent diseases[10] do not get rid of their feverishness very quickly, and in these cases cold does not accompany the heat." Now I consider

and uncertain: I have followed Jouanna's interpretation without being totally convinced of its correctness.   [10] Two of these diseases belong to the group of four acute diseases specified in *Regimen in Acute Diseases* 5 and *Affections* 6.

δέ τοῦτό μοι μέγιστον τεκμήριον ἡγεῦμαι εἶναι, ὅτι
οὐ διὰ τὸ θερμὸν ἁπλῶς πυρεταίνουσιν οἱ ἄνθρωποι,
οὐδὲ τοῦτ' εἴη τὸ αἴτιον τῆς κακώσιος μοῦνον, ἀλλ'
ἔστι καὶ πικρὸν καὶ θερμὸν τὸ αὐτό, καὶ ὀξὺ καὶ θερμόν, καὶ ἁλμυρὸν καὶ θερμόν, καὶ ἄλλα μυρία—καὶ
πάλιν γε ψυχρὸν μετὰ δυναμίων ἑτέρων. τὰ μὲν οὖν
λυμαινόμενα ταῦτ' ἐστί· συμπάρεστι δὲ καὶ τὸ θερμόν
ῥώμης μετέχον ὡς ἂν τὸ ἡγεύμενον καὶ παροξυνόμε-
νον καὶ αὐξόμενον ἅμα κείνῳ, δύναμιν δὲ οὐδεμίαν
πλείω τῆς προσηκούσης.

18. Δῆλα δὲ ταῦτα ὅτι ὧδε ἔχει ἐπὶ τῶνδε τῶν ση-
μείων· πρῶτον μὲν ἐπὶ τὰ φανερώτατα,[43] ὧν πάντες
614 ἔμπειροι πολλάκις ἐσμέν | τε καὶ ἐσόμεθα. τοῦτο μὲν
γάρ, ὅσοισιν ἂν ἡμέων κόρυζα ἐγγένηται καὶ ῥεῦμα
κινηθῇ διὰ τῶν ῥινῶν, τοῦτο ὡς τὸ πολὺ δριμύτερον
τοῦ πρότερον γινομένου τε καὶ ἰόντος ἐκ τῶν ῥινῶν
καθ' ἑκάστην ἡμέρην· καὶ οἰδέειν μὲν ποιέει τὴν ῥῖνα
καὶ συγκαίειν[44] θερμήν τε καὶ διάπυρον ἐσχάτως· ἢν
δὲ τὴν χεῖρα προσφέρῃς καὶ πλείω χρόνον παρῇ, καὶ
ἐξελκοῦται τὸ χωρίον ἄσαρκόν τε καὶ σκληρὸν ἐόν.
παύεται δέ πως τό γε καῦμα ἐκ τῆς ῥινός; οὐχ ὅταν
τὸ ῥεῦμα γίνηται καὶ ἡ φλεγμονὴ ᾖ, ἀλλ' ἐπειδὰν
παχύτερόν τε καὶ ἧσσον δριμὺ ῥέῃ, καὶ πέπον καὶ
μεμιγμένον μᾶλλον τῷ πρότερον γινομένῳ, τότε δὲ
ἤδη καὶ τὸ καῦμα πέπαυται. ἀλλ' οἷσι δὲ ὑπὸ ψύχεος
φανερῶς αὐτοῦ μόνου γίνεται μηδενὸς ἄλλου συμ-

43 -ώτατα M: -ώτερα A   44 συγκαίειν M: συγκαίει A

44

this my strongest evidence that patients do not become feverish merely from heat, and that heat is not the sole cause of their harm, and also that one and the same thing can be both bitter and hot, or acid and hot, or salty and hot, with numerous other combinations, and that cold, for its part, also combines with other potencies. In fact, it is these concomitant powers which really cause the harm. Admittedly, heat, too, is present, but merely as a subsidiary, having strength as it is directed, aggravated, and increased in combination with the other factor, but possessing no potency of its own greater than what properly belongs to it.

18. That this is so is plain if we consider the following pieces of evidence. First we have the most obvious things, which all of us often experience and will continue to experience. As a first example, those of us who suffer from a coryza, with a discharge through the nostrils, generally find this discharge more acrid than that which previously formed there and passed each day from the nostrils; this makes the nose swell and become inflamed with an extremely fiery heat. If you put your hand on it, and the disease is present for an unusually long time, the part actually becomes ulcerated, being fleshless and hard. But how does the heat of the nostril cease? Not when a discharge is taking place and inflammation is present, but when the discharge becomes thicker and less acrid, and concocted and more mixed than what was produced before, it is then that the heat finally ceases. Now in cases where the condition has clearly come from cold itself

παραγενομένου, πᾶσιν αὕτη ἡ ἀπαλλαγή· ἐκ μὲν τῆς
ψύξιος διαθερμανθῆναι, ἐκ δὲ τοῦ καύματος δια-
ψυχθῆναι, καὶ ταῦτα ταχέως παραγίνεται καὶ πέψιος
616 οὐδεμιῆς | προσδεῖται. τὰ δ' ἄλλα πάντα, ὅσα διὰ
χυμῶν δριμύτητας καὶ ἀκρησίας φημὶ ἔγωγε γίνεσθαι, τὸν αὐτὸν τρόπον ἀποκαθίσταται⁴⁵ πεφθέντα
καὶ κρηθέντα.

19. Ὅσα τε αὖ ἐπὶ τοὺς ὀφθαλμοὺς τρέπεται τῶν
ῥευμάτων, ὡς⁴⁶ ἰσχυρὰς καὶ παντοίας δριμύτητας
ἔχοντα, ἑλκοῖ μὲν βλέφαρα, κατεσθίει δ' ἐνίων γνά-
θους τε καὶ τὰ ὑπὸ τοῖσιν ὀφθαλμοῖσιν, ἐφ' ὅ τι ἂν
ἐπιρρυῆ, ῥήγνυσι δὲ καὶ διεσθίει τὸν ἀμφὶ τὴν ὄψιν
χιτῶνα. ὀδύναι δὲ καὶ καῦμα καὶ φλογμὸς ἔσχατος
κατέχει μέχρι τινός; μέχρι ἂν τὰ ῥεύματα πεφθῆ καὶ
γένηται παχύτερα καὶ λήμη ἀπ' αὐτῶν ᾖ. τὸ δὲ πε-
φθῆναι γίνεται ἐκ τοῦ μιχθῆναι καὶ κρηθῆναι ἀλλήλοισι καὶ συνεψηθῆναι. τοῦτο δέ, ὅσα ἐς τὴν
φάρυγγα, ἀφ' ὧν βράγχοι γίνονται καὶ κυνάγχαι,
ἐρυσιπέλατά τε καὶ περιπλευμονίαι, πάντα ταῦτα τὸ
μὲν πρῶτον ἁλμυρά τε καὶ ὑγρὰ καὶ δριμέα ἀφίει—
καὶ ἐν τοῖσι τοιούτοισιν ἔρρωται τὰ νοσήματα—ὅταν
δὲ παχύτερα καὶ πεπαίτερα γένηται καὶ πάσης δρι-
μύτητος ἀπηλλαγμένα, τότ' ἤδη καὶ οἱ πυρετοὶ παύονται καὶ τἄλλα τὰ λυπέοντα τὸν ἄνθρωπον. δεῖ δὲ
δήπου ταῦτα αἴτια ἑκάστου ἡγεῖσθαι εἶναι, ὧν παρε-
όντων μὲν τοιουτότροπον ἀνάγκη γίνεσθαι,⁴⁷ μετα-
618 βαλλόντων δ' ἐς ἄλλην κρῆσιν | παύεσθαι. ὁπόσα τε

alone, unaccompanied by any other factor, in every case recovery occurs by the cold being heated, and what is hot being cooled, and these changes take place rapidly and require no concoction. In all other diseases, those which I contend arise from acridness and the separation of the humors, resolution always comes the same way, through concoction and compounding (sc. of the humors).

19. In the second place, discharges that turn toward the eyes, possessing powerful, acrid humors of all sorts, ulcerate the eyelids, and in some cases corrode the parts to which they run, the cheeks and under the eyes; and they also rupture and eat through the tunic of the eyeball. Pains, burning, and intense inflammation prevail and for how long? Until the discharges are concocted and become thicker, so that rheum is formed from them. This concoction is the result of mixture, compounding, and digestion. Another example: fluxes to the throat from which sore throats, anginas, erysipelas and pneumonias are engendered all first produce discharges that are salty, fluid, and acrid, from which these diseases draw their strength. But when the discharges become thicker and more mature, and throw off all trace of their acridity, then the fevers too subside with the other symptoms that have distressed the patient. We must surely consider the cause of each complaint to be those substances which, when present, necessarily provoke the appearance of that particular form of disease, but which, on changing into a different mixture (sc. of potencies), make it remit. All conditions, then, re-

---

45 ἀποκαθίσταται M: καὶ ἀποκαθίστασθαι A
46 ὡς M: om. A
47 ἀνάγκη γίνεσθαι M: γενέσθαι ἀνάγκη A

οὖν ἀπ' αὐτῆς τῆς θέρμης εἰλικρινέος ἢ ψύξιος γίνεται καὶ μὴ μετέχει ἄλλης δυνάμιος μηδεμιῆς, οὕτω
παύοιτ' ἄν, ὅταν μεταβάλλῃ ἐκ τοῦ θερμοῦ ἐς τὸ ψυχρὸν καὶ ἐκ τοῦ ψυχροῦ ἐς τὸ θερμόν. μεταβάλλει δὲ
ὅνπερ προείρηταί μοι τρόπον.

Ἔτι τοίνυν τἆλλα ὅσα κακοπαθέει ὤνθρωπος
πάντα ἀπὸ δυναμίων γίνεται. τοῦτο μὲν γάρ, ὅταν
πικρότης τις ἀποχυθῇ, ἣν δὴ χολὴν ξανθὴν καλέο-
μεν, οἷαι ἄσαι καὶ καύματα καὶ ἀδυναμίαι κατέχου-
σιν· ἀπαλλασσόμενοι δὲ τούτου, ἐνίοτε καὶ καθαιρό-
μενοι, ἢ αὐτόματοι ἢ ὑπὸ φαρμάκου, ἢν ἐν καιρῷ τι
αὐτῶν γίνηται, φανερῶς καὶ τῶν πόνων καὶ τῆς θέρμης ἀπαλλάσσονται. ὅσον δ' ἂν χρόνον ταῦτα μετέωρα ᾖ καὶ ἄπεπτα καὶ ἄκρητα, μηχανὴ οὐδεμία
οὔτε τῶν πόνων παύεσθαι οὔτε τῶν πυρετῶν. καὶ
ὅσοισι δὲ ὀξύτητες προσίστανται δριμεῖαί τε καὶ ἰώ-
δεες, οἷαι λύσσαι καὶ δήξιες σπλάγχνων καὶ θώρηκος
καὶ ἀπορίη· οὐ παύεταί τε τούτου πρότερον, πρὶν ἢ
ἀποκαθαρθῇ τε καὶ καταστορεσθῇ καὶ μιχθῇ τοῖσιν
ἄλλοισιν. πέσσεσθαι δὲ καὶ μεταβάλλειν καὶ λεπτύ-
νεσθαί τε καὶ παχύνεσθαι ἐς χυμῶν εἶδος διὰ πολλῶν
εἰδέων καὶ παντοίων—διὸ καὶ κρίσιες καὶ ἀριθμοὶ τῶν
χρόνων ἐν τοῖσι τοιούτοισι μέγα δύνανται—πάντων
δὴ τούτων ἥκιστα προσήκει θερμῷ ἢ ψυχρῷ πάσχειν·
οὔτε γὰρ ἂν τοῦτό γε σαπείη οὔτε παχυνθείη. †τί γὰρ
αὐτὸ φήσωμεν εἶναι; κρήσιας αὐτῶν ἄλλην πρὸς ἄλ-
620 ληλα ἐχούσας δύναμιν.†[48] ἐπεὶ | ἄλλῳ γε οὐδενὶ τὸ

48

sulting from heat or cold pure and simple, and with no admixture of any other potency, will cease when the heat changes into cold or the cold into heat: this change takes place in the manner I have previously described.

Otherwise, all the other complaints to which a person is liable arise from potencies. Thus, for example, when there is an outpouring of the bitter principle, which we call yellow bile, great nausea, burning, and weakness prevail; but when the patient is relieved from this, sometimes by purgation—either spontaneous or induced by medicine—if one of these operations takes place at the proper time, he is clearly seen to be relieved of both the pains and the heat. But as long as these bitter fluids remain stirred up, unconcocted and uncompounded, there is no way for the pains and fevers to be relieved. And those who are attacked by pungent and acrid acids suffer greatly from frenzy, from gnawing in the viscera and the chest, and from distress. No relief from the condition arrives until (sc. the acidity) is purged away, or calmed down and mixed with the other humors. But least susceptible of all the things (sc. in the body) to be concocted or changed or thinned or thickened into some form of humors, by passing through many and various other forms—for which reason the crises and the numbering of the time periods take on great importance in these cases—are surely heat and cold, since neither of them could ever either ferment or thicken. †For what we shall say is this: combinations of these (i.e., hot and cold) that exhibit a different potency†

---

48 τί δ' ἂν αὐτὸ φαίημεν εἶαι κρῆσίς τε αὐτέων ἐστὶ πλὴν πρὸς ἄλληλα ἔχουσα δύναμιν M: τί γὰρ αὐτὸ φίσωμεν εἶναι; κρήσιας αὐτῶν ἄλλην πρὸς ἄλληλα ἐχούσας δύναμιν A

θερμὸν μιχθὲν παύσεται τῆς θέρμης ἢ τῷ ψυχρῷ οὐδέ
γε τὸ ψυχρὸν ἢ τῷ θερμῷ. τὰ δ᾽ ἄλλα πάντα τὰ περὶ
τὸν ἄνθρωπον, ὅσῳ ἂν πλείοσι μίσγηται, τοσούτῳ
ἠπιώτερα καὶ βελτίονα. πάντων δ᾽ ἄριστα διάκειται
ὥνθρωπος, ὅταν πέσσηται καὶ ἐν ἡσυχίῃ ᾖ, μηδεμίαν
δύναμιν ἰδίην ἀποδεικνύμενα.[49]

20. Περὶ μὲν οὖν τουτῶν ἱκανῶς μοι ἡγεῦμαι ἐπι-
δεδεῖχθαι. λέγουσι δέ τινες ἰητροὶ καὶ σοφισταί, ὡς
οὐκ εἴη δυνατὸν ἰητρικὴν εἰδέναι ὅστις μὴ οἶδεν ὅ τί
ἐστιν ἄνθρωπος, ἀλλὰ τοῦτο δεῖ καταμαθεῖν τὸν μέλ-
λοντα ὀρθῶς θεραπεύσειν τοὺς ἀνθρώπους. τείνει τε
αὐτοῖσιν ὁ λόγος ἐς φιλοσοφίην, καθάπερ Ἐμπεδο-
κλῆς ἢ ἄλλοι οἱ περὶ φύσιος γεγράφασιν ἐξ ἀρχῆς ὅ
τί ἐστιν ἄνθρωπος, καὶ ὅπως ἐγένετο πρῶτον καὶ
ὁπόθεν συνεπάγη. ἐγὼ δὲ τοῦτο μέν, ὅσα τινὶ εἴρηται
ἢ σοφιστῇ ἢ ἰητρῷ ἢ γέγραπται περὶ φύσιος, ἧσσον
νομίζω τῇ ἰητρικῇ τέχνῃ προσήκειν ἢ τῇ γραφικῇ.
622 νομίζω δὲ περὶ φύσιος | γνῶναί τι σαφὲς οὐδαμόθεν
ἄλλοθεν εἶναι ἢ ἐξ ἰητρικῆς· τοῦτο δὲ οἷόν τε καταμαθεῖν, ὅταν αὐτήν τις τὴν ἰητρικὴν ὀρθῶς πᾶσαν[50]
περιλάβῃ—μέχρι δὲ τούτου πολλοῦ μοι δοκέει δεῖν—
λέγω δὲ ταύτην τὴν ἱστορίην εἰδέναι, ἄνθρωπος τί
ἐστι καὶ δι᾽ οἵας αἰτίας γίνεται καὶ τἆλλα ἀκριβέως.

Ἐπεὶ τοῦτό γέ μοι δοκεῖ ἀναγκαῖον εἶναι ἰητρῷ
περὶ φύσιος εἰδέναι καὶ πάνυ σπουδάσαι ὡς εἴσεται,

49 -κνύμενα Ermerins: -κνύμενον ΜΑ
50 πᾶσαν Μ: om. Α

. . . since the hot will give up its heat only when mixed with cold, and the cold can be neutralized only by hot, whereas all the other components of the human body become milder and better, the greater the number of other components with which they are mixed: a person is in the best possible condition when his components are concocted and at rest, displaying no specific potency at all.

20. About this I think that I have given a full explanation. But certain physicians and philosophers assert that no one can know medicine who does not know what the human being is, that anyone intending properly to treat patients must become thoroughly knowledgeable about this: these arguments tend to move in the direction of philosophy, the way Empedocles and others have written about natural science—what a human being is from his beginning, how he first came into existence, and out of what material his body (sc. first) congealed. My own view, however, is first that everything philosophers or physicians have said or written on natural science no more pertains to medicine than it does to painting! Indeed, I believe that any precise knowledge about nature can come from nowhere else than from medicine. This knowledge it is possible to acquire when a person correctly takes up the whole study of medicine itself—but before that it seems to me very inadequate—I mean to possess this information accurately, what a human being is, by what causes he comes into being, and similar points.

The following, at least, I think a physician must know, and be at great pains to know, about natural science, if he

εἴπερ τι μέλλει τῶν δεόντων ποιήσειν—ὅ τί τέ ἐστιν
ἄνθρωπος πρὸς τὰ ἐσθιόμενά τε καὶ πινόμενα καὶ ὅ
τι πρὸς τὰ ἄλλα ἐπιτηδεύματα, καὶ ὅ τι ἀφ᾿ ἑκάστου
ἑκάστῳ συμβήσεται, καὶ μὴ ἁπλῶς οὕτως «πονηρόν
ἐστιν βρῶμα τυρός· πόνον γὰρ παρέχει τῷ πληρω-
θέντι αὐτοῦ», ἀλλὰ τίνα τε πόνον καὶ διὰ τί καὶ τίνι
τῶν ἐν τῷ ἀνθρώπῳ ἐνεόντων ἀνεπιτήδειον. ἔστι γὰρ
καὶ ἄλλα πολλὰ βρώματα καὶ πόματα πονηρά, ἃ δια-
τίθησι τὸν ἄνθρωπον οὐ τὸν αὐτὸν τρόπον. οὕτως οὖν
μοι ἔστω οἷον· οἶνος ἄκρητος πολλὸς ποθεὶς δια-
τίθησί πως τὸν ἄνθρωπον· καὶ πάντες ἂν οἱ ἰδόντες[51]
τοῦτο γνοίησαν, ὅτι αὕτη ἡ δύναμις οἴνου καὶ αὐτὸς
αἴτιος· καὶ οἷσί γε τῶν ἐν τῷ ἀνθρώπῳ τοῦτο δύναταί
γε μάλιστα, οἴδαμεν. τοιαύτην δὴ βούλομαι ἀληθείην
624 καὶ περὶ τῶν | ἄλλων φανῆναι. τυρὸς γάρ, ἐπειδὴ
τούτῳ σημείῳ ἐχρησάμην, οὐ πάντας ἀνθρώπους
ὁμοίως λυμαίνεται, ἀλλ᾿ εἰσὶν οἵτινες αὐτοῦ πληρού-
μενοι οὐδ᾿ ὁτιοῦν βλάπτονται, ἀλλὰ καὶ ἰσχύν οἷσιν
ἂν συμφέρῃ θαυμασίως παρέχεται. εἰσὶ δ᾿ οἳ χαλε-
πῶς ἀπαλλάσσουσι. διαφέρουσιν οὖν τούτων αἱ φύσιες. διαφέρουσι δὲ κατὰ τοῦτο, ὅπερ ἐν τῷ σώματι
ἔνεστι πολέμιον τυρῷ καὶ ὑπὸ τούτου ἐγείρεταί τε καὶ
κινέεται· οἷσιν ὁ τοιοῦτος χυμὸς τυγχάνει πλείων
ἐνεὼν καὶ μᾶλλον ἐνδυναστεύων ἐν τῷ σώματι, τούτους μᾶλλον κακοπαθέειν εἰκός. εἰ δὲ πάσῃ τῇ ἀνθρω-
πίνῃ φύσει ἦν κακόν, πάντας ἂν ἐλυμήνατο. ταῦτα δ᾿
εἴ τις εἰδείη, οὐκ ἂν πάσχοι.

21. Τὰ δ᾿ ἐν τῇσιν ἀνακομιδῇσι τῇσιν ἐκ τῶν νού-

is going to perform anything of his duty: what a person is in relation to foods and drinks, and to habits generally, and what effects each of these will have on any individual. It is not sufficient to learn simply: "Cheese is a bad food, since it gives a pain to anyone who eats too much of it." We must also know what the pain is, the reasons for it, and to which constituent of the person the cheese is uncongenial. For there are many other troublesome foods and drinks, which affect a person in different ways. I would therefore have the point put thus, for example: "Undiluted wine, drunk in large quantity, produces a certain effect upon a person." All who know this would realize this to be a potency of wine, and that wine itself is to blame; we know, too, through which parts of a person wine chiefly exerts this power. Just such precision of truth I would wish to be evident in all other instances. To return to my former example, cheese does not harm all people alike; some can eat their fill of it without the least harm, indeed those with whom it agrees are wonderfully strengthened by it. Others have difficulty tolerating it: thus the constitutions of these people differ, and the difference lies in the constituent of the body that is hostile to cheese, and is roused and stirred into action under its influence. Those in whom a humor of this kind is present in greater quantity, and exercising a greater control over their body, naturally suffer more severely. But if cheese were bad for the human constitution without exception, it would hurt everyone. Whoever knows the above truths will avoid suffering.

21. In convalescences from illness, and especially in

---

51 ἰδόντες M: οἱ ἰδότες A

σων, ἔτι δὲ καὶ ἐν τῆσι νούσοισι τῆσι μακρῆσι γίνονται πολλαὶ συνταράξιες, αἱ μὲν ἀπὸ ταὐτομάτου, αἱ
δὲ καὶ ἀπὸ τῶν προσενεχθέντων τῶν τυχόντων. οἶδα
δὲ τοὺς πολλοὺς ἰητρούς, ὥσπερ τοὺς ἰδιώτας, ἢν τύ-
χωσι περὶ τὴν ἡμέρην ταύτην τι κεκαινουργηκότες—ἢ
λουσάμενοι ἢ περιπατήσαντες ἢ φαγόντες τι ἑτεροῖον—ταῦτα δὲ πάντα βελτίω προσενηνεγμένα ἢ μή,
οὐδὲν ἧσσον τὴν αἰτίην τούτων τινὶ ἀνατιθέντας[52] τὸ
μὲν αἴτιον ἀγνοεῦντας, τὸ δὲ συμφορώτατον, ἢν οὕτω
τύχῃ, ἀφαιρεῦντας. δεῖ δὲ οὔ, ἀλλ' εἰδέναι, τί λουτρὸν
626 ἀκαίρως | προσγενόμενον ἐργάσεται ἢ τί κόπος. οὐδέ-
ποτε γὰρ ἡ αὐτὴ κακοπάθεια τούτων οὐδετέρου, οὐδέ
γε ἀπὸ πληρώσιος οὐδ' ἀπὸ βρώματος τοίου ἢ τοίου.
ὅστις οὖν ταῦτα μὴ εἴσεται ὡς ἕκαστα ἔχει πρὸς τὸν
ἄνθρωπον, οὔτε γινώσκειν τὰ γινόμενα ἀπ' αὐτῶν δυνήσεται οὔτε χρῆσθαι ὀρθῶς.

22. Δεῖν δέ μοι δοκέει καὶ ταῦτ' εἰδέναι, ὅσα τῷ
ἀνθρώπῳ παθήματα ἀπὸ δυναμίων γίνεται καὶ ὅσα
ἀπὸ σχημάτων. λέγω δέ τί τοῦτο;[53] δύναμιν μὲν εἶναι
τῶν χυμῶν τὰς ἀκρότητάς τε καὶ ἰσχύν, σχήματα δὲ
λέγω ὅσα ἔνεστιν ἐν τῷ ἀνθρώπῳ, τὰ μὲν κοιλά τε
καὶ ἐξ εὐρέος ἐς στενὸν συνηγμένα, τὰ δὲ καὶ ἐκπεπταμένα, τὰ δὲ στερεά τε καὶ στρογγύλα, τὰ δὲ
πλατέα τε καὶ ἐπικρεμάμενα, τὰ δὲ διατεταμένα, τὰ δὲ
μακρά, τὰ δὲ πυκνά, τὰ δὲ μανά τε καὶ τεθηλότα, τὰ
δὲ σπογγοειδέα τε καὶ ἀραιά. τοῦτο μὲν οὖν, ἑλκύσαι
ἐφ' ἑωυτὸ καὶ ἐπισπᾶσθαι ὑγρότητα ἐκ τοῦ ἄλλου
σώματος, πότερον τὰ κοιλά τε καὶ ἐκπεπταμένα ἢ τὰ

54

protracted illnesses, many disturbances occur, some spontaneously and others from things casually prescribed. I am
aware that most physicians, like laymen, when a patient
has done something unusual near the day of a disturbance—taken a bath or a walk, or eaten some strange food,
whether these things happen to provide some benefit
when administered or when not administered—assign the
cause to one of them, and, while ignorant of the real cause,
stop doing what may have been of the greatest value. This
they should not do, but rather be aware what the result of
a bath unseasonably taken will be, or of (sc. an unusual)
fatigue. For the trouble caused by each of these things is
peculiar to it, and the same with overfilling or with one
particular kind of food or other. Whoever therefore fails
to appreciate how each of these particular measures affects a person will be able neither to discover their consequences nor to use them properly.

22. I hold that it is also necessary to know which diseased states arise in patients from potencies, and which
from structures. What do I mean by this? A "potency" is
the intensity and strength of the humors; with "structures"
I mean the parts inside a person, some of which are hollow
and tapering from wide to narrow, others expanded, others solid and rounded, others broad and suspended, others
stretched, others long, others dense in texture, others
loose in texture and fleshy, and others spongy and porous.
Now for example which structure would be best able to
draw and attract to itself fluid from the rest of the body,

---

52 καὶ add. A
53 τοῦτο M: τοιοῦτον A

στερεά τε καὶ στρογγύλα ἢ τὰ κοῖλά τε καὶ ἐς στενὸν
ἐξ εὐρέος συνηγμένα δύναιτ᾽ ἂν μάλιστα; οἶμαι μὲν
τὰ τοιαῦτα, τὰ ἐς στενὸν συνηγμένα ἐκ κοίλου τε καὶ
εὐρέος. καταμανθάνειν δὲ δεῖ ταῦτα ἔξωθεν ἐκ τῶν
φανερῶν.

Τοῦτο μὲν γάρ, τῷ στόματι κεχηνὼς ὑγρὸν οὐδὲν
ἀνασπάσαις· προμυλλήνας δὲ καὶ συστείλας, πιέσας
τε τὰ χείλεα ἀνασπάσεις[54] καὶ ἐπί τε αὐλὸν προσθέ-
μενος ῥηϊδίως ἀνασπάσαις ἂν ὅ τι θέλοις. τοῦτο δέ,
628 αἱ σικύαι προσβαλλόμεναι ἐξ | εὐρέος ἐς στενότερον
συνηγμέναι πρὸς τοῦτο τετέχνέαται, πρὸς τὸ ἕλκειν
ἐκ τῆς σαρκὸς καὶ ἐπισπᾶσθαι, ἄλλα τε πολλὰ τοι-
ουτότροπα. τῶν δὲ ἔσω τοῦ ἀνθρώπου φύσις καὶ[55]
σχῆμα τοιοῦτον· κύστις τε καὶ κεφαλή, καὶ ὑστέραι
γυναιξίν· καὶ φανερῶς ταῦτα μάλιστα ἕλκει καὶ
πλήρεά ἐστιν ἐπακτοῦ ὑγρότητος αἰεί. τὰ δὲ κοῖλα καὶ
ἐκπεπταμένα ἐπεσρυεῖσαν[56] μὲν ἂν ὑγρότητα μάλιστα
δέξαιτο πάντων, ἐπισπάσαιτο δ᾽ ἂν οὐχ ὁμοίως. τὰ δέ
γε στερεὰ καὶ στρογγύλα οὔτ᾽ ἂν ἐπισπάσαιτο οὔτ᾽
ἂν ἐπεσρυεῖσαν δέξαιτο· περιολισθάνοι τε γὰρ ἂν καὶ
οὐκ ἔχοι ἕδρην, ἐφ᾽ ἧς μένοι. τὰ δὲ σπογγοειδέα τε
καὶ ἀραιά, οἷον σπλήν τε καὶ πνεύμων καὶ μαζοί,
προσκαθεζόμενα μάλιστα ἀναπίοι καὶ σκληρυνθείη
ἂν καὶ αὐξηθείη ὑγρότητος προσγενομένης ταῦτα μά-
630 λιστα. οὐ γὰρ ἂν | ὥσπερ ἦν[57] ἐν κοιλίῃ, ἐν ᾗ τὸ
ὑγρόν, ἔξω τε περιέχῃ αὕτη ἡ κοιλίη, καὶ ἐξαλίζοιτ᾽
ἂν καθ᾽ ἑκάστην ἡμέρην· ἀλλ᾽ ὅταν πίῃ καὶ δέξηται
αὐτὸς ἐς ἑωυτὸν[58] τὸ ὑγρόν, τὰ κενὰ καὶ ἀραιὰ ἐπλη-

the hollow and expanded, the solid and rounded, or the hollow and tapering? I think it would be the broad and hollow one that tapers. One should carefully deduce this from objects that can be observed.

For example, with your mouth wide open you cannot draw fluid in; but if you protrude and contract it, compressing your lips, you will draw some up; and if you insert a tube besides, you can easily draw up any liquid you wish. Again, cupping instruments applied to the skin, which are broad and tapering, are constructed in this way on purpose, in order to draw and attract fluid from the flesh, and there are many other instruments of a similar nature. The material and form of some of the parts inside the human being are also like this: the bladder, the head, and the uterus in women. These obviously attract powerfully, and are always full of fluids drawn in from outside. Hollow and expanded parts accept liquid flowing into them best of all, but do not attract it so well. Solid and rounded parts will neither attract fluid nor hold it when it flows to them, since the fluid will slip off them and find no place in which to rest. Spongy, porous parts, like the spleen, lungs and breasts, readily absorb what is in contact with them, and these parts tend most to harden and to enlarge on the addition of fluid; for they cannot, as in the case where the fluid is inside a cavity and enclosed on the outside by the cavity wall, even be expelled on a daily basis. But when one of these parts drinks up fluid and takes it into itself,

---

[54] τὰ χείλεα ἀνασπάσεις Jouanna: τὰ χείλεα M: ἀνασπάσειε τὰ χείλεα A   [55] τοῦ ἀνθρώπου φύσις καὶ M: φύσει τοῦ ἀνθρώπου A   [56] ἐπεσρυείσαν Kühlewein: ἐπιρρυεῖσαν M: ἐπεισρυνῆσαν A   [57] ἦν om. A   [58] αὐτὸς ἐς ἑωυτὸν Aldina: αὐτὸς ἑωυτὸν M: αὐτὸ ἐς ἑωυτὸ A

ρώθη καὶ τὰ σμικρὰ πάντη καὶ ἀντὶ μαλθακοῦ τε καὶ
ἀραιοῦ σκληρός τε καὶ πυκνὸς ἐγένετο καὶ οὔτ᾽ ἐκ-
πέσσει οὔτ᾽ ἀφίησι. ταῦτα δὲ πάσχει διὰ τὴν φύσιν
τοῦ σχήματος.

Ὅσα δὲ φῦσάν τε καὶ ἀνειλήματα ἀπεργάζονται
ἐν τῷ σώματι, προσήκει ἐν μὲν τοῖσι κοίλοισί τε καὶ
εὐρυχωρέσιν, οἷον κοιλίη τε καὶ θώρηκι, ψόφον τε καὶ
632 | πάταγον ἐμποιέειν. ὅ τι γὰρ ἂν μὴ ἀποπληρώσῃ
οὕτως ὥστε στῆναι, ἀλλ᾽ ἔχῃ μεταβολάς τε καὶ κινήσιας, ἀνάγκη ὑπ᾽ αὐτῶν καὶ ψόφον καὶ καταφανέας
κινήσιας γίνεσθαι· ὅσα δὲ σαρκώδεά τε καὶ μαλθακά,
ἐν τοῖσι τοιούτοισι νάρκας τε καὶ πληρώματα οἷα ἐν
τοῖσιν ἀποφραγεῖσι[59] γίνεται. ὅταν δ᾽ ἐγκύρσῃ πλατεῖ
τε καὶ ἀντικειμένῳ, καὶ πρὸς αὐτὸ ἀντιπέσῃ, καὶ φύσει τοῦτο τύχῃ ἐὸν μήτε ἰσχυρόν, ὥστε δύνασθαι
ἀνέχεσθαι τὴν βίην καὶ μηδὲν κακὸν παθεῖν, μήτε
μαλθακόν τε καὶ ἀραιόν, ὥστ᾽ ἐκδέξασθαί τε καὶ ὑπεῖ-
ξαι, ἁπαλὸν δὲ καὶ τεθηλὸς καὶ ἔναιμον καὶ πυκνόν,
οἷον ἧπαρ, διὰ μὲν τὴν πυκνότητα καὶ πλατύτητα ἀν-
θέστηκέ τε καὶ οὐχ ὑπείκει—φῦσα δ᾽ ἐπιχεομένη[60]
αὔξεταί τε καὶ ἰσχυροτέρη γίνεται καὶ ὁρμᾷ μάλιστα
πρὸς τὸ ἀντιπαῖον—διὰ δὲ τὴν ἁπαλότητα καὶ τὴν
ἐναιμότητα οὐ δύναται ἄνευ πόνων εἶναι, καὶ διὰ
ταύτας τὰς προφάσιας ὀδύναι τε ὀξύαται καὶ πυκνόταται πρὸς τοῦτο τὸ χωρίον γίνονται ἐμπυήματά τε
634   καὶ φύματα | πλεῖστα. γίνεται δὲ καὶ ὑπὸ φρένας
ἰσχυρῶς, ἧσσον δὲ πολλόν. διάτασις μὲν γὰρ φρενῶν
πλατείη καὶ ἀντικειμένη, φύσις δὲ νευρωδεστέρη τε

what is empty and porous fills up completely, even at the minutest level, what was soft and porous becomes hard and dense, and neither concoction nor expulsion occurs. This happens because of the nature of their structure.

Whatever produces wind and colic in the body naturally gives rise in the hollow, wide open parts, such as the cavity and the chest, to noise and rumbling. For when wind fails to fill a part up and come to rest, but keeps moving and changing its position, this must provoke noise and perceptible movements. In the parts that are soft and fleshy, numbness and obstruction (sc. result), like what happens with blockages.[11] But when flatulence meets a broad, resistant body, and rushes against it, and this body happens by nature to be neither strong enough to withstand the violence without suffering harm, nor soft and porous enough to admit it and recede, but tender, fleshy, full of blood, and dense, like the liver, it resists without yielding. As the flatulence being poured against it increases and becomes stronger, dashing violently against the obstacle resisting it, the part, owing to its tenderness and the blood it contains, cannot avoid pain. For these reasons, the sharpest and most frequent pains occur in this region, and abscesses and growths too are very common. These symptoms are also felt strongly beneath the diaphragm, but much less severely, since the extended surface of the diaphragm is broad, resistant, of a more sinewy

---

[11] The manuscript readings signify "those with cut throats," which Cornarius renders in *trucidatis* and Foes *jugulatis*: Coray conjectures "blockages" and Littré "apoplexies."

---

[59] -φραγεῖσι Coray (Jouanna): -σφαγεῖσι M: -σφαγίσι A: -πληγεῖσι Littré   [60] ἐπιχεομένη A: ἐπιδεχομένη M

καὶ ἰσχυροτέρη, διὸ ἧσσον ἐπώδυνά ἐστιν. γίνεται δὲ
καὶ περὶ ταῦτα καὶ πόνοι καὶ φύματα.

23. Πολλὰ δὲ καὶ ἄλλα καὶ ἔσω καὶ ἔξω τοῦ σώματος εἴδεα σχημάτων, ἃ μεγάλα ἀλλήλων διαφέρει
πρὸς τὰ παθήματα καὶ νοσέοντι καὶ ὑγιαίνοντι, οἷον
κεφαλαὶ σμικραὶ ἢ μεγάλαι, τράχηλοι λεπτοὶ ἢ παχέες, μακροὶ ἢ βραχέες, κοιλίαι μακραὶ ἢ στρογ-
γύλαι, θώρηκος καὶ πλευρέων πλατύτητες ἢ στενότη-
τες, ἄλλα μυρία, ἃ δεῖ πάντα εἰδέναι ᾗ διαφέρει, ὅπως
τὰ αἴτια ἑκάστων εἰδὼς ὀρθῶς φυλάσσηται.

24. Περὶ δὲ δυναμίων χυμῶν αὐτῶν τε ἕκαστος ὅ τι
δύναται ποιεῖν τὸν ἄνθρωπον ἐσκέφθαι, ὥσπερ καὶ
πρότερον εἴρηται, καὶ τὴν συγγένειαν ὡς ἔχουσι πρὸς
ἀλλήλους. λέγω δὲ τὸ τοιοῦτον· εἰ γλυκὺς χυμὸς ἐὼν
μεταβάλλοι ἐς ἄλλο εἶδος, μὴ ἀπὸ συγκρήσιος, ἀλλὰ
αὐτὸς ἐξιστάμενος, ποῖός τις ἂν πρῶτος γένοιτο, πι-
κρὸς ἢ ἁλμυρὸς ἢ στρυφνὸς ἢ ὀξύς; οἶμαι μέν, ὀξύς.
636 ὁ ἄρα ὀξὺς χυμὸς ἂν ἐπιτήδειος | προσφέρειν[61] ἂν τῶν
λοιπῶν εἴη μάλιστα, εἴπερ ὁ γλυκύς γε τῶν πάντων
ἀνεπιτηδειότατος.[62]

Οὕτως εἴ τις δύναιτο ζητέων ἔξωθεν ἐπιτυγχάνειν,
καὶ δύναιτ᾽ ἂν πάντων ἐκλέγεσθαι αἰεὶ τὸ βέλτιστον.
βέλτιστον δέ ἐστι αἰεὶ τὸ προσωτάτω τοῦ ἀνεπιτη-
δείου ἀπέχον.

[61] προσφέρειν Kühlewein: προσφέρων M: προσφορῶν A
[62] ἀνεπιτηδ. M: ἐπιτηδ. A

and stronger texture, and so less subject to pain: but here too pains and growths occur.

23. There are many other structural forms, both on the inside and on the outside of the body, which differ greatly from one another with regard to disordered states in both the ill and the healthy, such as whether the head is large or small, the neck thin or thick, long or short, the cavities long or rounded, the chest and ribs broad or narrow; and there are very many other things, the differences among which must all be known, so that knowledge of the causes of each thing may ensure that the proper precautions are taken.

24. As far as the potencies are concerned, each of the individual humors must be investigated with regard to what effect it can have on a person, as I have indicated above, and the relationships that exist between them. My opinion is something like this: if a humor that is sweet changes to another form, not through compounding (sc. with another one) but through its own alteration, which humor would it first become, sharp or salty or astringent or acid? I believe acid. Therefore the acid humor would be the most appropriate of the other humors to prescribe in a situation where the sweet humor was the most inappropriate of all the humors.[12]

If a man can in this way conduct with success inquiries outside the human body, he will also be able to select the very best treatments. Best is always the thing farthest removed from what is inappropriate.

[12] Neither the text nor the sense of this sentence is secure.

AIRS WATERS PLACES

INTRODUCTION

Erotian includes the title *On Places and Seasons* (περὶ τόπων καὶ ὡρῶν) in his introductory census of Hippocratic works, in the category "etiological and natural" (αἰτιολογικὰ δὲ καὶ φυσικά), and assigns one gloss (O 11) specifically to *On Seasons and Places*; altogether about twenty words originating certainly or probably from the treatise are included in his glossary, one of which contains a reference to Epicles of Crete, who is reported to have excerpted material from the third-century BC glossator Bacchius of Tanagra.[1]

Galen includes ten words from *Airs Waters Places* in his glossary, attributing one (σ 59) to *On Seasons Places Waters* (περὶ ὡρῶν καὶ τόπων καὶ ὑδάτων),[2] he mentions in his *Commentary on Hippocrates' Aphorisms* and in *On His Own Books*[3] that he wrote a commentary in three books to περὶ τόπων, ἀέρων, ὑδάτων—which is lost in the original Greek, but survives in a ninth-century Arabic translation in the manuscript Cairensis Ṭal'aṭ ṭibb 550 (dated 887 Hegira = AD 1483) (= Galen [Ar. comm.])[4]—

1 Nachmanson, pp. 318–24.
2 See Perilli, p. 408, for references.
3 Cf. Boudon, p, 160.
4 See M. Ullmann, "Galens Kommentar zu der Schrift *De aere*

and he quotes several passages from the text in his other writings.[5] Numerous Greek and Latin writers express ideas similar to those found in *Airs Waters Places*, others include direct quotations from the text, and yet others name the treatise as their source explicitly.[6]

A fifth- or sixth-century Latin translation from the Greek (= Lat.) is transmitted in the three ninth-century manuscripts, Parisinus Latinus 7027, Glasgow Hunter 96, and Ambrosianus Latinus G 108 inf., while the twelfth-century manuscript, Monacensis Latinus 23535, contains a different Latin translation of chapter 1—chapter 6 (καὶ δρόσοι πίπτουσι) from the Greek (= Lat. 2), dated by Grensemann to the second half of the eleventh century as the work of Alfanus of Salerno or possibly some translator in his circle.[7] An Arabic translation of the text made by extracting the lemmata from Galen's commentary (= Gal. [Ar.]) is edited and translated into English in J. N. Mattock and M. C. Lyons. *Kitāb Buqrāt fi'l-amrād al-bilādiyya. Hippocrates: On Endemic Diseases (Airs Waters Places)*[8] (Cambridge, 1969). For an extensive collection of cita-

---

aquis locis," in Joly, pp. 353–65. Jouanna *Airs* (pp. 139–48) gives a comprehensive description of Galen (*Ar. comm.*) based on his access to an unpublished German translation of the entire *Commentary* made by G. Strohmaier.

[5] See *Testimonien* vol. II,1, pp. 25–51, and vol. II,2, pp. 30–36.

[6] See *Testimonien* vol. I, pp. 23–39.

[7] See H. Grensemann, *Hippokrates. De aeribus aquis locis. Interlineare Ausgabe der spätlateinischen Übersetzung und des Fragments einer hochmittelalterlichen Übersetzung* (Bonn, 1996).

[8] See Ullmann, pp. 27f; Sezgin, pp. 36f.

tions from the text of *Airs Waters Places* in medieval Byzantine, Latin, and Arabic writers, see *Testimonien* vol. III, pp. 12–39.

The two parts of *Airs Waters Places* are different in their subject matter and purpose, although they share many assumptions, methods, and expressions.

The first, medical, section gives an account of how environmental factors such as a town's prevalent winds, waters available for drinking, position with respect to the daily course of the sun, and terrain, as well as the particular climatic characteristics of a specific year, determine the diseases likely to arise in its inhabitants, and how an itinerant physician arriving in an unfamiliar town can use such understanding to make accurate prognoses and apply successful therapies.

The chapters are:

1–2	General introduction
3	Winds in towns facing south
4	Winds in towns facing north
5	Winds in towns facing east
6	Winds in towns facing west
7	Waters from marshes and springs
8	Waters from rain and snow
9	Mixed waters
10	The effects of particular seasons
11	The effects of seasons in general

The second, ethnographic, section of *Airs Waters Places*, which shows affinities with the geographical sections of Herodotus' *Histories*, describes the specific physical and mental characteristics of various peoples in Asia and Europe,[9] and explains how these are caused by natural ($\phi\acute{\upsilon}\sigma\iota\varsigma$) and cultural ($\nu\acute{o}\mu o\varsigma$) influences. The chapters are:

12–13   General introduction

14	The "long-heads"	
15	dwellers on the Phasis River	— Asia
16	Comparative mental traits	

17	The *Sauromatae* women	
18	Lifestyle of the Scythian nomads	
19	Effects of the extreme cold on the Scythians	
20	Physique of the Scythians and its causes	— Europe
21	Impotence and infertility of the Scythian men	
22	The Scythian eunuch-like "un-manlies"	

23	Summary of climatic causes and effects
24	Examples of human types and their causes

The Greek text of *Airs Waters Places* is transmitted in its entirety in only one independent manuscript, V, in which it has suffered a serious derangement: due to the accidental inversion of a bifolium in a quire of one of V's ancestors, two extensive pieces of text (ch. 9 ὅτι τὸ παχύτατον

---

[9] Cf. Jouanna *Airs* (pp. 299f.) on the role Egypt and Libya may have played in the treatise originally.

to ch. 10 ἐγγένηται, and ch. 3 τοῦ δὲ χειμῶνος to ch. 9 λιθιῶντες) have disappeared from their correct place in *Airs Waters Places* (between ch. 3 τοῦ δὲ χειμῶνος ψυχρά and ch. 10 καὶ λειεντερίαι) and reappeared in this inverted order in the text of the preceding treatise, *Wounds in the Head*, in chapter 21 after τῷ ἕλκει.[10]

Three other Greek sources also contribute directly to our knowledge of the text:

1. The manuscript Parisinus Graecus 2047 A (XIII c.) contains a text of chapter 24 from ὁκόσοι μὲν χώρην to οὐχ ἁμαρτήσεις that differs significantly from V's and represents an equally important strand of the transmission. (= P)
2. A Greek manuscript once in the possession of A. Gadaldini (1515–1575), but apparently no longer extant, contained a number of variants of the *Airs Waters Places* text, reports of which have been preserved in four sources:

    a. in the manuscript Barberinianus Graecus 5 (1558) (= Barb.), which gives evidence of containing readings directly from Gadaldini's "extremely old copy."[11]
    b. as readings attributed by Gadaldini to "an extremely old copy" (*antiquissimi exemplaris*) and "my most ancient Greek codex" (*in nostro codice graeco vetustissimo*), and printed on folios 2–6 of the second section of the *Galeni opera omnia latine in septem*

10 See Jouanna *Airs*, pp. 85f.

11 See J. Ilberg, "Zur Überlieferungsgeschichte des Hippokrates," *Philologus* 52 (1894): 426–28; Jouanna *Airs*, pp. 104f.

classes digesta. Ex Juntarum quarta editione, edited by Gadaldini and published in Venice in 1565 (= Gad. [J]).

c. as marginal notes in two Greek Hippocrates editions once in the possession of Gadaldini, the Aldine *Omnia opera Hippocratis* (Venice, 1526) (= Gad. [A]) and the Froben *Hippocratis . . . libri omnes* (Basel, 1538) (= Gad. [B]), the first of which was once kept in the Ambrosian Library in Milan but is now missing, and the second of which is still in Milan.

d. as approximately thirty textual variants printed in Baldinius, which the commentator tells us originated in an "ancient codex" (*antiquo codice*) and were sent to him from Padua.

The variant readings presented in these sources influenced by Gadaldini's "ancient codex" are sometimes the same but in other cases are in only one or more witness. From marginal notes in Gad. (A), we also know that the text of Gadaldini's "ancient codex" had been rearranged intentionally by some ancient or medieval scribe, who moved chapters 7–11 from after chapter 6 to after chapter 24, as well as making other more minor changes of order.

3. Significant contributions to our knowledge of the original *Airs Waters Places* text are also made by Galen's commentary now extant only in Arabic translation as mentioned above, which often discusses textual variants of the Hippocratic text he cites, as well as commenting on the text's interpretation.

69

Airs Waters Places is printed in all the collected editions and translations, including Zwinger, Adams, Kühlewein, Heiberg, and Chadwick. The text has also received much attention in the scholarly literature, e.g.:

Cornarius, I. *Hippocratis Coi De aere, aquis et locis libellus*. Basel, 1529.

Baldinius, B. *In librum Hyppocratis De Aquis, Aere et Locis Commentaria*. Florence, 1586. (= Baldinius)

Septalius, L. *In Librum Hippocratis Coi de Aeribus, Aquis, Locis Commentarii V.* Köln, 1590. (= Septalius)

Coray, A. *Traité d' Hippocrate. Des airs, des eaux et des lieux*. Paris, 1800. (= Coray); 2nd ed., 1816.

Jurk, J. *Ramenta Hippocratea*. Diss., Berlin, 1900.

Gundermann, G. *Hippocratis* De aere aquis loci *mit der alten lateinischen Übersetzung*. Bonn, 1911.

Jacoby, F. "Zu Hippokrates' ΠΕΡΙ ΑΕΡΩΝ ΥΔΑΤΩΝ ΤΟΠΩΝ." *Hermes* 46 (1911): 518–67.

Edelstein, L. ΠΕΡΙ ΑΕΡΩΝ *und die Sammlung der hippokratischen Schriften*. Berlin, 1931.

Diller, H. *Die Überlieferung der hippokratischen Schrift* ΠΕΡΙ ΑΕΡΩΝ ΥΔΑΤΩΝ ΤΟΠΩΝ. Leipzig, 1932.

———. *Wanderarzt und Aitiologe*. Leipzig, 1934.

Lesky, E. "Zur Lithiasisbeschreibung in περὶ ἀέρων ὑδάτων τόπων." *Wiener Studien* 63 (1948): 69–83. (= Lesky)

The two most important recent editions of *Airs Waters Places* are:

Diller, H. *Hippokrates. Über die Umwelt.* CMG I 1,2. Berlin, 1970.

Jouanna, J. *Hippocrate. Airs, Eaux, Lieux.* Budé II(2). Paris, 1996.

The present edition takes account of the scholarship cited but depends in particular on Jouanna's edition.

ΠΕΡΙ ΑΕΡΩΝ ΥΔΑΤΩΝ ΤΟΠΩΝ

II 12
Littré

1. Ἰητρικὴν ὅστις βούλεται ὀρθῶς ζητεῖν, τάδε χρὴ
ποιέειν· πρῶτον μὲν ἐνθυμέεσθαι τὰς ὥρας τοῦ ἔτεος,
ὅ τι δύναται ἀπεργάζεσθαι ἑκάστη· οὐ γὰρ ἐοίκασιν
οὐδέν, ἀλλὰ πολὺ διαφέρουσιν αὐταί τε ἐφ᾽ ἑωυτῶν
καὶ ἐν τῇσι μεταβολῇσιν· ἔπειτα δὲ τὰ πνεύματα τὰ
θερμά τε καὶ τὰ ψυχρά, μάλιστα μὲν τὰ κοινὰ πᾶσιν
ἀνθρώποισιν, ἔπειτα δὲ καὶ τὰ[1] ἐν ἑκάστῃ χώρῃ ἐπιχώρια ἐόντα. δεῖ δὲ καὶ τῶν ὑδάτων ἐνθυμέεσθαι τὰς
δυνάμιας· ὥσπερ γὰρ ἐν τῷ στόματι διαφέρουσι καὶ
ἐν τῷ σταθμῷ, οὕτω καὶ ἡ δύναμις διαφέρει πολὺ
ἑκάστου. ὥστε ἐς πόλιν ἐπειδὰν ἀφίκηταί τις, ἧς
ἄπειρός ἐστι, διαφροντίσαι χρὴ τὴν θέσιν αὐτῆς,
ὅκως κέεται καὶ πρὸς τὰ πνεύματα καὶ πρὸς τὰς ἀνατολὰς τοῦ ἡλίου. οὐ γὰρ τὠυτὸ δύναται ἥτις πρὸς
βορέην κέεται καὶ ἥτις πρὸς νότον οὐδ᾽ ἥτις πρὸς
ἥλιον ἀνίσχοντα οὐδ᾽ ἥτις πρὸς δύνοντα. ταῦτα δὲ
χρὴ ἐνθυμέεσθαι ὡς κάλλιστα καὶ τῶν ὑδάτων πέρι
ὡς ἔχουσι, καὶ πότερον ἑλώδεσι χρέονται καὶ μαλα-
κοῖσιν ἢ σκληροῖσί τε καὶ ἐκ μετεώρων καὶ ἐκ πετρω-
δέων εἴτε ἁλυκοῖσι καὶ ἀτεράμνοισι· καὶ τὴν γῆν,
πότερον ψιλή τε καὶ ἄνυδρος ἢ δασεῖα καὶ ἔφυδρος

AIRS WATERS PLACES

1. Anyone who wishes to investigate medicine correctly must do the following: first consider the seasons of the year and the effects they can have, for they are not at all the same, but differ greatly both among themselves and during the changes (sc. from one season to another); then the winds, both hot and cold, mainly the ones that are common among all people, but also those that are prevalent in each particular location. He must also consider the potencies of the waters, for just as these are different to the tongue and in their weight, their effects too are very different. Thus, when a person arrives in a town with which he is unfamiliar, he should examine the position in which it lies in relationship both to the winds and to the risings of the sun. For a town that faces north cannot have the same effect as one that faces south, or one that faces the rising of the sun, or one that faces the setting of the sun. These things must be considered with the greatest care, and also the waters available, whether people employ such as are marshy and soft, or those from high places and rocks which are hard, or those that are salty and refractory.[1] The terrain too, whether it is exposed and water-

---

[1] *Refractory*: see Note on Technical Terms.

---

[1] τὰ Gad. (B): om. V

καὶ εἴτε ἐν κοίλῳ ἐστὶ καὶ πνιγηρὴ εἴτε μετέωρος καὶ
ψυχρή· καὶ τὴν δίαιταν τῶν ἀνθρώπων, ὁκοίη ἥδον-
ται, πότερον φιλοπόται καὶ ἀριστηταὶ καὶ ἀταλαίπωροι ἢ φιλογυμνασταί τε καὶ φιλόπονοι καὶ ἐδωδοὶ καὶ
ἄποτοι. |

14   2. Καὶ ἀπὸ τούτων χρὴ ἐνθυμέεσθαι ἕκαστα. εἰ γὰρ
ταῦτα εἰδείη τις καλῶς, μάλιστα μὲν πάντα, εἰ δὲ μή,
τά γε πλεῖστα, οὐκ ἂν αὐτὸν λανθάνοι ἐς πόλιν ἀφ-
ικνεόμενον, ἧς ἂν ἄπειρος ᾖ, οὔτε νοσήματα ἐπιχώρια
οὔτε τῶν κοιλιῶν ἡ φύσις, ὁκοίη τίς ἐστιν· ὥστε μὴ
ἀπορέεσθαι ἐν τῇ θεραπείῃ τῶν νούσων μηδὲ διαμαρ-
τάνειν· ἃ εἰκός ἐστι γίνεσθαι, ἢν μή τις ταῦτα πρότερον εἰδὼς προφροντίσῃ περὶ ἑκάστου· τοῦ δὲ χρόνου
προϊόντος καὶ τοῦ ἐνιαυτοῦ λέγοι ἄν, ὁκόσα τε νοσήματα μέλλει πάγκοινα τὴν πόλιν κατασχήσειν ἢ
θέρεος ἢ χειμῶνος, ὅσα τε ἴδια ἑκάστῳ κίνδυνος γίνεσθαι ἐκ μεταβολῆς τῆς διαίτης. εἰδὼς γὰρ τῶν
ὡρέων τὰς μεταβολὰς καὶ τῶν ἄστρων ἐπιτολάς τε
καὶ δύσιας, καθότι ἕκαστον τούτων γίνεται, προειδείη
ἂν τὸ ἔτος ὁκοῖόν τι μέλλει γίνεσθαι. οὕτως ἄν τις
ἐρευνώμενος[2] καὶ προγινώσκων τοὺς καιροὺς μάλιστ᾽
ἂν εἰδείη περὶ ἑκάστου καὶ τὰ πλεῖστα τυγχάνοι τῆς
ὑγιείης καὶ κατ᾽ ὀρθὸν φέροιτο οὐκ ἐλάχιστα ἐν τῇ
τέχνῃ. εἰ δὲ δοκέοι τις ταῦτα μετεωρολόγα εἶναι, εἰ
μετασταίη τῆς γνώμης, μάθοι ἄν, ὅτι οὐκ ἐλάχιστον
μέρος συμβάλλεται ἀστρονομίη ἐς ἰητρικήν, ἀλλὰ

----

[2] ἐρευνώμενος V: ἐννοούμενος Gad. (B)

less, or bushy and well-watered, or in a hollow and suffocating, or high and dry. The mode of life, too, that the
inhabitants prefer, whether they are given to drink, they
take lunch, and they are adverse to exertion, or they like
exercise and effort, they eat heartily, and they drink little.

2. On the basis of this evidence, each situation must
be considered, and if a person understands these matters
well—best all of them, or, if not that, then at least most of
them—it will not escape him on arrival in an unfamiliar
town what the local diseases are or the state of the cavities[2]
of its inhabitants. Thus, he will not be at a loss with regard
to treatment, nor will he make mistakes, which is likely to
happen if he does not know this from earlier experience,
and does not consider each case in advance. Then, as time
passes through the year, he will be able to say which epidemic diseases will befall the town in summer or winter,
as well as the individual ones a person runs the risk of
contracting through some change in his regimen. For by
knowing the changes of the seasons, as well as the risings
and settings of the stars that determine these changes,
he will know in advance how any year is going to proceed.
By investigating in this way and recognizing the critical
times in advance, he will generally understand each situation, and in most cases secure health for the patient and
win great success in his profession. If someone thinks
that all this belongs rather to meteorology,[3] by keeping
his mind open he will learn that astronomy makes no
small contribution to the art of medicine, but a very great

[2] *Cavity*: see Note on Technical Terms, vol. 8, p. 10f.
[3] *Meteorology*: see Note on Technical Terms.

πάνυ πλεῖστον. ἅμα γὰρ τῇσιν ὥρῃσι καὶ αἱ κοιλίαι
μεταβάλλουσιν τοῖσιν ἀνθρώποισιν.

3. Ὅκως δὲ χρὴ ἕκαστα τῶν προειρημένων σκοπεῖν
καὶ βασανίζειν, ἐγὼ φράσω σαφέως. ἥτις μὲν πόλις
πρὸς τὰ πνεύματα κεῖται τὰ θερμά—ταῦτα δ᾽ ἐστὶ
16 μεταξὺ τῆς τε χειμερινῆς ἀνατολῆς | τοῦ ἡλίου καὶ
τῶν δυσμέων τῶν χειμερινῶν—καὶ αὐτῇ ταῦτα τὰ
πνεύματά ἐστι σύννομα, τῶν δὲ ἀπὸ τῶν ἄρκτων
πνευμάτων σκέπη, ἐν ταύτῃ τῇ πόλει ἐστὶ τά τε ὕδατα
πολλὰ καὶ ὑφαλυκά, καὶ[3] ἀνάγκη εἶναι μετέωρα, τοῦ
μὲν θέρεος θερμά, τοῦ δὲ χειμῶνος ψυχρά· τούς τε
ἀνθρώπους τὰς κεφαλὰς ὑγρὰς ἔχειν καὶ φλεγματώ-
δεας, τάς τε κοιλίας αὐτῶν πυκνὰ ἐκταράσσεσθαι
ἀπὸ τῆς κεφαλῆς τοῦ φλέγματος ἐπικαταρρέοντος· τά
τε εἴδεα ἐπὶ τὸ πλῆθος αὐτῶν ἀτονώτερα εἶναι· ἐσθίειν
δ᾽ οὐκ ἀγαθοὺς εἶναι οὐδὲ πίνειν. ὁκόσοι μὲν γὰρ κε-
18 φαλὰς ἀσθενέας ἔχουσιν, οὐκ ἂν εἴησαν ἀγαθοὶ | πί-
νειν· ἡ γὰρ κραιπάλη μᾶλλον πιέζει. νοσήματά τε
τάδε ἐπιχώρια εἶναι· πρῶτον μὲν τὰς γυναῖκας νο-
σερὰς καὶ ῥοώδεας εἶναι· ἔπειτα πολλὰς ἀτόκους ὑπὸ
νούσου καὶ οὐ φύσει τιτρώσκεσθαί τε πυκνά· τοῖσί
τε παιδίοισιν ἐπιπίπτειν σπασμούς τε καὶ ἄσθματα
καὶ ἃ νομίζουσι τὸ παιδίον ποιεῖν καὶ ἱερὴν νοῦσον
εἶναι· τοῖσι δὲ ἀνδράσι δυσεντερίας καὶ διαρροίας καὶ
ἠπιάλους καὶ πυρετοὺς πολυχρονίους χειμερινοὺς καὶ
ἐπινυκτίδας πολλὰς καὶ αἱμορροΐδας ἐν τῇ ἕδρῃ.

---

[3] ὑφαλυκὰ Gad. (B): ὕφαλοι καὶ V

one: for it is in time with the seasons that the cavities in humans also change.

3. How each of the matters mentioned should be investigated and tested, I will now explain in detail. A town that lies facing the hot winds—these winds will be between the winter rising of the sun and its winter setting (i.e., south)—and in which these are prevalent, while at the same time being sheltered from winds coming from (sc. the constellations of) the Bears (i.e., north), must have waters that are plentiful, somewhat salty, and near the surface, hot in the summer and cold in the winter. The heads of the inhabitants are moist and subject to phlegm, and their cavities are frequently disturbed by phlegm that runs down into them from the head. Their appearance is for the most part rather flabby, and they do not eat or drink very well, since people with weak heads avoid drinking, since the aftereffects are very distressing to them. The common diseases in such a place are the following. First, the women are sickly and subject to fluxes; then many are childless as the result of some disease—not by nature— while abortions are frequent. The children (sc. in this town) are subject to convulsions and shortness of breath, which in the local opinion constitutes "the childhood condition," and is equivalent to the sacred disease.[4] The men suffer from dysenteries, diarrheas, agues, chronic fevers in winter, many attacks of epinyctides,[5] and from hemor-

[4] *Sacred Disease*: see Note on Technical Terms.
[5] Epinyctides: see vol. 8, p. 245, n. 8.

πλευρίτιδες δὲ καὶ περιπνευμονίαι καὶ καῦσοι καὶ
ὁκόσα ὀξέα νοσήματα νομίζονται εἶναι οὐκ ἐγγίνονται πολλά. οὐ γὰρ οἷόν τε, ὅκου ἂν κοιλίαι ὑγραὶ
ἔωσι, τὰς νούσους ταύτας ἰσχύειν. ὀφθαλμίαι τε ἐγγίνονται ὑγραὶ καὶ οὐ χαλεπαί, ὀλιγοχρόνιοι, ἢν μή
τι κατάσχῃ νόσημα πάγκοινον ἐκ μεταβολῆς μεγάλης.[4] καὶ ὁκόταν τὰ πεντήκοντα ἔτεα ὑπερβάλλωσι,
κατάρροοι ἐπιγενόμενοι ἐκ τοῦ ἐγκεφάλου παραπλη-
κτικοὺς ποιέουσι τοὺς ἀνθρώπους, ὁκόταν ἐξαίφνης
ἡλιωθέωσι τὴν κεφαλὴν ἢ ριγώσωσι. ταῦτα μὲν τὰ
νοσήματα αὐτοῖσιν ἐπιχώριά ἐστι. χωρὶς δέ, ἤν τι
πάγκοινον κατάσχῃ νόσημα ἐκ μεταβολῆς τῶν
ὡρέων, καὶ τούτου μετέχουσιν.

4. Ὁκόσαι δ᾽ ἀντικέονται τούτων πρὸς τὰ πνεύματα
τὰ ψυχρὰ μεταξὺ τῶν δυσμέων τῶν θερινῶν τοῦ ἡλίου
καὶ τῆς ἀνατολῆς τῆς θερινῆς, καὶ αὐτῇσι ταῦτα τὰ
πνεύματα ἐπιχώριά ἐστι, τοῦ δὲ νότου καὶ τῶν θερμῶν πνευμάτων σκέπη, ὧδε ἔχει περὶ τῶν πόλεων
τούτων· πρῶτον μὲν τὰ ὕδατα σκληρά τε καὶ ψυχρὰ
ὡς ἐπὶ τὸ πλῆθος γλυκέα τε.[5] τοὺς δὲ ἀνθρώπους εὐ-
τόνους τε καὶ σκελιφροὺς ἀνάγκη | εἶναι, τούς τε
πλείους τὰς κοιλίας ἀτεράμνους ἔχειν καὶ σκληρὰς
τὰς κάτω, τὰς δὲ ἄνω εὐρωτέρας· χολώδεάς τε μᾶλλον ἢ φλεγματίας εἶναι. τὰς δὲ κεφαλὰς ὑγιηρὰς
ἔχουσι καὶ σκληράς, ῥηγματίαι τέ εἰσιν ἐπὶ τὸ
πλῆθος. νοσεύματα δὲ αὐτοῖσιν ἐπιδημεῖ τάδε· πλευ-
ρίτιδές τε πολλαὶ αἵ τε ὀξεῖαι νομιζόμεναι νοῦσοι.
ἀνάγκη δὲ ὧδε ἔχειν, ὁκόταν αἱ κοιλίαι σκληραὶ ἔω-

20

rhoids in the seat. Pleurisies, pneumonias, ardent fevers, and any diseases classified as "acute" rarely occur, since it is impossible for such diseases to be severe in persons whose cavities are moist. Ophthalmias with watering of the eyes occur, but are not serious; they are of short duration, unless a general epidemic takes place due to some abrupt change. When the people are past fifty years of age, downward fluxes supervening from the brain cause paralyses, due to their having been suddenly struck on the head by the sun, or chilled. These are the common diseases in the place: furthermore, if some epidemic disease attacks as the result of a change of the seasons, people suffer from that too.

4. Towns with the opposite situation, facing cold winds coming from between the summer setting of the sun and the summer rising (i.e., north) which prevail there, while at the same time being sheltered from the southern, hot winds are as follows. First, the waters of the region are for the most part hard and cold, and also sweet. The people must be vigorous and lean, and most have refractory lower cavities tending to constipation, with the upper cavities being more fluent; they incline more toward bile than toward phlegm. They have healthy dry heads, but are generally subject to internal ruptures. The common diseases in the place are the following. Many pleurisies and what are classified as "acute" diseases are common; this must be so, since their cavities are constipated. Also the

---

σιν· ἔμπυοί τε πολλοὶ γίνονται ἀπὸ πάσης προφάσιος. τούτου δὲ αἴτιόν ἐστι τοῦ σώματος ἡ ἔντασις
καὶ ἡ σκληρότης τῆς κοιλίης. ἡ γὰρ ξηρότης ῥηγμα-
τίας ποιέει εἶναι καὶ τοῦ ὕδατος ἡ ψυχρότης. ἐδωδοὺς
δὲ ἀνάγκη τὰς τοιαύτας φύσιας εἶναι καὶ οὐ πολυ-
πότας· οὐ γὰρ οἷόν τε ἅμα πολυβόρους τε εἶναι καὶ
πολυπότας. ὀφθαλμίας τε γίνεσθαι μὲν διὰ χρόνου,
γίνεσθαι δὲ σκληρὰς καὶ ἰσχυράς, καὶ εὐθέως ῥήγνυ-
σθαι τὰ ὄμματα· αἱμορροίας δὲ ἐκ τῶν ῥινῶν τοῖσι
νεωτέροισι τριήκοντα ἐτέων γίνεσθαι ἰσχυρὰς τοῦ
θέρεος· τά τε ἱερὰ νοσεύματα καλεύμενα, ὀλίγα μὲν
ταῦτα, ἰσχυρὰ δέ. μακροβίους δὲ τοὺς ἀνθρώπους
τούτους μᾶλλον εἰκὸς εἶναι τῶν ἑτέρων· τά τε ἕλκεα
οὐ φλεγματώδεα ἐγγίνεσθαι οὐδὲ ἀγριοῦσθαι· τά τε
22   ἤθεα ἀγριώτερα ἢ ἡμερώτερα. | τοῖσι μὲν ἀνδράσι
ταῦτα τὰ νοσήματα ἐπιχώριά ἐστι· καὶ χωρίς, ἤν τι
πάγκοινον κατάσχῃ ἐκ μεταβολῆς τῶν ὡρέων· τῆσι
δὲ γυναιξί· πρῶτον μὲν στέριφαι[6] πολλαὶ γίνονται διὰ
τὰ ὕδατα ἐόντα σκληρά τε καὶ ἀτέραμνα καὶ ψυχρά.
αἱ γὰρ καθάρσιες οὐκ ἐπιγίνονται τῶν ἐπιμηνίων ἐπι-
τήδειαι, ἀλλὰ ὀλίγαι καὶ πονηραί. ἔπειτα τίκτουσι χα-
λεπῶς· τιτρώσκουσι δὲ οὐ σφόδρα. ὁκόταν δὲ τέκωσι,
τὰ παιδία ἀδύνατοι τρέφειν εἰσί· τὸ γὰρ γάλα ἀπο-
σβέννυνται ἀπὸ τῶν ὑδάτων τῆς σκληρότητος καὶ ἀτε-
ραμνίης· φθισιές τε γίνονται συχναὶ ἀπὸ τῶν τοκε-
τῶν. ὑπὸ γὰρ βίης ῥήγματα ἴσχουσι καὶ σπάσματα.
τοῖσί τε παιδίοισιν ὕδρωπες ἐγγίνονται ἐν τοῖσιν ὄρ-
χεσιν, ἕως μικρὰ ᾖ· ἔπειτα προϊούσης τῆς ἡλικίης

slightest cause produces internal suppurations in many patients, due to the tautness of their body and the constipation of their cavity: for their dryness, combined with the coldness of the water, makes them liable to internal ruptures. Such natures must eat well and limit their drinking, since it is impossible at the same time both to eat heartily and to drink copious. Opththalmias occur all the time, they are indurated and severe, and the eyes soon rupture. Violent hemorrhages from the nostrils occur in people under thirty years of age in the summer; cases of the so-called sacred disease are infrequent, but severe when they do occur. These persons are more likely to be long-lived than others. Ulcers that develop do not have phlegm, nor do they become malignant, although they tend more toward malignancy than toward quiescence. These are the common diseases of the men, and besides some epidemic disease may attack as the result of a change in the seasons. As for the women, first many of them become barren through the waters being hard, refractory, and cold, since their cleanings do not follow regularly according to the months, but are scanty and troublesome. Then childbirth is difficult, although abortion is rare. When they do give birth, they are unable to nourish their children, since their milk dries up because of the hardness and refractoriness of the water. Consumption often comes on after their deliveries, for due to the straining they have ruptures and tears. In the boys, as long as they are little, edemas occur in their testicles, but then as their age advances

---

^{6} στέριφαι Diller: στερίφαι Coray: στεριφναὶ V

ἀφανίζονται· ἠβῶσί τε ὀψὲ ἐν ταύτῃ τῇ πόλει. περὶ
μὲν οὖν τῶν θερμῶν πνευμάτων καὶ τῶν ψυχρῶν καὶ
τῶν πόλεων τούτων ὧδε ἔχει ὡς προείρηται.

5. Ὁκόσαι δὲ κέονται πρὸς τὰ πνεύματα μεταξὺ
τῶν θερινῶν ἀνατολέων τοῦ ἡλίου καὶ τῶν χειμερινῶν
καὶ ὁκόσαι τὸ ἐναντίον τούτων, ὧδε ἔχει περὶ αὐτῶν·
ὁκόσαι μὲν πρὸς τὰς ἀνατολὰς τοῦ ἡλίου κέονται,
ταύτας εἰκὸς εἶναι ὑγιεινοτέρας τῶν πρὸς τὰς ἄρκτους
ἐστραμμένων καὶ τῶν πρὸς τὰ θερμά <πνεύματα>,[7] ἢν
καὶ στάδιον τὸ μεταξὺ ᾖ. πρῶτον[8] μὲν γὰρ μετριώτε-
ρον ἔχει τὸ θερμὸν καὶ τὸ ψυχρόν· ἔπειτα τὰ ὕδατα,
ὁκόσα πρὸς τὰς τοῦ ἡλίου ἀνατολάς ἐστι, ταῦτα λαμ-
πρά τε εἶναι ἀνάγκη καὶ εὐώδεα καὶ μαλθακά, ἠέρα
τε μὴ[9] ἐγγίνεσθαι ἐν ταύτῃ τῇ πόλει· ὁ γὰρ ἥλιος
κωλύει ἀνίσχων καὶ καταλάμπων. †τὸ γὰρ ἑωθινὸν
ἑκάστοτε αὐτὸς [ὁ ἠὴρ][10] ἐπέχει ὡς ἐπὶ τὸ πολύ.† τά
24 τε εἴδεα τῶν ἀνθρώπων | εὔχροά τε καὶ ἀνθηρά ἐστι
μᾶλλον, ἢν μή τις νοῦσος ἄλλη κωλύῃ. λαμπρόφωνοί
τε οἱ ἄνθρωποι ὀργήν τε καὶ ξύνεσιν βελτίους εἰσὶ
τῶν πρὸς βορέην,[11] ᾗπερ[12] καὶ τὰ ἄλλα τὰ ἐμφυόμενα
ἀμείνω ἐστίν. ἔοικέ τε μάλιστα ἡ οὕτω κειμένη πόλις
ἦρι κατὰ τὴν μετριότητα τοῦ θερμοῦ καὶ τοῦ ψυχροῦ·
τά τε νοσεύματα ἐλάσσω μὲν γίνεται καὶ ἀσθενέστερα, ἔοικε δὲ τοῖσι ἐν τῇσι πόλεσι γινομένοισι νο-
σεύμασι τῇσι πρὸς τὰ θερμὰ πνεύματα ἐστραμμένῃ-
σιν. αἵ τε γυναῖκες αὐτόθι ἀρικύμονές εἰσι σφόδρα
καὶ τίκτουσι ῥηϊδίως. περὶ μὲν τούτων ὧδε ἔχει.

these disappear. Puberty comes late in such a town. And so, with regard to warm and cold winds affecting towns, things are as I have described.

5. Towns that lie facing winds that come from between the summer and winter risings of the sun (i.e., east), and those opposite to these are as follows. Those that lie facing the risings of the sun are likely to be healthier than those turned toward the Bears and those facing the hot winds (i.e., south), even though (sc. only) a *stadion* separates them. First, the heat and the cold are more moderate; then the waters in such a town that faces the risings of the sun must be clear, sweet-smelling, and soft, and free of mist, since the sun prevents its formation by shining over the town as it comes up, generally dominating it all through the morning. The people's appearance is characterized by a better complexion and more florid than elsewhere, unless some disease otherwise prevents this. They have clear voices, and a better temper and intelligence than those who face the north, just as all things growing toward the east are better. A town so situated most resembles the spring, because of the moderation of its heat and cold; diseases are less frequent and milder, although they resemble those in towns turned toward the warm winds. The women there are very prolific and give birth easily. So it is in these matters.

---

[7] πνεύματα add. Diller after *ad calidiores flatus* Lat. *versus calidos ventos* Lat. 2 "facing hot winds" Gal. (Ar.): om. V

[8] πρῶτον Coray after *primo* Lat. *primum* Lat. 2: πρότερον V

[9] ἠέρα τε μὴ Jurk (p. 61): ἢ ἐρατεινὰ V: καὶ ἔρα τε μὴ Gad. (B)   [10] ὁ ἠὴρ V: del. Diller, but the text and its sense are by no means clear: cf. Jouanna's discussion, pp. 264f.   [11] πρὸς βορέην recc.: προσβορέων V   [12] ἧπερ Coray: εἴπερ V

6. Ὁκόσαι δὲ πρὸς τὰς δύσιας κεῖνται καὶ αὐτῇσίν
ἐστι σκέπη τῶν πνευμάτων τῶν ἀπὸ τῆς ἠοῦς πνεόν-
των τά τε θερμὰ πνεύματα παραρρεῖ καὶ τὰ ψυχρὰ
ἀπὸ τῶν ἄρκτων, ἀνάγκη ταύτας τὰς πόλιας θέσιν
κέεσθαι νοσερωτάτην. πρῶτον μὲν γὰρ τὰ ὕδατα οὐ
λαμπρά· αἴτιον δέ, ὅτι ὁ ἠὴρ τὸ ἑωθινὸν κατέχει ὡς
ἐπὶ τὸ πολύ, ὅστις τῷ ὕδατι ἐγκαταμιγνύμενος τὸ
λαμπρὸν ἀφανίζει· ὁ γὰρ ἥλιος πρὶν ἄνω ἀρθῆναι οὐκ
ἐπιλάμπει. τοῦ δὲ θέρεος ἕωθεν μὲν αὖραι ψυχραὶ
πνέουσι καὶ δρόσοι πίπτουσι· τὸ δὲ λοιπὸν ἥλιος
ἐγκαταδύνων ὥστε μάλιστα διέψει τοὺς ἀνθρώπους·
διὸ καὶ ἀχρόους τε εἰκὸς εἶναι καὶ ἀρρώστους, τῶν
τε νοσευμάτων πάντων μετέχειν μέρος τῶν προειρη-
μένων· οὐδὲν αὐτοῖσιν ἀποκέκριται. βαρυφώνους τε
26 εἰκὸς | εἶναι καὶ βραγχώδεας διὰ τὸν ἠέρα, ὅτι
ἀκάθαρτος ὡς ἐπὶ τὸ πολὺ αὐτόθι γίνεται καὶ νοτώ-
δης·[13] οὔτε γὰρ ὑπὸ τῶν βορείων ἐκκρίνεται σφόδρα·
οὐ γὰρ προσέχουσι τὰ πνεύματα· ἅ τε προσέχουσιν
αὐτῇσι[14] καὶ πρόσκεινται ὑδατεινότατά ἐστιν· ἐπεὶ
<τοιαῦτα>[15] τὰ ἐπὶ τῆς ἑσπέρης πνεύματα· ἔοικέν τε
μετοπώρῳ μάλιστα ἡ θέσις ἡ τοιαύτη τῆς πόλιος
κατὰ τὰς τῆς ἡμέρης μεταβολάς, ὅτι πολὺ τὸ μέσον
γίνεται τοῦ τε ἑωθινοῦ καὶ τοῦ πρὸς τὴν δείλην. περὶ
μὲν πνευμάτων, ἅ τέ ἐστιν ἐπιτήδεια καὶ ἀνεπιτήδεια,
ὧδε ἔχει.

---

[13] νοτώδης Jacoby (p. 560): νοσώδης V   [14] αὐτῇσι Diller:
αὐταίησι V   [15] τοιαῦτα add. Foes n. 20

6. Towns that lie facing the settings (i.e., west), that are sheltered from the winds blowing from the east, but that the hot winds blow past on the one side, as do the cold ones from the Bears on the other side—these cities must be lying in a most unhealthy position. For first their waters are not clear, because the morning mist continues for a long time so that it becomes mixed with the water and removes its clearness, since the sun does not shine upon the town before it is high in the sky. In the summer cold breezes blow in the morning and there are heavy dews; for the rest of the day the sun as it progresses toward the west scorches the inhabitants to the last degree, so that they are likely to be of a pale color and ailing, and to share all the diseases mentioned above; none is particular to them.[6] They are likely to have deep voices and a tendency to sore throats because of the mist, since it continues for a long time there without being purified, and is moist. For it is not dispersed very much by the north winds, since they lack any effect there, whereas the winds that do have an effect and prevail in these towns are very moist, such being the nature of winds from the west. In its daily changes a town in this position mainly resembles the autumn, since between the temperature in the morning and the temperature toward evening there is a great difference. So it is regarding the winds that are favorable and unfavorable.

[6] Jones *Loeb*: "αὐτοῖσιν may be either a dative of advantage or one of disadvantage. There can thus be two meanings: (1) 'for none are isolated to their advantage,' *i.e.* they are exempt from none; (2) 'for none are isolated to their disadvantage,' *i.e.* they have no disease peculiar to themselves."

7. Περὶ δὲ τῶν λοιπῶν ὑδάτων βούλομαι διηγήσα-
σθαι, ἅ τέ ἐστι νοσώδεα καὶ ἃ ὑγιεινότατα καὶ ὁκόσα
ἀφ᾿ ὕδατος κακὰ εἰκὸς γίνεσθαι καὶ ὅσα ἀγαθά. πλεῖστον γὰρ μέρος ξυμβάλλεται ἐς τὴν ὑγιείην. ὁκόσα
μὲν οὖν ἐστιν ἑλώδεα καὶ στάσιμα καὶ λιμναῖα, ταῦτα
ἀνάγκη τοῦ μὲν θέρεος εἶναι θερμὰ καὶ παχέα καὶ
ὀδμὴν ἔχοντα, ἅτε οὐκ ἀπόρρυτα ἐόντα· ἀλλὰ τοῦ τε
ὀμβρίου ὕδατος ἐπιτρεφομένου[16] ἀεὶ νέου τοῦ τε ἡλίου
καίοντος ἀνάγκη ἄχροά τε εἶναι καὶ πονηρὰ καὶ χολώδεα, τοῦ δὲ χειμῶνος παγετώδεά τε καὶ ψυχρὰ καὶ
τεθολωμένα ὑπό τε χιόνος καὶ παγετῶν, ὥστε φλεγ-
ματωδέστατα εἶναι καὶ βραγχωδέστατα. τοῖσι δὲ πίνουσι σπλῆνας μὲν ἀεὶ μεγάλους εἶναι καὶ μεμνω-
μένους καὶ τὰς γαστέρας σκληράς τε καὶ λεπτὰς
καὶ θερμάς, τοὺς δὲ ὤμους καὶ τὰς κληῗδας καὶ τὸ
πρόσωπον καταλελεπτύνθαι καὶ κατισχάνθαι·[17] ἐς
γὰρ τὸν σπλῆνα αἱ σάρκες ξυντήκονται· διότι ἰσχνοὶ
28 | εἰσιν· ἐδωδούς τε εἶναι τοὺς τοιούτους καὶ διψηρούς·
τάς τε κοιλίας ξηροτάτας [τε][18] καὶ τὰς ἄνω καὶ τὰς
κάτω ἔχειν, ὥστε τῶν φαρμάκων ἰσχυροτέρων δεῖσθαι. τοῦτο μὲν τὸ νόσημα αὐτοῖσι ξύντροφόν ἐστι
καὶ θέρεος καὶ χειμῶνος. πρὸς δὲ τούτοισιν οἱ ὕδρωπες [τε][19] καὶ πλεῖστοι γίνονται καὶ θανατωδέστατοι.
τοῦ γὰρ θέρεος δυσεντερίαι τε πολλαὶ ἐμπίπτουσι καὶ
διάρροιαι καὶ πυρετοὶ τεταρταῖοι πολυχρόνιοι. ταῦτα
δὲ τὰ νοσεύματα μηκυνθέντα τὰς τοιαύτας φύσιας ἐς
ὕδρωπας καθίστησι καὶ ἀποκτείνει. ταῦτα μὲν αὐτοῖσι
τοῦ θέρεος γίνεται. τοῦ δὲ χειμῶνος τοῖσι νεωτέροισι

7. For the rest, I would like to expound upon waters that bring disease or excellent health, and on the evil or good things that are likely to arise from water, for they contribute a large part to health. Now such waters as are marshy, standing and stagnant must in summer be hot, thick and odorous, inasmuch as they have no outflow, but are constantly being added to by fresh rain water, and heated by the sun: these must be of bad color, troublesome and bilious. In winter they must be frosty, cold and turbid through the snow and ice, so as to be very conducive to phlegm and sore throats. In people who drink such waters, the spleen invariably becomes large and indurated, and their bellies are hard, narrow and hot, while their shoulders, collarbones and faces are lean and emaciated; the fact is that their flesh is melting and passing to their spleen, which causes them to become emaciated: such people eat and drink heavily, and their cavities, both the upper and the lower, are extremely dry, so that they require more powerful laxatives. This malady is endemic in both summer and winter. In addition, the dropsies that occur are very numerous and very deadly: in the summer there are dysenteries, diarrheas and chronic quartan fevers, which as they stretch out in time bring natures of this kind into dropsies, and kill them. These are the diseases suffered in summer. In winter, pneumonias and delirious states occur

---

¹⁶ ἐπιτρεφομένου V: ἐπιφερομένου Gad. (B)

¹⁷ καὶ κατισχάνθαι Jouanna: om. V: καὶ κατασχάνθαι Gad. (B): καὶ κατισχνάνθαι Diller

¹⁸ τε V: del. Coray

¹⁹ τε V: del. Diller

μὲν περιπνευμονίαι τε καὶ μανιώδεα νοσεύματα, τοῖσι
δὲ πρεσβυτέροισι καῦσοι διὰ τὴν τῆς κοιλίης σκλη-
ρότητα. τῆσι δὲ γυναιξὶν οἰδήματα ἐγγίνεται καὶ
φλέγμα λευκόν, καὶ ἐν γαστρὶ ἴσχουσι μόλις καὶ
τίκτουσι χαλεπῶς· μεγάλα τε τὰ ἔμβρυα καὶ οἰδέ-
οντα. ἔπειτα ἐν τῆσι τροφῆσι φθινώδεά τε καὶ πονηρὰ
γίνεται· ἥ τε κάθαρσις τῆσι γυναιξὶν οὐκ ἐπιγίνεται
χρηστὴ μετὰ τὸν τόκον. τοῖσι δὲ παιδίοισι κῆλαι ἐπιγίνονται μάλιστα καὶ τοῖσιν ἀνδράσι κιρσοὶ καὶ ἕλκεα ἐν τῆσι κνήμῃσιν, ὥστε τὰς τοιαύτας φύσιας οὐχ
οἷόν τε μακροβίους εἶναι, ἀλλὰ προγηράσκειν τοῦ
χρόνου τοῦ ἱκνευμένου. ἔτι δὲ αἱ γυναῖκες δοκέουσιν
ἔχειν ἐν γαστρί, καὶ ὁκόταν ὁ τόκος ᾖ, ἀφανίζεται τὸ
πλήρωμα τῆς γαστρός. τοῦτο δὲ γίνεται [ὑπὸ ὑδέ-
ρου][20] ὁκόταν ὑδρωπιήσωσιν αἱ ὑστέραι.

Τὰ μὲν τοιαῦτα ὕδατα νομίζω μοχθηρὰ εἶναι πρὸς
ἅπαν χρῆμα· δεύτερα δὲ ὅσων[21] εἶεν αἱ πηγαὶ ἐκ πε-
τρέων—σκληρὰ γὰρ ἀνάγκη εἶναι—ἢ ἐκ γῆς,[22] ὅκου
30 θερμὰ | ὕδατά ἐστιν, ἢ σίδηρος γίνεται ἢ χαλκὸς ἢ
ἄργυρος ἢ χρυσὸς ἢ θεῖον ἢ στυπτηρίη ἢ ἄσφαλτον
ἢ νίτρον. ταῦτα γὰρ πάντα ὑπὸ βίης γίνονται τοῦ
θερμοῦ. οὐ τοίνυν οἷόν τε ἐκ τοιαύτης γῆς ὕδατα
ἀγαθὰ γίνεσθαι, ἀλλὰ σκληρὰ καὶ καυσώδεα διουρεῖν τε[23] χαλεπὰ καὶ πρὸς τὴν διαχώρησιν ἐναντία
εἶναι. ἄριστα δὲ ὁκόσα ἐκ μετεώρων χωρίων ῥέει καὶ
λόφων γεηρῶν· αὐτά τε γάρ ἐστι γλυκέα καὶ λευκὰ
καὶ τὸν οἶνον φέρειν ὀλίγον οἷά τέ ἐστιν· τὸν δὲ χειμῶνα θερμὰ γίνεται, τοῦ δὲ θέρεος ψυχρά. οὕτω γὰρ

in young people, while in older ones there is ardent fever, caused by the hardness of their cavity. In the women there are swellings and white phlegm; they rarely conceive and they are delivered with difficulty. The babies are large and edematous: then, as they are nursed, they become consumptive and miserable. The (sc. lochial) cleaning after childbirth is not successful. The children have many hernias and the men suffer frequently from varicose veins and ulcers on their lower legs, so that people with such natures cannot be long-lived, but must grow old before their time comes. Furthermore, the women appear to be pregnant, but when their delivery should arrive, the fullness of their belly simply disappears, this condition arising when the uterus suffers from dropsy.

Such waters I hold to be bad for any purpose. The next worst waters are springs coming from rocks—for such water must be hard—or from the earth where there are hot waters, or where iron is produced, or copper, or silver, or gold, or sulfur, or alum, or asphalt, or soda, since all these arise from the violent action of heat. From such earth, then, good waters cannot come, but hard, burning waters, difficult to pass, and causing constipation. Best are waters that flow from high places and earthy hills: for these are sweet and limpid of themselves, and they are drinkable even with only a little wine. In winter they become warm, and in summer cold, because they originate from very

---

[20] ὑπὸ ὑδέρου V: del. Gad. (B)

[21] ὅσων Aldina: ὅσον V

[22] ἢ ἐκ γῆς Gad. (B): ἢ εἴτε V

[23] διουρεῖν τε Jouanna: διουρέεται V: διουρέεσθαι Gad. (B)

ἂν εἴη ἐκ βαθυτάτων πηγέων· μάλιστα δὲ ἐπαινέω[24]
ὧν τὰ ῥεύματα πρὸς τὰς ἀνατολὰς τοῦ ἡλίου ἐρρώ-
γασι καὶ μᾶλλον πρὸς τὰς θερινάς. ἀνάγκη γὰρ λαμ-
πρότερα εἶναι καὶ εὐώδεα καὶ κοῦφα. ὁκόσα δέ ἐστιν
ἀλυκὰ καὶ ἀτέραμνα καὶ σκληρά, τὰ[25] μὲν πάντα πίνειν οὐκ ἀγαθά· εἰσὶ δ' ἔνιαι φύσιες καὶ νοσεύματα,
ἐς ἃ ἐπιτήδειά ἐστι τὰ τοιαῦτα ὕδατα πινόμενα, περὶ
ὧν φράσω αὐτίκα. ἔχει δὲ καὶ περὶ τούτων ὧδε· ὁκόσων μὲν αἱ πηγαὶ πρὸς τὰς ἀνατολὰς ἔχουσι, ταῦτα
μὲν ἄριστα αὐτὰ ἑωυτῶν ἐστι· δεύτερα δὲ [τῶν][26] τὰ
μεταξὺ τῶν θερινῶν ἀνατολέων ἐστὶ τοῦ ἡλίου καὶ
δύσεων, καὶ μᾶλλον τὰ πρὸς τὰς ἀνατολάς· τρίτα δὲ
τὰ μεταξὺ τῶν δυσμέων τῶν θερινῶν καὶ τῶν χειμε-
ρινῶν· φαυλότατα δὲ τὰ πρὸς τὸν νότον τὰ μεταξὺ
τῆς χειμερινῆς[27] ἀνατολῆς καὶ δύσιος, καὶ ταῦτα
τοῖσι μὲν νοτίοισι πάνυ πονηρά, τοῖσι δὲ βορείοισιν
ἀμείνω.

32 Τούτοισι δὲ πρέπει ὧδε χέεσθαι· | ὅστις μὲν ὑγιαί-
νει τε καὶ ἔρρωται, μηδὲν διακρίνειν, ἀλλὰ πίνειν αἰεὶ
τὸ παρέον. ὅστις δὲ νούσου εἵνεκα βούλεται τὸ ἐπιτη-
δειότατον πίνειν, ὧδε ἂν ποιέων μάλιστα τυγχάνοι
τῆς ὑγιείης· ὁκόσων μὲν αἱ κοιλίαι σκληραί εἰσι καὶ
ξυγκαίειν ἀγαθαί [εἶναι],[28] τούτοισι μὲν τὰ γλυκύτατα
ξυμφέρει καὶ κουφότατα καὶ λαμπρότατα· ὁκόσων δὲ
μαλθακαὶ αἱ νηδύες καὶ ὑγραί εἰσι καὶ φλεγματώδεες,
τούτοισι δὲ τὰ σκληρότατα καὶ ἀτεραμνότατα καὶ τὰ

---

[24] ἐπαινέω Gad. (B): ἐπαινεῖν V   [25] τὰ Linden: τῷ V

deep springs.[7] I particularly commend springs whose water flows out in the direction of the rising—by preference the summer rising—of the sun. For they must be brighter, sweet-smelling and light; all, however, that are salty, refractory, and hard are not good to drink, although there are some constitutions and some diseases for which drinking such waters is advantageous: about these I will speak presently. Here are more details about spring waters: those whose sources face the risings of the sun are the very best; second are those between the summer risings of the sun and the summer settings, in particular those in the direction of the risings; third are those between the summer and winter settings; worst are those that face the south, that is, those between the winter rising and setting: these are very bad in conjunction with south winds, but somewhat better with north winds.

Waters from springs should be employed thus. A person in health and strength needs to make no choice, but can always drink whatever is available, but a person who because of disease wishes to drink the most suitable water will secure health best by doing the following. Persons whose cavities are constipated and liable to become inflamed benefit from the sweetest, lightest and most sparkling of waters. But those whose cavities are soft, moist, and phlegmatic, benefit from the hardest, most refractory,

[7] Cf. *Nature of the Child* 13–14; Loeb *Hippocrates*, vol. 10, pp. 67–75.

---

26 τῶν V: del. Littré
27 τῆς χειμερινῆς Gad. (B): θερινῆς V
28 εἶναι V: del. Linden

ὑφαλυκά· οὕτω γὰρ ἂν ξηραίνοιντο μάλιστα. ὁκόσα
γὰρ ὕδατά ἐστιν ἕψειν ἄριστα καὶ τακερώτατα, ταῦτα
καὶ τὴν κοιλίην διαλύειν εἰκὸς μάλιστα καὶ διατήκειν·
ὁκόσα δέ ἐστιν ἀτέραμνα καὶ σκληρὰ καὶ ἥκιστα
ἕψειν ἀγαθά, ταῦτα δὲ ξυνίστησι μάλιστα τὰς κοιλίας καὶ ξηραίνει. ἀλλὰ γὰρ ψευσάμενοι εἰσὶν οἱ ἄνθρωποι τῶν ἁλμυρῶν ὑδάτων πέρι δι' ἀπειρίην, καὶ
ὅτι νομίζεται διαχωρητικά· τὰ δὲ ἐναντιώτατά ἐστι
πρὸς τὴν διαχώρησιν· ἀτέραμνα γὰρ καὶ ἀνέψανα,
ὥστε καὶ τὴν κοιλίην ὑπ' αὐτῶν στύφεσθαι μᾶλλον ἢ
τήκεσθαι. καὶ περὶ μὲν τῶν πηγαίων ὑδάτων ὧδε ἔχει.

8. Περὶ δὲ τῶν ὀμβρίων καὶ ὁκόσα ἀπὸ χιόνος
φράσω ὅκως ἔχει. τὰ μὲν οὖν ὄμβρια κουφότατα καὶ
γλυκύτατά ἐστι καὶ λεπτότατα καὶ λαμπρότατα. τήν
τε γὰρ ἀρχὴν ὁ ἥλιος ἀνάγει καὶ ἀναρπάζει τοῦ ὕδατος τό τε λεπτότατον καὶ κουφότατον. δῆλον δὲ οἱ
ἅλες ποιέουσι. τὸ μὲν γὰρ ἁλμυρὸν λείπεται αὐτοῦ
ὑπὸ πάχεος καὶ βάρεος καὶ γίνεται ἅλες, τὸ δὲ λεπτότατον ὁ ἥλιος ἀναρπάζει ὑπὸ κουφότητος· ἀνάγει
34 δὲ τὸ τοιοῦτο οὐκ ἀπὸ τῶν ὑδάτων μόνον τῶν | λι-
μναίων, ἀλλὰ καὶ ἀπὸ τῆς θαλάσσης καὶ ἐξ ἁπάντων
ἐν ὁκόσοισιν ὑγρόν τι ἔνεστιν. ἔνεστι δὲ ἐν παντὶ
χρήματι. καὶ ἐξ αὐτῶν τῶν ἀνθρώπων ἄγει τὸ λεπτότατον τῆς ἰκμάδος καὶ κουφότατον. τεκμήριον δὲ
μέγιστον· ὅταν ἄνθρωπος ἐν ἡλίῳ βαδίζῃ ἢ καθίζῃ
ἱμάτιον ἔχων, ὁκόσα μὲν τοῦ χρωτὸς ὁ ἥλιος ἐφορᾷ,
οὐχ ἱδρώῃ ἄν· ὁ γὰρ ἥλιος ἀναρπάζει τὸ προφαινό-
μενον τοῦ ἱδρῶτος· ὁκόσα δὲ ὑπὸ τοῦ ἱματίου ἐσκέπα-

and salty of waters, for in this way they would best be dried up. For the waters that tend most to boil and melt are likely to relax the cavity and melt its contents away, whereas those that are refractory, hard and least suitable for boiling generally contract the cavities and dry them. In fact people are mistaken about saline waters through their inexperience, in thinking that salty waters are laxative: in reality they are the very opposite of laxative, being refractory and unsuitable for boiling, so that they congeal the cavities rather than melting them. Such are the facts about spring waters.

8. I will now explain how it is with rainwaters and waters from snow. Now rainwaters are the lightest, sweetest, finest, and brightest. To begin with, the sun lifts up and carries off the finest and lightest part of water, as the formation of salt makes clear: the salty element is left behind because of its thickness and weight, and becomes salt, whereas the finest part (sc. of the water) is carried off by the sun due to its lightness. The sun lifts up this element not only from pools, but also from the sea and from anything that has moisture in it—and there is moisture in everything. Even from human beings the sun draws up the finest and lightest part of their moistures. Here is a convincing proof. Whenever a person walks or sits in the sun while wearing a cloak, the parts of his skin the sun shines on will not (sc. be seen to) sweat, since the sun is carrying away the sweat as it is appearing. But those parts which

σται ἢ ὑπ' ἄλλου του, ἱδροῖ. ἐξάγεται μὲν γὰρ ὑπὸ
τοῦ ἡλίου καὶ βιάζεται, σῴζεται δὲ ὑπὸ τῆς σκέπης,
ὥστε μὴ ἀφανίζεσθαι ὑπὸ τοῦ ἡλίου. ὁκόταν δὲ ἐς
σκιὴν ἀφίκηται, ἅπαν τὸ σῶμα ὁμοίως ἰδίει·[29] οὐ γὰρ
ἔτι ὁ ἥλιος ἐπιλάμπει. διὰ ταῦτα δὲ καὶ σήπεται τῶν
ὑδάτων τάχιστα πάντων[30] καὶ ὀδμὴν ἴσχει πονηρὴν
τὸ ὄμβριον, ὅτι ἀπὸ πλείστων ξυνῆκται καὶ ξυμμέμι-
κται, ὥστε σήπεσθαι τάχιστα. ἔτι δὲ πρὸς τούτοισιν
ἐπειδὰν ἁρπασθῇ καὶ μετεωρισθῇ περιφερόμενον καὶ
καταμεμιγμένον ἐς τὸν ἠέρα, τὸ μὲν θολερὸν αὐτοῦ
καὶ νυκτοειδὲς ἐκκρίνεται καὶ ἐξίσταται καὶ γίνεται
ἠὴρ καὶ ὁμίχλη, τὸ δὲ λαμπρότατον καὶ κουφότατον
αὐτοῦ λείπεται καὶ γλυκαίνεται ὑπὸ τοῦ ἡλίου καιό-
μενόν τε καὶ ἑψόμενον. γίνεται δὲ καὶ τἆλλα πάντα τὰ
ἑψόμενα αἰεὶ γλυκέα. ἕως μὲν οὖν διεσκεδασμένον ᾖ
καὶ μήπω ξυνεστήκῃ, φέρεται μετέωρον. ὁκόταν δέ
κου ἀθροισθῇ καὶ ξυστραφῇ ἐς τὸ αὐτὸ ὑπὸ ἀνέμων
ἀλλήλοισιν ἐναντιωθέντων ἐξαίφνης, τότε καταρρήγνυται, ᾗ ἂν τύχῃ πλεῖστον ξυστραφέν. τότε γὰρ ἐοικὸς τοῦτο μᾶλλον γίνεσθαι, ὁκόταν τὰ νέφεα [μὴ][31]
ὑπὸ ἀνέμου ⟨σύ⟩στασιν ἔχοντα[32] ὡρμημένα ἐόντα καὶ
χωρέοντα ἐξαίφνης ἀντικόψῃ πνεῦμα ἐναντίον καὶ
36   ἕτερα νέφεα· ἐνταῦθα ⟨τὰ⟩[33] μὲν πρῶτον αὐτοῦ | ξυ-
στρέφεται, τὰ δὲ ὄπισθεν ἐπιφέρεται [τε][34] καὶ οὕτω
παχύνεται καὶ μελαίνεται καὶ ξυστρέφεται ἐς τὸ αὐτό·

---

29 ἰδίει Heringa (p. 45) from Erotian's gloss ἰδίειν: δίει V
30 πάντων A. Anastassiou (Jouanna): ταῦτα V

are covered by his cloak or by anything else do sweat, since the sweat, being forcibly drawn out by the sun is held back by the covering so that it cannot disappear through the sun's power. But when the person comes into the shade, his whole body sweats evenly, since the sun is no longer shining on him. For the same reason rainwater also becomes putrid most quickly of all waters, and has a fetid odor, because it is compounded and commingled from a great number of sources, so that it quickly putrefies. Furthermore, when this is carried away and raised upward, as it is tossed about and mixed into the air, the turbid and dark part separates out, is expelled, and becomes mist and fog, while the brightest and lightest part is left behind and sweetened, by being burned and boiled by the sun. All other things, too, that are boiled always become sweet. Now as long as this moisture is spread around and not yet collected together, it continues to move upward, but whenever it collects somewhere and condenses in one place, because of contrary winds suddenly pressing against one another, then it pours down its rain at the place where it happens to be most powerfully condensed. This is more likely to occur when the clouds have gained a certain consistency because of the wind, and being set in motion and advancing, they suddenly collide with an opposing wind and other clouds; thereupon the original clouds condense at that place, and the clouds added later become thick and dark, and condense in one location, and from their weight

---

31 μὴ V: del. Kühlewein after Latin and Arabic translations

32 συστάσιν ἔχοντα Kühlewein after Lat. *concursum habentes*: Gal. (Ar.). *"is firmly packed together"*: στάσιν ἔχοντος V

33 τὰ Gad. (B): om. V   34 τε V: del. Coray

καὶ ὑπὸ βάρεος καταρρήγνυται καὶ ὄμβροι γίνονται.
ταῦτα μέν ἐστιν ἄριστα κατὰ τὸ εἰκός. δεῖται δὲ ἀφέ-
ψεσθαι καὶ ἀποσήπεσθαι· εἰ δὲ μή, ὀδμὴν ἴσχει πονηρὴν καὶ βράγχοι καὶ βῆχες καὶ βαρυφωνίη[35] τοῖς
πίνουσι προσίσταται.

Τὰ δὲ ἀπὸ χιόνος καὶ κρυστάλλων πονηρὰ πάντα.
ὁκόταν γὰρ ἅπαξ παγῇ, οὐκέτι ἐς τὴν ἀρχαίην φύσιν
καθίσταται, ἀλλὰ τὸ μὲν αὐτοῦ λαμπρὸν καὶ κοῦφον
καὶ γλυκὺ ἐκπήγνυται[36] καὶ ἀφανίζεται, τὸ δὲ θολωδέστατον καὶ σταθμωδέστατον λείπεται. γνοίης δ' ἂν
ὧδε· εἰ γὰρ βούλει, ὅταν ᾖ χειμών, ἐς[37] ἀγγεῖον μέτρῳ
ἐγχέας ὕδωρ θεῖναι ἐς τὴν αἰθρίην, ἵνα πήξεται μάλιστα, ἔπειτα τῇ ὑστεραίῃ εἰσενεγκὼν εἰς ἀλέαν, ὅκου
χαλάσει μάλιστα ὁ παγετός, ὁκόταν δὲ λυθῇ, ἀνα-
μετρεῖν τὸ ὕδωρ, εὑρήσεις ἔλασσον συχνῷ. τοῦτο
τεκμήριον, ὅτι ὑπὸ τῆς πήξιος ἀφανίζεται καὶ ἀναξηραίνεται τὸ κουφότατον καὶ λεπτότατον, οὐ γὰρ τὸ
βαρύτατον καὶ παχύτατον· οὐ γὰρ ἂν δύναιτο. ταύτῃ
οὖν νομίζω πονηρότατα ταῦτα τὰ ὕδατα εἶναι τὰ ἀπὸ
χιόνος καὶ κρυστάλλου καὶ τὰ τούτοισιν ἑπόμενα
πρὸς ἅπαντα χρήματα. περὶ μὲν οὖν ὀμβρίων ὑδάτων
καὶ τῶν ἀπὸ χιόνος καὶ κρυστάλλων οὕτως ἔχει.

9. Λιθιῶσι δὲ μάλιστα ἄνθρωποι καὶ ὑπὸ νεφριτί-
δων καὶ στραγγουρίης ἁλίσκονται καὶ ἰσχιάδων, καὶ
κῆλαι γίνονται, ὅκου ὕδατα πίνουσι παντοδαπώτατα
καὶ ἀπὸ ποταμῶν μεγάλων, ἐς οὓς ποταμοὶ ἕτεροι ἐμ-
βάλλουσι, καὶ ἀπὸ λίμνης, ἐς ἣν ῥεύματα πολλὰ καὶ

water pours down and becomes rain. Such waters are naturally the best, but they do need to be boiled off and purified from putrefaction, if they are not to have a fetid odor, and provoke sore throats, coughs and hoarseness in the people who drink them.

Waters from snow and ice are all bad. For water, once frozen, never again recovers its original nature, but its clear, light, sweet part is frozen out of it and disappears, while the most turbid and heavy part remains. You may learn this from the following experiment. If you like, pour by measure, in winter, water into a vessel and set it in the open air, in order that it will best freeze; then on the next day bring it into a warm place, where the ice will best melt. When it has melted, on remeasuring the water you will find it to be much less than it was. This proves that on freezing water's lightest and finest part disappears and dries up, but naturally not its heaviest and thickest part, which could do no such thing. Therefore, I am of the opinion that such waters, derived from snow and ice, and waters similar to these, are the worst for all purposes. With regard to rainwaters, and of those from snow and ice, so it is.

9. People suffer from stone in particular, and are attacked by kidney diseases, strangury and sciaticas, and hernias occur, in places where they drink water compounded of very many different kinds, coming both from large rivers into which other rivers flow, or from a lake fed by many streams of various sorts, and also whenever they

---

[35] βράγχοι . . . βαρυφωνίη Gad. (B): βράγχος καὶ βαρυφωνίην V   [36] ἐκπήγνυται Gad. (B): ἐκκρίνεται V
[37] ἢ χειμών, ἐς Coray: οἱ χειμῶνες V: χειμὼν εἰς Gad. (B)

παντοδαπὰ ἀφικνεῦνται, καὶ ὁκόσοι ὕδασιν ἐπακτοῖσι
38 | χρέονται διὰ μακροῦ ἀγομένοισι καὶ μὴ ἐκ βραχέος.
οὐ γὰρ οἷόν τε ἕτερον ἑτέρῳ ἐοικέναι ὕδωρ, ἀλλὰ τὰ
μὲν γλυκέα εἶναι, τὰ δὲ ἁλυκά τε καὶ στυπτηριώδεα,
τὰ δὲ ἀπὸ θερμῶν ῥέειν. ξυμμισγόμενα δὲ ταῦτα ἐς
ταὐτὸν ἀλλήλοισι στασιάζει<ν>[38] καὶ κρατέειν ἀεὶ τὸ
ἰσχυρότατον. ἰσχύει δὲ οὐκ ἀεὶ τωὐτό, ἀλλὰ ἄλλοτε
ἄλλο κατὰ τὰ πνεύματα· τῷ μὲν γὰρ βορέης τὴν
ἰσχὺν παρέχεται, τῷ δὲ ὁ νότος, καὶ τῶν λοιπῶν πέρι
ωὑτὸς λόγος. ὑφίστασθαι οὖν τοῖσι τοιούτοισιν
ἀνάγκη ἐν τοῖσιν ἀγγείοισιν ἰλὺν καὶ ψάμμον· καὶ
ἀπὸ τούτων πινομένων τὰ νοσήματα γίνεται τὰ προ-
ειρημένα· ὅτι δὲ οὐχ ἅπασιν, ἑξῆς φράσω.

Ὁκόσων μὲν ἥ τε κοιλίη εὔροός τε καὶ ὑγιηρή ἐστι
καὶ ἡ κύστις μὴ πυρετώδης μηδὲ ὁ στόμαχος τῆς
κύστιος ξυμπέφρακται λίην, οὗτοι μὲν διουρεῦσι ῥηϊδίως, καὶ ἐν τῇ κύστει οὐδὲν ξυστρέφεται. ὁκόσων δὲ
ἂν ἡ κοιλίη πυρετώδης ᾖ, ἀνάγκη καὶ τὴν κύστιν
τωὐτὸ πάσχειν. ὁκόταν γὰρ θερμανθῇ μᾶλλον τῆς
φύσιος, ἐφλέγμηνεν αὐτῆς ὁ στόμαχος. ὁκόταν δὲ
ταῦτα πάθῃ, τὸ οὖρον οὐκ ἀφίησιν, ἀλλ᾿ ἐν ἑωυτῇ
ξυνέψει καὶ συγκαίει. καὶ τὸ μὲν λεπτότατον αὐτοῦ
[ἀποκρίνεται][39] καὶ τὸ καθαρώτατον διεῖ καὶ ἐξου-
ρεῖται, τὸ δὲ παχύτατον καὶ θολωδέστατον ξυστρέφεται καὶ συμπήγνυται. τὸ μὲν πρῶτον μικρόν, ἔπειτα
δὲ μεῖζον γίνεται. κυλινδεύμενον γὰρ ὑπὸ τοῦ οὔρου,
ὅ τι ἂν ξυνίστηται παχύ, ξυναρμόζει πρὸς ἑωυτό, καὶ
οὕτως αὔξεταί τε καὶ πωροῦται· καὶ ὁκόταν οὐρῇ,

employ imported waters led from a long distance rather than a short distance. For one water cannot be like another, but some are sweet, others are salty and astringent, and others flow from hot springs. As these are commingled, they struggle against one another, and it is always the strongest one that prevails. But it is not always the same one that dominates, for at one time one does, and at another time another does, according to the winds, since one water is given strength by the north wind, another by the south wind, and the rest according to the same principle. And so such waters must precipitate silt and sand in the vessels, and when these waters are drunk the diseases mentioned above arise. That there are exceptions, I will explain.

Persons whose cavity is fluent and healthy, whose bladder is not feverish, and the orifice of whose bladder is not abnormally contracted, pass urine easily, and nothing solid forms in their bladder. But those whose cavity is feverish must also have feverishness of the bladder, and when it becomes unnaturally heated its orifice becomes inflamed. When it suffers thus, it fails to expel its urine, but boils it and concocts it inside itself. The urine's purest part passes out and is excreted as urine, while the thickest and most turbid part is compressed and coagulated—at first in tiny pieces, and then they become larger. For as this sediment is rolled around by the urine, the part that has become solid adds matter to itself, and so it continues to grow and calcify. When the patient passes urine, this material falls

---

38 στασιάζειν Jouanna: στασιάζει V
39 ἀποκρίνεται V: del. Jouanna after Lat. Gal. (Ar.)

99

πρὸς τὸν στόμαχον τῆς κύστιος προσπίπτει ὑπὸ τοῦ
οὔρου βιαζόμενον καὶ κωλύει οὐρεῖν καὶ ὀδύνην παρέχει ἰσχυρήν· ὥστε τὰ αἰδοῖα τρίβουσι καὶ ἕλκουσι
τὰ παιδία τὰ λιθιῶντα· δοκεῖ γὰρ αὐτοῖσι τὸ αἴτιον
ἐνταῦθα εἶναι τῆς οὐρήσιος. τεκμήριον δέ, ὅτι οὕτως
ἔχει· τὸ γὰρ οὖρον λαμπρότατον οὐρέουσιν οἱ λιθιῶν-
40   τες, ὅτι | τὸ παχύτατον καὶ θολωδέστατον αὐτοῦ μένει
καὶ συστρέφεται. τὰ μὲν πλεῖστα οὕτω λιθιᾷ· γίνεται
δὲ πῶρος[40] καὶ ἀπὸ τοῦ γάλακτος, ἢν μὴ ὑγιηρὸν ᾖ,
ἀλλὰ θερμόν τε λίην καὶ χολῶδες. τὴν γὰρ κοιλίην
διαθερμαίνει καὶ τὴν κύστιν, ὥστε τὸ οὖρον ξυγκαιό-
μενον ταῦτα πάσχειν. καὶ φημι ἄμεινον εἶναι τοῖσι
παιδίοισι τὸν οἶνον ὡς ὑδαρέστατον διδόναι· ἧσσον
γὰρ τὰς φλέβας ξυγκαίει καὶ συναναίνει. τοῖσι δὲ
θήλεσι λίθοι οὐ γίνονται ὁμοίως· ὁ γὰρ οὐρητὴρ |
42   βραχύς ἐστιν ὁ τῆς κύστιος καὶ εὐρύς, ὥστε βιάζεται
τὸ οὖρον ῥηιδίως. οὔτε γὰρ τῇ χειρὶ τρίβει τὸ αἰδοῖον
ὥσπερ τὸ ἄρσεν, οὔτε ἅπτεται τοῦ οὐρητῆρος· ἐς γὰρ
τὰ αἰδοῖα ξυντέτρηνται, οἱ δὲ ἄνδρες οὐκ εὐθὺ τέτρην-
ται, διότι καὶ[41] οἱ οὐρητῆρες οὐκ εὐρεῖς· καὶ πίνουσι
πλεῖον ἢ οἱ παῖδες. περὶ μὲν οὖν τούτων ὧδε ἔχει ἢ
ὅτι τούτων ἐγγύτατα.

10. Περὶ δὲ τῶν ὡρέων ὧδε ἄν τις ἐνθυμεύμενος
διαγινώσκοι, ὁκοῖόν τι μέλλει ἔσεσθαι τὸ ἔτος, εἴτε
νοσερὸν εἴτε ὑγιηρόν· ἢν μὲν γὰρ κατὰ λόγον γένηται
τὰ σημεῖα ἐπὶ τοῖς ἄστροισι δύνουσί τε καὶ ἐπιτέλ-
λουσιν, ἔν τε τῷ μετοπώρῳ ὕδατα γένηται, καὶ ὁ χειμὼν μέτριος καὶ μήτε λίην εὔδιος μήτε ὑπερβάλλων

against the orifice of his bladder, being driven forward by the urine, blocks the passage of the urine, and provokes violent pain. Thus young children suffering from stone rub and pull at their genitalia, believing that the cause of their urination is there. Here is evidence for my account. Patients suffering from stone pass urine that is exceptionally clear, because its thickest and most turbid part has remained behind and solidified. This is so in most cases of stone, but a stone can also form from milk, if it is not healthy, but is too hot and bilious, for then it heats the cavity and the bladder, so that the urine is concocted and suffers the same things. Also I contend that it is better to give young children only very diluted wine, since this burns and dries the vessels less. In females stones do not form so frequently, since the urethra of their bladder is short and wide, so that urine is easily forced out; they do not rub their genitalia with their hand as the male does, nor do they touch their urethra. This is so because their urethra is connected directly into the genitalia—men do not have it connected straight like this, which is why their urethras are not wide—and also because they drink more than boys do. On this subject, so it is, or something very like this.

10. With regard to the seasons, a person considering the following points could distinguish how a year was going to be, whether unhealthy or healthy. For if the signs are regular at the times when the stars set and rise, if there are rains in the fall, if the winter is moderate (being neither too mild nor exceeding due measure with regard to

---

[40] πῶρος Heiberg: πρὸς V: πόρος Gad. (B)
[41] διότι καὶ Lesky (p. 75): καὶ διότι V

τὸν καιρὸν τῷ ψύχει, ἔν τε τῷ ἦρι ὕδατα γένηται
ὡραῖα καὶ ἐν τῷ θέρει, οὕτω τὸ ἔτος ὑγιεινότατον
εἰκὸς εἶναι.

Ἢν δὲ ὁ μὲν χειμὼν αὐχμηρὸς καὶ βόρειος γένηται, τὸ δὲ ἦρ ἔπομβρον καὶ νότιον, ἀνάγκη τὸ θέρος
πυρετῶδες γίνεσθαι καὶ ὀφθαλμίας καὶ δυσεντερίας
ἐμποιεῖν. ὁκόταν γὰρ τὸ πνῖγος ἐπιγένηται ἐξαίφνης
τῆς τε γῆς ὑγρῆς ἐούσης ὑπὸ τῶν ὄμβρων τῶν ἐαρι-
44   νῶν | καὶ ὑπὸ τοῦ νότου, ἀνάγκη διπλόον τὸ καῦμα
εἶναι, ἀπό τε τῆς γῆς διαβρόχου ἐούσης καὶ θερμῆς
καὶ ὑπὸ τοῦ ἡλίου καίοντος, τῶν τε κοιλιῶν μὴ ξυνε-
στηκυιῶν τοῖς ἀνθρώποις μήτε τοῦ ἐγκεφάλου ἀνεξη-
ρασμένου—οὐ γὰρ οἷόν τε τοῦ ἦρος τοιούτου ἐόντος
μὴ οὐ πλαδᾶν τὸ σῶμα καὶ τὴν σάρκα—ὥστε τοὺς
πυρετοὺς ἐπιπίπτειν ὀξυτάτους ἅπασι, μάλιστα δὲ
τοῖσι φλεγματίῃσι. τὰς δὲ δυσεντερίας εἰκός ἐστι γίνεσθαι καὶ τῇσι γυναιξὶ καὶ τοῖς εἴδεσι τοῖς ὑγρο-
τάτοισι. καὶ ἢν μὲν ἐπὶ κυνὸς ἐπιτολῇ ὕδωρ ἐπιγένηται καὶ χειμὼν καὶ οἱ ἐτησίαι πνεύσωσιν, ἐλπὶς
παύσασθαι καὶ τὸ μετόπωρον ὑγιηρὸν γενέσθαι· ἢν
δὲ μή, κίνδυνος θανάτους τε γενέσθαι τοῖσι παιδίοισι
καὶ τῇσι γυναιξί, τοῖσι δὲ πρεσβύτῃσιν ἥκιστα, τούς
τε περιγενομένους ἐς τεταρταίους ἀποτελευτᾶν καὶ ἐκ
τῶν τεταρταίων ἐς ὕδρωπας.

Ἢν δ' ὁ μὲν χειμὼν νότιος γένηται καὶ ἔπομβρος
καὶ εὔδιος, τὸ δὲ ἦρ βόρειόν τε καὶ αὐχμηρὸν καὶ
χειμέριον, πρῶτον μὲν τὰς γυναῖκας, ὁκόσαι ἂν τύχω-
σιν ἐν γαστρὶ ἔχουσαι καὶ ὁ τόκος αὐτῇσιν ᾖ πρὸς

cold) and if the rains are seasonable in spring and summer, the year is likely to be very healthy.

a) If, however, the winter is dry and dominated from the north, and the spring is rainy and southerly, the summer cannot fail to have fevers, and to bring ophthalmias and dysenteries. For when a suffocating heat suddenly comes on while the earth is soaked as the result of the spring rains and the south wind, the heat must be redoubled, coming both from the hot, sodden earth and from the burning sun, and also from people's cavities not being firm and their brain not being kept dry—for it is not possible, when the spring is like this, for the body and its flesh not to be swollen with fluid—so that the fevers that attack will be of the acutest kind in all cases, but especially among the phlegmatic. It is also likely that dysenteries will come on in the women, and in people with very humid constitutions. But if at the rising of Sirius rains occur and storms, and the Etesian winds blow, there is hope that the diseases will cease and that the fall will be healthy. Otherwise, there is a danger of deaths in the children and women—although very little among the old men—and for patients who survive to end up with quartan fevers, and that from the quartan fevers dropsies will follow.

b) But if the winter is southerly, rainy and mild, and the spring is northerly, dry and stormy, first women who happen to be pregnant and whose delivery is due toward

46 τὸ ἦρ, τιτρώσκεσθαι· | ὁκόσαι δ' ἂν καὶ τέκωσιν,
ἀκρατέα τὰ παιδία τίκτειν καὶ νοσώδεα, ὥστε ἢ
αὐτίκα ἀπόλλυσθαι, ἢ ζώειν[42] λεπτά τε ἐόντα καὶ
ἀσθενέα καὶ νοσώδεα. ταῦτα μὲν τῇσι γυναιξί· τοῖσι
δὲ λοιποῖσι δυσεντερίας καὶ ὀφθαλμίας ξηρὰς καὶ
ἐνίοισι κατάρρους ἀπὸ τῆς κεφαλῆς ἐπὶ τὸν πνεύμονα.
τοῖσι μὲν οὖν φλεγματίῃσι τὰς δυσεντερίας εἰκὸς γίνεσθαι καὶ τῇσι γυναιξὶ φλέγματος ἐπικαταρρυέντος
ἀπὸ τοῦ ἐγκεφάλου διὰ τὴν ὑγρότητα τῆς φύσιος·
τοῖσι δὲ χολώδεσιν ὀφθαλμίας ξηρὰς διὰ τὴν θερμότητα καὶ ξηρότητα τῆς σαρκός, τοῖσι δὲ πρεσβύ-
τῃσι κατάρρους διὰ τὴν ἀραιότητα καὶ τὴν ἔκτηξιν
τῶν φλεβῶν, ὥστε ἐξαίφνης τοὺς μὲν ἀπόλλυσθαι,
τοὺς δὲ παραπλήκτους γίνεσθαι τὰ δεξιά.[43] ὁκόταν
γὰρ τοῦ χειμῶνος ἐόντος νοτίου καὶ θερμοῦ τὸ σῶμα[44]
μὴ ξυνιστῆται[45] μηδὲ αἱ φλέβες, τοῦ ἦρος ἐπιγενομένου βορείου καὶ αὐχμηροῦ καὶ ψυχροῦ ὁ ἐγκέφαλος,
ὁπηνίκα αὐτὸν ἔδει ἅμα τῷ ἦρι διαλύεσθαι καὶ καθ-
αίρεσθαι ὑπό τε κορύζης καὶ βράγχου,[46] τηνικαῦτα |
48 πήγνυταί τε καὶ συνίσταται, ὥστε ἐξαίφνης τοῦ
θέρεος ἐπιγενομένου καὶ τοῦ καύματος καὶ <μεγά-
λης>[47] τῆς μεταβολῆς ἐπιγινομένης ταῦτα τὰ νοσεύματα ἐπιπίπτειν. καὶ ὁκόσαι μὲν τῶν πόλεων κέονταί
τε καλῶς τοῦ ἡλίου καὶ τῶν πνευμάτων ὕδασί τε
χρέονται ἀγαθοῖσιν, αὗται μὲν ἧσσον αἰσθάνονται
τῶν τοιούτων μεταβολέων· ὁκόσαι δὲ ὕδασί τε ἑλείοισι
χρέονται καὶ λιμνώδεσι κέονταί τε μὴ καλῶς τῶν
πνευμάτων καὶ τοῦ ἡλίου, αὗται δὲ μᾶλλον. κἢν μὲν

spring are likely to have an abortion, or if they do give birth, it will be to children that are weak and sickly, so that these either die at once, or they live but are puny, weak and sickly. So much for the women: the rest of the population will have dysenteries and dry ophthalmias, and in some cases fluxes descending from the head to the lung. Now whereas in phlegmatics dysenteries are likely to occur, and also in women from phlegm running down from the brain because of the humidity of their nature, in the bilious there are dry ophthalmias because of the hotness and dryness of their flesh, and in old men catarrhs caused by their flabbiness and the attenuation of their vessels, so that some suddenly die, while others become paralyzed on the right side. For when, owing to the winter being southerly and warm, neither the body nor the vessels become firm, and then a northerly, dry, cold spring supervenes, the brain—just at the time when it ought to have been relaxed along with the spring and purged by a cold in the head and sore throats—congeals and hardens, so that when the summer and its heat suddenly supervene, and a great change occurs, these diseases befall. Towns that are well situated with regard to sun and winds, and use good waters, are less sensitive to such changes; but those that use marshy and stagnant waters, and are not well situated with regard to the winds and the sun, are more affected. And

---

42 ζώειν Coray: ζῶσι V

43 Add. ἢ τὰ ἀριστερὰ Gad. (B)

44 τὸ σῶμα Zwinger[marg]: τοῦ σώματος V

45 ὁ ἐγκέφαλος add. Gad. (B)

46 βράγχου Jouanna: βράγχων V

47 μεγάλης add. Jouanna after Gal. (Ar.)

τὸ θέρος αὐχμηρὸν γένηται, θᾶσσον παύονται αἱ νοῦ-
σοι· ἢν δὲ ἔπομβρον, πολυχρόνιοι γίνονται· καὶ φα-
γεδαίνας κίνδυνος ἐγγίνεσθαι ἀπὸ πάσης προφάσιος,
ἢν ἕλκος ἐγγένηται· καὶ λειεντερίαι καὶ ὕδρωπες τε-
λευτῶσι τοῖσι νοσεύμασιν ἐπιγίνονται· οὐ γὰρ ἀπο-
ξηραίνονται αἱ κοιλίαι ῥηϊδίως.

    Ἢν δὲ τὸ θέρος ἔπομβρον γένηται καὶ νότιον καὶ
τὸ μετόπωρον, τὸν χειμῶνα ἀνάγκη νοσερὸν εἶναι καὶ
50 | τοῖς φλεγματίῃσι καὶ τοῖς γεραιτέροισι τεσσαρά-
κοντα ἐτέων καύσους γίνεσθαι εἰκός, τοῖσι δὲ χολώδεσι πλευρίτιδας καὶ περιπνευμονίας.

    Ἢν δὲ τὸ θέρος αὐχμηρὸν γένηται καὶ νότιον, τὸ
δὲ μετόπωρον ἔπομβρον καὶ βόρειον, κεφαλαλγίας ἐς
τὸν χειμῶνα καὶ σφακέλους τοῦ ἐγκεφάλου εἰκὸς γίνεσθαι, καὶ προσέτι βῆχας καὶ βράγχους καὶ κορύ-
ζας, ἐνίοισι δὲ καὶ φθίσιας.

    Ἢν δὲ βόρειόν τε ᾖ καὶ ἄνυδρον καὶ μήτε ἐπὶ τῷ
κυνὶ γένηται ὕδωρ[48] μήτε ἐπὶ τῷ ἀρκτούρῳ, τοῖσι μὲν
φλεγματίῃσι φύσει οὕτως ἄν[49] ξυμφέροι μάλιστα καὶ
τοῖς ὑγροῖς τὰς φύσιας καὶ τῇσι γυναιξί, τοῖσι δὲ
χολώδεσι τοῦτο πολεμιώτατον γίνεται· λίην γὰρ ἀνα-
ξηραίνονται· καὶ ὀφθαλμίαι αὐτοῖσιν ἐπιγίνονται ξη-
ραί, καὶ πυρετοὶ ὀξέες καὶ πολυχρόνιοι, ἐνίοισι δὲ καὶ
μελαγχολίαι· τῆς γὰρ χολῆς τὸ μὲν ὑγρότατον καὶ
ὑδαρέστατον ἀναλοῦται, τὸ δὲ παχύτατον καὶ δριμύ-
τατον λείπεται καὶ τοῦ αἵματος κατὰ τὸν αὐτὸν λόγον·
ἀφ᾽ ὧν ταῦτα τὰ νοσεύματα αὐτοῖσι γίνεται. τοῖσι δὲ
φλεγματίῃσι πάντα ταῦτα ἀρωγά ἐστιν. ἀποξηραίνον-

if the summer is dry, the diseases will cease more quickly, but if it is rainy, they persist for a long time, and there is a danger that erosions will develop from many causes if there is an ulcer; lienteries and dropsies supervene when the diseases end, since the cavities are not easily dried up.

c) If the summer is rainy and southerly, and the fall as well, the winter must be unhealthy: in both phlegmatics and men over forty years of age ardent fevers are likely to arise, and in the bilious, pleurisies and pneumonias

d) If the summer is dry and southerly, and the autumn rainy and northerly, there are likely to be headaches in winter and sphacelus of the brain, and also coughs, sore throats, and coryzas, and in some cases consumptions as well.

e) But if the (sc. summer) weather is northerly and dry, with no rain falling at the rising of either Sirius or Arcturus, for those who are naturally phlegmatic, this would be very beneficial—and to those who are humid in their natures, and for women—but it is very harmful to the bilious: for these become excessively dry, and are attacked by dry ophthalmias and by chronic acute fevers, in some cases by melancholies too. This is because the most humid and watery part of the bile is used up, while the thickest and most acrid part is left behind, and similarly with the blood: from these causes such diseases befall the bilious. But for the phlegmatic, all these (sc. weather conditions) are beneficial, since they thoroughly dry the patients, who

---

[48] μήτε ἐπὶ τῷ κυνὶ γένηται ὕδωρ Jouanna after Gad. (B): om. V

[49] οὕτως ἂν Gad. (B): om. V

ται γὰρ καὶ ἐς τὸν χειμῶνα ἀφικνέονται οὐ πλαδῶν-
τες, ἀλλὰ ἀναξηραινόμενοι.

[Ἢν δὲ ὁ χειμὼν βόρειος γένηται καὶ ξηρός, τὸ δὲ
ἦρ νότιον καὶ ἔπομβρον, κατὰ τὸ θέρος ὀφθαλμίαι
γίνονται ἰσχυραί, τοῖσι δὲ παισὶ καὶ γυναιξὶ πυρετοί.]⁵⁰

11. Κατὰ ταῦτά τις ἐννοεύμενος καὶ σκοπεύμενος
προειδείη ἂν τὰ πλεῖστα τῶν μελλόντων ἔσεσθαι ἀπὸ
τῶν μεταβολέων. φυλάσσεσθαι δὲ χρὴ μάλιστα τὰς
μεταβολὰς τῶν ὡρέων τὰς μεγίστας καὶ μήτε φάρμα-
κον διδόναι ἑκόντα μήτε καίειν ὅτι ἐς κοιλίην μήτε |
52 τάμνειν, πρὶν παρέλθωσιν ἡμέραι δέκα ἢ καὶ πλείο-
νες· μέγισται δέ εἰσιν αἵδε καὶ ἐπικινδυνόταται· ἡλίου
τροπαὶ ἀμφότεραι καὶ μᾶλλον θεριναὶ καὶ ἰσημερίαι
νομιζόμεναι εἶναι ἀμφότεραι, μᾶλλον δὲ αἱ μετοπωρι-
ναί· δεῖ δὲ καὶ τῶν ἄστρων τὰς ἐπιτολὰς φυλάσσεσθαι καὶ μάλιστα τοῦ κυνός, ἔπειτα ἀρκτούρου, καὶ
ἔτι πληϊάδων δύσιν. τά τε γὰρ νοσεύματα μάλιστα ἐν
ταύτῃσι τῇσιν ἡμέρῃσι κρίνεται· καὶ τὰ μὲν ἀποφθί-
νει, τὰ δὲ λήγει, τὰ δὲ ἄλλα πάντα μεθίσταται ἐς
ἕτερον εἶδος καὶ ἑτέρην κατάστασιν. περὶ μὲν τούτων
οὕτως ἔχει.

12. Βούλομαι δὲ περὶ τῆς Ἀσίης καὶ τῆς Εὐρώπης
δεῖξαι ὁκόσον διαφέρουσιν ἀλλήλων ἐς τὰ πάντα καὶ
περὶ τῶν ἐθνέων τῆς μορφῆς, ὅτι⁵¹ διαλλάσσει καὶ
μηδὲν ἔοικεν ἀλλήλοισι. περὶ μὲν οὖν ἀπάντων πολὺς

---

⁵⁰ ἦν δὲ ὁ χειμὼν . . . πυρετοί V Gad. (B): del. Cornarius

then arrive at winter without edematous swellings and in a well dried state.

f) [If the winter is northerly and dry, and the spring southerly and rainy, in the summer there will be severe ophthalmias, and in the children and women fevers.][8]

11. By investigating and observing such matters, a person will be able to foresee most of the consequences of the changes. One must be especially on one's guard against the most important changes of the seasons, and avoid giving a purgative medication or applying any cautery at all for the cavity, or incising, before ten days are past, or even more. Most important (sc. of the changes) are the following, and most dangerous: both solstices, especially the summer one, and both equinoxes (as they are reckoned), especially the autumnal. One must also pay attention to the risings of the stars, especially of Sirius , then of Arcturus, and also to the setting of the Pleiades: for diseases generally have their crises on these days, some proving fatal, others coming to an end, and all the rest changing to a different form and to a different constitution. So much on this subject.

12. It is my intention to demonstrate, regarding Asia and Europe, how they differ from one another in every respect, and that the peoples in the two regions are different in their bodily form and completely dissimilar. Now

[8] See paragraph a), where this passage first appears.

Zwinger: Foes (n. 66), *Quae ante haec leguntur,* ἦν δὲ ὁ χειμὼν βόρειος, *suspecta sunt, cum paulo ante iam scripta sint, ideoque, cum Interpretibus ea reiecimus*

[51] ὅτι Coray: τί V: ὡς Gad. (B)

ἂν εἴη λόγος, περὶ δὲ τῶν μεγίστων καὶ πλεῖστον δια-
φερόντων ἐρέω ὥς μοι δοκεῖ ἔχειν. τὴν Ἀσίην πλεῖστον διαφέρειν φημὶ τῆς Εὐρώπης ἐς τὰς φύσιας τῶν
ξυμπάντων τῶν τε ἐκ τῆς γῆς φυομένων καὶ τῶν ἀνθρώπων. πολὺ γὰρ καλλίονα καὶ μείζονα πάντα γίνεται ἐν τῇ Ἀσίῃ, ἥ τε χώρη τῆς χώρης ἡμερωτέρη καὶ
τὰ ἤθεα τῶν ἀνθρώπων ἠπιώτερα καὶ εὐοργητότερα.[52]
τὸ δὲ αἴτιον τούτων ἡ κρᾶσις τῶν ὡρέων, ὅτι τοῦ
54 ἡλίου ἐν μέσῳ τῶν | ἀνατολέων κεῖται πρὸς τὴν ἠῶ
τοῦ τε ψυχροῦ πορρωτέρω. τὴν δὲ αὔξησιν καὶ ἡμε-
ρότητα παρέχει πλεῖστον ἁπάντων, ὁκόταν μηδὲν ᾖ
ἐπικρατέον βιαίως, ἀλλὰ παντὸς ἰσομοιρίη δυνα-
στεύῃ. ἔχει δὲ κατὰ τὴν Ἀσίην οὐ πανταχῇ ὁμοίως,
ἀλλ' ὅση μὲν τῆς χώρης ἐν μέσῳ κεῖται τοῦ θερμοῦ
καὶ τοῦ ψυχροῦ, αὕτη μὲν εὐκαρποτάτη ἐστὶ καὶ εὐ-
δενδροτάτη καὶ εὐδιεστάτη καὶ ὕδασι καλλίστοισι
κέχρηται τοῖσί τε οὐρανίοισι καὶ τοῖς ἐκ τῆς γῆς. οὔτε
γὰρ ὑπὸ τοῦ θερμοῦ ἐκκέκαυται λίην οὔτε ὑπὸ αὐχμῶν καὶ ἀνυδρίης ἀναξηραίνεται, οὔτε ὑπὸ ψύχεος
βεβιασμένη <οὔτε>[53] νοτίη τε καὶ διάβροχός ἐστιν
ὑπό τε ὄμβρων πολλῶν καὶ χιόνος· τά τε ὡραῖα αὐτόθι πολλὰ ἐοικὸς γίνεσθαι, ὅσα τε ἀπὸ σπερμάτων
καὶ ὁκόσα αὐτὴ ἡ γῆ ἀναδιδοῖ φυτά, ὧν τοῖς καρ-
ποῖσι χρέονται ἄνθρωποι, ἡμεροῦντες ἐξ ἀγρίων καὶ
ἐς ἐπιτήδειον μεταφυτεύοντες· τά τε ἐντρεφόμενα κτή-
νεα εὐθηνεῖν εἰκός, καὶ μάλιστα τίκτειν τε πυκνότατα
καὶ ἐκτρέφειν κάλλιστα· τούς τε ἀνθρώπους εὐτρα-
φέας εἶναι καὶ τὰ εἴδεα καλλίστους καὶ μεγέθει μεγί-

since to encompass all the nations would require an exceedingly long explanation, I shall limit my exposition to the greatest and most extreme contrasts. I hold that Asia differs very greatly from Europe in the natures of everything, of both the things that grow from the earth, and of the human inhabitants. For everything grows to be much more beautiful and grand in Asia; this region is more cultivated than the other one, and the peoples in it are gentler and more cultivated. The cause of this is the blending of the seasons, which occurs because Asia lies toward the east, between the risings of the sun, and farther away from the cold. Growth and cultivation are best furthered when no single force predominates, but an equilibrium of each exists. Although uniformity does not exist all through Asia, the region lying in the middle between the heat and the cold is very fruitful, well-wooded and mild, and it has the very best water, whether coming from the heavens or from the earth. This region is not excessively scorched by heat, nor dried out by drought and lack of water, nor is it oppressed by cold, nor yet damp and soaked with excessive rains and snow. The harvests there are likely to be plentiful, both those from seed and those which the earth brings forth spontaneously, the fruit of which men then take over, domesticating them from wild plants, and transplanting them to a suitable soil. The cattle too raised there are likely to flourish, and especially to give birth frequently and to rear these to be fine specimens. The people will be well nourished, of very fine form and the greatest size, differ-

---

52 εὐοργητότερα Heringa (p. 49): εὐεργότερα V
53 Add. Coray

στους καὶ ἥκιστα διαφόρους ἐς τά τε εἴδεα αὐτοὺς
ἑωυτῶν[54] καὶ τὰ μεγέθεα· εἰκός τε τὴν χώρην ταύτην
τοῦ ἦρος ἐγγύτατα εἶναι κατὰ τὴν φύσιν καὶ τὴν με-
56 τριότητα | τῶν ὡρέων. τὸ δὲ ἀνδρεῖον καὶ τὸ ταλαίπω-
ρον καὶ τὸ ἔμπονον καὶ τὸ θυμοειδὲς οὐκ ἂν δύναιτο
ἐν τοιαύτῃ φύσει ἐγγίνεσθαι . . .[55] μήτε ὁμοφύλου
μήτε ἀλλοφύλου, ἀλλὰ τὴν ἡδονὴν ἀνάγκη κρατέειν·
διότι πολύμορφα γίνεται τὰ ἐν τοῖς θηρίοις. περὶ μὲν
οὖν Αἰγυπτίων καὶ Λιβύων οὕτως ἔχειν μοι δοκεῖ.

13. Περὶ δὲ τῶν ἐν δεξιᾷ τοῦ ἡλίου ἀνατολέων τῶν
θερινῶν[56] μέχρι Μαιώτιδος λίμνης—οὗτος γὰρ ὅρος
τῆς Εὐρώπης καὶ τῆς Ἀσίης—ὧδε ἔχει περὶ αὐτῶν· τὰ
δὲ ἔθνεα ταῦτα ταύτῃ διάφορα αὐτὰ ἑωυτῶν μᾶλλόν
ἐστι τῶν προδιηγημένων διὰ τὰς μεταβολὰς τῶν
ὡρέων καὶ τῆς χώρης τὴν φύσιν. ἔχει δὲ καὶ κατὰ τὴν
γῆν ὁμοίως ἅπερ καὶ κατὰ τοὺς ἄλλους ἀνθρώπους.
ὅκου γὰρ αἱ ὧραι μεγίστας μεταβολὰς ποιέονται καὶ
58 πυκνοτάτας, ἐκεῖ καὶ ἡ χώρη ἀγριωτάτη | καὶ ἀνωμα-
λωτάτη ἐστί, καὶ εὑρήσεις ὄρεά τε πλεῖστα καὶ δάσεα

---

[54] αὐτοὺς ἑωυτῶν Kühlewein: αὐτοῦ V: σφίσιν ἑωυτοῖσιν
Gad. (B)

[55] A lacuna is postulated at this point by Galen Zwinger, etc.

[56] θερινῶν Gal. (Ar. comm.) Septalius Gadaldini's manuscript
as noted in Foes (n. 73): χειμερινῶν V

---

[9] As was remarked by Galen in his commentary, the narrative
jumps at this point from Asia to the end of an account of Egypt
and Libya, presumably an extensive loss of text.

ing little from one another in bodily form or size. It is obvious that this region will be most like the spring in its nature and in the mildness of its seasons. Bravery, endurance, industry and impetuosity could never arise from a nature like this . . .[9] neither of the same species nor of a foreign one, but pleasure must hold the greatest sway; this is why so many different forms exist in the wild animals.[10] Such in my opinion is the condition of the Egyptians and Libyans.

13. As to the dwellers on the right side of the summer risings of the sun (i.e., southeast) as far as Lake Maeotis, which is the boundary between Europe and Asia,[11] their condition is as follows. These peoples have more variation among themselves than the ones I have already described, due to the changes of the seasons and the nature of their country; just as these factors affect the soil, they affect the human beings too. For where the seasons have the greatest and most frequent changes, the land will be the most savage and irregular, and you will find the most wooded

[10] Cf. Aristotle, *Generation of Animals*, 746b 7–11: "And the proverb about Libya, that 'Libya is always producing something new,' is said to have originated from animals of different species uniting with one another in that country, for it is said that because of the want of water all meet at the few places where springs are to be found, and that even different kinds unite in consequence."

[11] The boundary between Europe and Asia is drawn here along the north side of Lake Maeotis (Sea of Azov) and along the Tanaïs River (Don), which flows into its northeast corner. According to Herodotus (4, 45), other writers placed the border more to the south along the Phasis River (Rioni).

καὶ πεδία καὶ λειμῶνας ὄντας. ὅκου δὲ αἱ ὧραι μὴ
μεγάλα[57] ἀλλάσσουσιν, ἐκείνοις ἡ χώρη ὁμαλωτάτη
ἐστίν. οὕτω δὲ ἔχει καὶ περὶ τῶν ἀνθρώπων, εἴ τις
βούλεται ἐνθυμεῖσθαι. εἰσὶ γὰρ φύσιες αἱ μὲν ὄρεσιν
ἐοικυῖαι δενδρώδεσί τε καὶ ἐφύδροισιν, αἱ δὲ λεπτοῖσί
τε καὶ ἀνύδροις, αἱ δὲ λειμακεστέροις τε καὶ ἑλώδε-
σιν, αἱ δὲ πεδίῳ τε καὶ ψιλῇ καὶ ξηρῇ γῇ. αἱ γὰρ ὧραι
αἱ μεταλλάσσουσαι τῆς μορφῆς τὴν φύσιν εἰσὶ διάφοροι. ἢν δὲ διάφοροι ἔωσι μέγα[58] σφέων αὐτέων, διαφοραὶ καὶ πλείονες γίνονται τοῖς εἴδεσι.

14. Καὶ ὁκόσα μὲν ὀλίγον διαφέρει τῶν ἐθνέων
παραλείψω, ὁκόσα δὲ μεγάλα ἢ φύσει ἢ νόμῳ, ἐρέω
περὶ αὐτῶν ὡς ἔχει. καὶ πρῶτον περὶ τῶν Μακροκε-
φάλων. τούτων γὰρ οὐκ ἔστιν ἄλλο ἔθνος ὁμοίας τὰς
κεφαλὰς ἔχον οὐδέν· τὴν μὲν γὰρ ἀρχὴν ὁ νόμος
αἰτιώτατος ἐγένετο τοῦ μήκεος τῆς κεφαλῆς, νῦν δὲ
καὶ ἡ φύσις ξυμβάλλεται τῷ νόμῳ. τοὺς γὰρ μακρο-
τάτην ἔχοντας τὴν κεφαλὴν γενναιοτάτους ἡγέονται.
ἔχει δὲ περὶ νόμου ὧδε· τὸ παιδίον ὁκόταν γένηται
τάχιστα, τὴν κεφαλὴν αὐτοῦ ἔτι ἁπαλὴν ἐοῦσαν μαλθακοῦ ἐόντος ἀναπλάσσουσι τῇσι χερσὶ καὶ ἀναγκά-
ζουσιν ἐς τὸ μῆκος αὔξεσθαι δεσμά τε προσφέροντες
καὶ τεχνήματα ἐπιτήδεια, ὑφ' ὧν τὸ μὲν σφαιροειδὲς
τῆς κεφαλῆς κακοῦται, τὸ δὲ μῆκος αὔξεται. οὕτω τὴν
ἀρχὴν ὁ νόμος κατειργάσατο, ὥστε ὑπὸ βίης τοιαύτην
τὴν φύσιν γενέσθαι· τοῦ δὲ χρόνου προϊόντος ἐν φύ-
60 σει ἐγένετο, ὥστε τὸν νόμον μηκέτι | ἀναγκάζειν. ὁ

mountains, plains, and meadows. But where the seasons do not change much, the land will be very consistent for its inhabitants. So it is with human beings too, if someone were willing to think the matter over: some humans have natures that resemble wooded, well-watered mountains, others are like thin, waterless soil, others like marshy meadows, and others like a plain with bare, dry earth. In fact, the seasons that modify the nature of bodily form are the ones that vary: if the variations among them are great, all the greater will be the differences in the shapes they produce.

14. The nations that differ only a little from one another I will omit, and describe the situation of those that differ greatly, either in their nature or in their law. First, the long-heads, for there is no other people with heads like theirs. At the beginning, it was their law that had the greatest influence on the length of their heads, but now nature too contributes to what law started. The individuals among them that have the longest heads they consider to be the noblest; their law is as follows: as soon as a child is born, they remodel its head with their hands as long as it is still soft and the child is tender, and force it to increase in length by applying bandages and suitable appliances, which overcome the roundness of the head and increase its length. In this way law takes the first step, so that by its compulsion a new nature is called into being; then, as time passes, what has been achieved by law becomes natural, so that law no longer needs to exert its force. For the

---

[57] μεγάλα Barb.: μεγάλαι V
[58] μέγα Coray: μετὰ V

γὰρ γόνος πανταχόθεν ἔρχεται τοῦ σώματος,[59] ἀπό τε
τῶν ὑγιηρῶν ὑγιηρὸς ἀπό τε τῶν νοσερῶν νοσερός. εἰ
οὖν γίνονται ἔκ τε τῶν φαλακρῶν φαλακροὶ καὶ ἐκ
τῶν γλαυκῶν γλαυκοὶ καὶ ἐκ διεστραμμένων στρε-
βλοὶ ὡς ἐπὶ τὸ πλῆθος, καὶ περὶ τῆς ἄλλης μορφῆς
ὁ αὐτὸς λόγος, τί κωλύει καὶ ἐκ μακροκεφάλου μακρο-
κέφαλον γίνεσθαι; νῦν δ᾽ ὁμοίως οὐκέτι γίνονται ἦ[60]
πρότερον· ὁ γὰρ νόμος οὐκέτι ἰσχύει διὰ τὴν ὁμιλίην
τῶν ἀνθρώπων. περὶ μὲν οὖν τούτων οὕτως ἔχειν μοι
δοκεῖ.

15. Περὶ δὲ τῶν ἐν Φάσει, ἡ χώρη ἐκείνη ἐλώδης
ἐστὶ καὶ θερμὴ καὶ ὑδατεινὴ καὶ δασεῖα, ὄμβροι τε
αὐτόθι γίνονται πᾶσαν ὥρην πολλοί τε καὶ ἰσχυροί·
ἥ τε δίαιτα τοῖς ἀνθρώποις ἐν τοῖς ἕλεσίν ἐστι, τά τε
οἰκήματα ξύλινα καὶ καλάμινα ἐν τοῖς ὕδασι μεμηχα-
νημένα· ὀλίγῃ τε χρέονται βαδίσει κατὰ τὴν πόλιν
καὶ τὸ ἐμπόριον, ἀλλὰ μονοξύλοις διαπλέουσιν ἄνω
καὶ κάτω· διώρυγες γὰρ πολλαί εἰσι. τὰ δὲ ὕδατα
θερμὰ καὶ στάσιμα πίνουσιν ὑπό τε τοῦ ἡλίου σηπό-
μενα καὶ ὑπὸ τῶν ὄμβρων ἐπαυξόμενα, αὐτός τε ὁ
Φᾶσις στασιμώτατος πάντων τῶν ποταμῶν καὶ ῥέων
ἠπιώτατα. οἵ τε καρποὶ ⟨οἱ⟩[61] γινόμενοι αὐτόθι πάντες
ἀναλδέες εἰσι καὶ τεθηλυσμένοι καὶ ἀτελέες ὑπὸ πο-
λυπληθείης τοῦ ὕδατος· διὸ καὶ οὐ πεπαίνονται. ἠήρ
τε πολὺς κατέχει τὴν χώρην ἀπὸ τῶν ὑδάτων. διὰ

---

[59] τοῦ σώματος Gad. (B): om. V, add. καὶ τοῦ σώματος after
ὑγιηρὸς   [60] ἦ Coray: ἢ V   [61] οἱ add. Coray

generative seed comes from all parts of (sc. the parent's) body, healthy seed from healthy parts, diseased seed from diseased parts. If, therefore, from bald parents children likely to become bald are born, from gray-eyed parents gray-eyed children, from parents with strabismus children usually having strabismus, and so on according to the same principle with any other feature, what is to prevent a long-headed parent also having a long-headed child?[12] At the present time, this does not happen so much as it did before, since the law is no longer as strict as it once was, owing to the long-heads' intercourse with other peoples. Such in my opinion is the case of the long-heads.

15. The dwellers on the river Phasis: the land there is marshy, hot, humid, and well-wooded; frequent, violent rains fall in every season. The inhabitants live in the marshes; their dwellings are made of wood and reeds, and constructed in the water; they make little use of walking in the city and the harbor, but paddle up and down in dug-out canoes, for there are many canals. They drink from waters that are hot and stagnant, being putrefied by the sun and swollen by the rains; the Phasis itself is the slowest and most stagnant of all rivers. The fruits that grow in this country are all stunted, soft and defective, owing to the excess of water, and for this reason they do not ripen. Much fog arising from the waters envelops the land. For

[12] Cf. *Sacred Disease* 5. For a modern formulation of the theory of the inheritance of characters acquired during the lifetime of an individual, see J. B. P. A. Lamarck, *Animaux sans vertèbres*, Introduction (Paris, 1815).

62 ταύτας δὴ τὰς | προφάσιας τὰ εἴδεα ἀπηλλαγμένα
τῶν λοιπῶν ἀνθρώπων ἔχουσιν οἱ Φασιηνοί· τά τε
γὰρ μεγέθεα μεγάλοι, τὰ πάχεα δ᾽ ὑπερπάχητες, ἄρ-
θρον τε κατάδηλον οὐδὲν οὐδὲ φλέψ· τήν τε χροιὴν
ὕπωχρον[62] ἔχουσιν ὥσπερ ὑπὸ ὑδέρου[63] ἐχόμενοι·
φθέγγονταί τε βαρύτατον ἀνθρώπων, τῷ ἠέρι χρεώμενοι οὐ λαμπρῷ, ἀλλὰ νοτώδει τε καὶ λιβρῷ· πρός
τὸ ταλαιπωρεῖν τὸ σῶμα ἀργότεροι πεφύκασιν. αἵ τε
ὧραι οὐ πολὺ μεταλλάσσουσιν οὔτε πρὸς τὸ πνῖγος
οὔτε πρὸς τὸ ψῦχος. τά τε πνεύματα τὰ πολλὰ νότια
πλὴν αὔρης μιῆς ἐπιχωρίης. αὕτη δὲ πνεῖ ἐνίοτε
βίαιος καὶ χαλεπὴ καὶ θερμή· καὶ κέγχρονα ὀνομά-
ζουσι τοῦτο τὸ πνεῦμα. ὁ δὲ βορέης οὐ σφόδρα ἀφ-
ικνεῖται· ὁκόταν δὲ πνέῃ, ἀσθενὴς καὶ βληχρός. καὶ
περὶ μὲν τῆς φύσιος τῆς διαφορῆς καὶ τῆς μορφῆς
τῶν ἐν τῇ Ἀσίῃ καὶ τῇ Εὐρώπῃ οὕτως ἔχει.

16. Περὶ δὲ τῆς ἀθυμίης τῶν ἀνθρώπων καὶ τῆς
ἀνανδρείης, ὅτι ἀπολεμώτατοί εἰσι τῶν Εὐρωπαίων οἱ
Ἀσιηνοὶ καὶ ἡμερώτεροι τὰ ἤθεα αἱ ὧραι αἴτιαι μάλιστα, οὐ μεγάλας τὰς μεταβολὰς ποιεύμεναι οὔτε ἐπὶ
τὸ θερμὸν οὔτε ἐπὶ τὸ ψυχρόν, ἀλλὰ παραπλήσιαι. οὐ
γὰρ γίνονται ἐκπλήξιες τῆς γνώμης οὔτε μετάστασις
64 ἰσχυρὴ | τοῦ σώματος, ἀφ᾽ ὅτων εἰκὸς τὴν ὀργὴν
ἀγριοῦσθαί τε καὶ τοῦ ἀγνώμονος καὶ θυμοειδέος μετέχειν μᾶλλον ἢ ἐν τῷ αὐτῷ ἀεὶ ἐόντα. αἱ γὰρ μεταβολαί εἰσι τῶν πάντων ⟨αἱ⟩ αἰεί τ᾽[64] ἐγείρουσαι τὴν
γνώμην τοῦ ἀνθρώπου καὶ οὐκ ἐῶσαι ἀτρεμίζειν. διὰ
ταύτας ἐμοὶ δοκέει τὰς προφάσιας ἀναλκὲς εἶναι τὸ

these reasons, therefore, the bodily form of the Phasians is different from that of other peoples. In size, they are large, and in fullness exceedingly fat, so that in their body no joint or vein is visible. Their complexion is slightly yellowish, looking as if they were suffering from dropsy. They have the deepest voice of any people, because the air they breathe is not clear, but moist and hazy. They are by nature disinclined to physical exertion. The seasons there do not vary much, either toward suffocating heat or toward cold. The winds are mostly moist, except one breeze peculiar to the country which sometimes blows forcefully, violent and hot; this wind they call *cenchron*. The north wind rarely comes up, and when it does blow, it is weak and gentle. Such are the differences in the nature and bodily form of the inhabitants of Asia and Europe.

16. With regard to the lack of spirit and bravery of the people, the main reason why Asiatics are so unwarlike in comparison to Europeans, and more gentle in character, is the seasons, which bring no great changes toward heat or cold, but are always much the same. Thus there are no mental shocks or radical alterations of the body, from which anger is likely to be provoked, with a greater share of cruelty and hot temper, than being in a monotonous sameness would. For it is changes of all kinds that perpetually rouse the temper of a person and do not leave him in peace. It is for these reasons, I think, that the Asiatic race

---

62 ὕπωχρον Gad. (B): ὠχρὴν V
63 ὑδέρου Jouanna after Lat.: ἰκτέρου V
64 ⟨αἱ⟩ αἱεί τ᾽ Jouanna: ἀεὶ τὲ V

γένος τὸ Ἀσιηνὸν καὶ προσέτι διὰ τοὺς νόμους. τῆς γὰρ Ἀσίης τὰ πολλὰ βασιλεύεται. ὅκου δὲ μὴ αὐτοὶ ἑωυτῶν εἰσὶ καρτεροὶ οἱ ἄνθρωποι μηδὲ αὐτόνομοι, ἀλλὰ δεσπόζονται, οὐ περὶ τούτου αὐτοῖσιν ὁ λόγος ἐστίν, ὅπως τὰ πολέμια ἀσκήσωσιν, ἀλλ' ὅκως μὴ δόξωσι μάχιμοι εἶναι. οἱ γὰρ κίνδυνοι οὐχ ὅμοιοί εἰσι. τοὺς μὲν γὰρ στρατεύεσθαι εἰκὸς καὶ ταλαιπω-ρέειν καὶ ἀποθνῄσκειν ἐξ ἀνάγκης ὑπὲρ τῶν δεσπο-τέων ἀπό τε παιδίων καὶ γυναικὸς ἐόντας καὶ τῶν λοιπῶν φίλων. καὶ ὁκόσα μὲν ἂν χρηστὰ καὶ ἀνδρεῖα ἐργάσωνται, οἱ δεσπόται ἀπ' αὐτῶν αὔξονταί τε καὶ ἐκφύονται, τοὺς δὲ κινδύνους καὶ θανάτους αὐτοὶ καρ-ποῦνται. ἔτι δὲ πρὸς τούτοισι τῶν τοιούτων ἀνθρώπων ἀνάγκη ἐρημοῦσθαι τὴν γῆν ὑπό τε πολεμίων καὶ ἀργίης. ὥστε καὶ εἴ τις φύσει πέφυκεν ἀνδρεῖος καὶ εὔψυχος, ἀποτρέπεσθαι τὴν γνώμην ὑπὸ τῶν νόμων. μέγα δὲ τεκμήριον τούτων· ὁκόσοι γὰρ ἐν τῇ Ἀσίῃ Ἕλληνες ἢ βάρβαροι μὴ δεσπόζονται, ἀλλ' αὐτόνο-μοί εἰσι καὶ ἑωυτοῖσι ταλαιπωρεῦσιν, οὗτοι μαχι-μώτατοί εἰσι πάντων· τοὺς γὰρ κινδύνους ἑωυτῶν πέρι κινδυνεύουσι, καὶ τῆς ἀνδρείης αὐτοὶ τὰ ἆθλα φέρον-ται καὶ τῆς δειλίης τὴν ζημίην ὡσαύτως. εὑρήσεις δὲ καὶ τοὺς Ἀσιηνοὺς διαφέροντας αὐτοὺς ἑωυτῶν, τοὺς 66 μὲν βελτίονας, | τοὺς δὲ φαυλοτέρους ἐόντας. τούτων δὲ αἱ μεταβολαὶ αἴτιαι τῶν ὡρέων, ὥσπερ μοι εἴρηται ἐν τοῖς προτέροισι. καὶ περὶ μὲν τῶν ἐν τῇ Ἀσίῃ οὕτως ἔχει.

17. Ἐν δὲ τῇ Εὐρώπῃ ἔστιν ἔθνος Σκυθικόν, ὃ περὶ

is cowardly, and also because of their laws, the greater part of Asia being governed by kings. Now where people are not their own masters and independent, but are subject to despots, their consideration is not how to conduct war, but how to appear unwarlike. For the risks they run are not compensated: subjects are regularly compelled to serve in the army, to endure hardships, and to risk death in the interest of their rulers, and to be parted from their wife, their children, and their friends: but from their worthy, brave deeds it is the rulers that increase their power and wealth, while the subjects reap only the dangers and the deaths. Moreover, the land of a people like this must become a wasteland as a result of the hostilities and from neglect. Thus, even if some naturally brave and spirited man is born, his temper will be subverted by their laws. Here is a capital proof of the fact. All the inhabitants of Asia, whether Greek or non-Greek, who are not subject to a ruler, but are independent and exert themselves for their own advantage, are the most warlike of all. For they risk danger for their own sakes, and for their bravery they themselves carry off the prizes, as likewise the penalty if they are craven. In fact you will find that Asiatics also differ from one another, some being braver, and others more cowardly; the reason for this, as I have said above, is the changes of the seasons. Such is the situation in Asia.

17. In Europe there is a Scythian people, dwelling

τὴν λίμνην οἰκεῖ τὴν Μαιῶτιν διαφέρον τῶν ἐθνέων
τῶν ἄλλων. Σαυρομάται καλεῦνται. τούτων αἱ γυναῖκες ἱππάζονταί τε καὶ τοξεύουσι καὶ ἀκοντίζουσιν
ἀπὸ τῶν ἵππων καὶ μάχονται τοῖς πολεμίοις, ἕως ἂν
παρθένοι ἔωσιν. οὐκ ἀποπαρθενεύονται δέ, μέχρις ἂν
τῶν πολεμίων τρεῖς ἀποκτείνωσι, καὶ οὐ πρότερον
συνοικέουσιν ἤπερ τὰ ἱερὰ θύσωσιν τὰ ἐν νόμῳ.[65] ἣ
δ᾽ ἂν ἄνδρα ἑωυτῇ ἄρηται, παύεται ἱππαζομένη, ἕως
ἂν μὴ ἀνάγκη καταλάβῃ παγκοίνου στρατείης. τὸν
δεξιὸν δὲ μαζὸν οὐκ ἔχουσι. παιδίοις γὰρ ἐοῦσιν ἔτι
νηπίοις αἱ μητέρες χαλκίον τετεχνημένον [ἢ][66] ἐπ᾽
αὐτῷ τούτῳ διάπυρον ποιέουσαι πρὸς τὸν μαζὸν τι-
θέασι τὸν δεξιὸν καὶ ἐπικαίεται, | ὥστε τὴν αὔξησιν
φθείρεσθαι, ἐς δὲ τὸν δεξιὸν ὦμον καὶ βραχίονα πᾶσαν τὴν ἰσχὺν καὶ τὸ πλῆθος ἐκδιδόναι.

18. Περὶ δὲ τῶν λοιπῶν Σκυθέων τῆς μορφῆς, ὅτι
αὐτοὶ ἑωυτοῖσιν[67] ἐοίκασι καὶ οὐδαμῶς ἄλλοις, ωὑτὸς
λόγος καὶ περὶ τῶν Αἰγυπτίων, πλὴν ὅτι οἱ μὲν ὑπὸ
τοῦ θερμοῦ εἰσι βεβιασμένοι, οἱ δὲ ὑπὸ τοῦ ψυχροῦ.
ἡ δὲ Σκυθέων ἐρημίη καλευμένη πεδιάς ἐστι καὶ λει-
μακώδης καὶ ὑψηλὴ καὶ ἔνυδρος μετρίως. ποταμοὶ
γάρ εἰσι μεγάλοι, οἳ ἐξοχετεύουσι τὸ ὕδωρ ἐκ τῶν
πεδίων. ἐνταῦθα καὶ οἱ Σκύθαι διαιτεῦνται, νομάδες δὲ
καλεῦνται, ὅτι οὐκ ἔστιν οἰκήματα, ἀλλ᾽ ἐν ἁμάξῃσιν
οἰκεῦσιν. αἱ δὲ ἅμαξαί εἰσιν αἱ μὲν ἐλάχισται τετρά-
κυκλοι, αἱ δὲ ἑξάκυκλοι· αὗται δὲ πίλοισι[68] περιπε-
φραγμέναι ⟨εἰσίν⟩·[69] εἰσὶ δὲ καὶ τετεχνασμέναι ὥσπερ

around Lake Maeotis, who differ from other nations: their name is *Sauromatae*. The women, as long as they are unmarried, ride horseback, shoot with the bow, throw the javelin from their horses, and fight with their enemies. They do not give up the unmarried state until they have killed three enemies, and they do not cohabit before they have performed sacrificial rites in accordance with their law. A woman who has won a husband gives up horseback riding, unless necessity requires that she take part in some general expedition. These women have no right breast, for while they are still infants their mothers heat a bronze instrument constructed for this purpose red-hot, and apply it to their right breast to cauterize it, so that its growth is arrested, and all its strength and bulk are diverted to the right shoulder and arm.

18. As for the bodily form of the rest of the Scythians, why they are like one another and not at all like other people, the same explanation applies as for the Egyptians, except that the latter are determined by heat, whereas the former are determined by cold. What is called the Scythian wilderness is a high, grassy, moderately well-watered plain, since there are large rivers draining the water from the plains. It is there that the Scythians pass their life, and they are called nomads, because they have no houses, but live in wagons. The wagons have at the least four wheels, but others six wheels. These are covered over with felt and set up like houses inside, some with one room, and others

---

^{65} τὰ ἐν νόμῳ A. Portus in Foes: τῷ ἐννόμῳ V
^{66} ἢ del. Heringa (p. 51) ^{67} ἑωυτοῖσιν Zwinger: αὐτοῖσιν V ^{68} πίλοισι Erotian: πηλοῖς V ^{69} Add. Jouanna

οἰκήματα τὰ μὲν ἁπλᾶ, τὰ δὲ καὶ τριπλᾶ. ταῦτα δὲ
καὶ στεγνὰ πρὸς ὕδωρ καὶ πρὸς χιόνα καὶ πρὸς τὰ
πνεύματα. τὰς δὲ ἁμάξας ἕλκουσι ζεύγεα τὰς μὲν δύο,
τὰς δὲ τρία βοῶν κέρως ἄτερ. οὐ γὰρ ἔχουσι κέρατα
ὑπὸ τοῦ ψύχεος. ἐν ταύτῃσι μὲν οὖν τῇσιν ἁμάξῃσιν
⟨αἱ⟩[70] γυναῖκες διαιτεῦνται. αὐτοὶ δ' ἐφ' ἵππων ὀχεῦν-
ται οἱ ἄνδρες. ἕπονται δὲ αὐτοῖς καὶ τὰ πρόβατα
⟨τὰ⟩[71] ἐόντα καὶ αἱ βόες καὶ οἱ ἵπποι. μένουσι δ' ἐν
τῷ αὐτῷ τοσοῦτον χρόνον, ὅσον ἂν ἀποχρῇ αὐτοῖσι
τοῖς κτήνεσιν ὁ χόρτος· ὁκόταν δὲ μηκέτι, ἐς ἑτέρην
χώρην μετέρχονται. αὐτοὶ δ' ἐσθίουσι κρέα ἑφθὰ[72] καὶ
πίνουσι γάλα ἵππων, καὶ ἱππάκην τρώγουσι· τοῦτο δ'
70 ἐστὶ | τυρὸς ἵππων. τὰ μὲν ἐς τὴν δίαιταν αὐτῶν
οὕτως ἔχει καὶ τοὺς νόμους.

19. Περὶ δὲ τῶν ὡρέων καὶ τῆς μορφῆς, ὅτι πολὺ
ἀπήλλακται τῶν λοιπῶν ἀνθρώπων τὸ Σκυθικὸν γένος
καὶ ἔοικεν αὐτὸ ἑωυτῷ ὥσπερ τὸ Αἰγύπτιον· καὶ ἥκιστα πολύγονόν ἐστι, καὶ ἡ χώρη ἐλάχιστα θηρία
τρέφει κατὰ μέγεθος καὶ πλῆθος. κεῖται γὰρ ὑπ' αὐτῇσι τῇσιν ἄρκτοις καὶ τοῖς ὄρεσι τοῖς Ῥιπαίοισιν,
ὅθεν ὁ βορέης πνέει. ὅ τε ἥλιος τελευτῶν ἐγγύτατα
γίνεται, ὁκόταν ἐπὶ τὰς θερινὰς ἔλθῃ περιόδους, καὶ
τότε ὀλίγον χρόνον θερμαίνει καὶ οὐ σφόδρα· τὰ δὲ
πνεύματα τὰ ἀπὸ τῶν θερμῶν πνέοντα ⟨οὐκ⟩[73] ἀφ-
ικνεῖται, ἢν μὴ ὀλιγάκις καὶ ἀσθενέα, ἀλλ' ἀπὸ τῶν
ἄρκτων ἀεὶ πνέουσι πνεύματα ψυχρὰ ἀπό τε χιόνος
καὶ κρυστάλλου καὶ ὑδάτων πολλῶν. οὐδέποτε δὲ τὰ
ὄρεα ἐκλείπει· ἀπὸ τούτων δὲ ἀοίκητά[74] ἐστιν. ἠήρ τε

with three; they are impenetrable to rain, snow, and wind. Two or three yoke of hornless oxen—they lack horns because of the cold—draw each wagon. Now in these wagons the women pass their life, while the men ride alone on horseback, followed by the sheep they have, their cattle and their horses. They remain in the same place as long as it has sufficient fodder for their animals; when this gives out, they move on. The people eat boiled meats and drink mares' milk, and also consume *hippace*, a cheese made from mares' milk. Such are their way of living and their laws.

19. As to their seasons and their bodily form, the Scythians race is as different as possible from the rest of humanity, and all the same, each resembling the other, like the Egyptian race; it is the least prolific, and the country supports the least game, in both size and number. For it lies under (sc. the constellations of) the Bears and the Rhipaean mountains, out of which the north wind blows. The sun comes very close to them only at the end of its course, when it reaches the summer solstice, and for a short time it warms them, although not very much. The winds blowing from hot regions do not reach them, except rarely, and with little force; but from the Bears cold winds constantly blow, chilled by snow, ice, and many waters, which are never absent from the mountains, making them uninhab-

---

70 Add. Linden
71 Add. Coray
72 κρέα ἑφθὰ Aldina: κρέδεφθα. V
73 Add. Littré
74 ἀοίκητά Coray: διοικητά V

κατέχει πολὺς τῆς ἡμέρης τὰ πεδία, καὶ ἐν αὐτοῖσι[75]
72 | διαιτεῦνται· ὥστε τὸν μὲν χειμῶνα ἀεὶ εἶναι, τὸ δὲ
θέρος ὀλίγας ἡμέρας καὶ ταύτας μὴ λίην. μετέωρα
γὰρ τὰ πεδία καὶ ψιλὰ καὶ οὐκ ἐστεφάνωνται ὄρεσιν,
ἀλλ' ἢ <τ>αύτη[76] ἀπὸ τῶν ἄρκτων· αὐτόθι καὶ τὰ θηρία οὐ γίνεται μεγάλα, ἀλλ' οἷά τέ ἐστιν ὑπὸ γῆν
σκεπάζεσθαι. ὁ γὰρ χειμὼν κωλύει καὶ τῆς γῆς ἡ
ψιλότης, ὅτι οὐκ ἔστιν ἀλέη οὐδὲ σκέπη. αἱ γὰρ μεταβολαὶ τῶν ὡρέων οὔκ εἰσι μεγάλαι οὐδὲ ἰσχυραί,
ἀλλ' ὅμοιαι καὶ ὀλίγον μεταλλάσσουσαι· διότι καὶ τὰ
εἴδεα ὅμοιοι αὐτοὶ[77] ἑωυτοῖς εἰσι σίτῳ τε χρεώμενοι
ἀεὶ ὁμοίῳ ἐσθῆτί τε τῇ αὐτῇ καὶ θέρεος καὶ χειμῶνος,
τόν τε ἠέρα ὑδατεινὸν ἕλκοντες καὶ παχύν, τά τε
ὕδατα πίνοντες ἀπὸ χιόνος καὶ παγετῶν, τοῦ τε τα-
λαιπώρου ἀπεόντες. οὐ γὰρ οἷόν τε τὸ σῶμα ταλαι-
πωρέεσθαι οὐδὲ τὴν ψυχήν, ὅκου μεταβολαὶ μὴ γίνονται ἰσχυραί. διὰ ταύτας τὰς ἀνάγκας τὰ εἴδεα
αὐτῶν παχέα ἐστὶ καὶ σαρκώδεα καὶ <ἄν>αρθρα[78] καὶ
ὑγρὰ καὶ ἄτονα, αἵ τε κοιλίαι ὑγρόταται πασέων κοιλιῶν αἱ κάτω. οὐ γὰρ οἷόν τε νηδὺν ἀναξηραίνεσθαι
ἐν τοιαύτῃ χώρῃ καὶ φύσει καὶ ὥρης καταστάσει,
ἀλλ' ἀίδια[79] πιμελέα[80] τε καὶ ψιλὴν τὴν σάρκα τά τε
εἴδεα ἔοικεν ἀλλήλοισι τά τε ἄρσενα τοῖς ἄρσεσι καὶ
τὰ θήλεα τοῖς θήλεσι. τῶν γὰρ ὡρέων παραπλησίων
ἐουσέων φθοραὶ οὐκ ἐγγίνονται οὐδὲ κακώσιες ἐν τῇ
τοῦ γόνου ξυμπήξει, ἢν μή τινος ἀνάγκης βιαίου
τύχῃ ἢ νούσου.

[75] αὐτοῖσι Ermerins: νότοισι V

itable. A dense fog envelops the plains by day, in which the Scythians pass their life, so that winter is perennial, while summer lasts only a few days and is feeble. The plains are high and bare, and are not encircled by mountains, though they slope from the Bears. There the wild animals do not become very large, but only such as are able to find shelter under the ground; it is the winter that inhibits their growth, as does the bareness of the land, which provides neither warmth nor shelter. The changes of the seasons are neither great nor violent, but the seasons are uniform and hardly alter. For this reason, too, the people are like one another in bodily form, since they always use similar foods and the same clothing, both summer and winter, they breathe moist, thick air, they drink waters from ice and snow, and they lack physical exertion. For neither can the body be strained, nor for that matter the mind, where there are no significant changes (sc. of the circumstances). These causes make their bodily form thick, fleshy, showing no joints, moist, and flaccid, and their lower cavities the most moist of cavities. For the lower belly cannot possibly become dry in a land like this, with such a nature and such a seasonal order, but their flesh must always be fat and smooth, and their physical forms similar, men's to men's and women's to women's. For as the seasons are uniform there are no corruptions or aberrations in the coagulation of the seed, except as the result of some violent cause or disease.

---

76 ἢ ταύτῃ Jouanna: ἡ αὐτῃ V
77 ὁμοῖοι αὐτοὶ Coray: ὅμοια αὐτὰ V
78 ἄναρθρα Zwinger^marg.: ἄρθρα V
79 ἀλλ' ἀίδια Jouanna after Lat. *sed semper*: ἀλλὰ διὰ V
80 πιμελέα Coray: πιμελήν V

ΠΕΡΙ ΑΕΡΩΝ ΥΔΑΤΩΝ ΤΟΠΩΝ

20. Μέγα δὲ τεκμήριον ἐς τὴν ὑγρότητα παρέξομαι.
74 Σκυθέων | γὰρ τοὺς πολλούς, ἅπαντας ὅσοι νομάδες,
εὑρήσεις κεκαυμένους τούς τε ὤμους καὶ τοὺς βραχίο-
νας καὶ τοὺς καρποὺς τῶν χειρῶν καὶ τὰ στήθεα
[ἰσχία]⁸¹ καὶ τὴν ὀσφῦν δι᾿ ἄλλ᾿ οὐδὲν ἢ διὰ τὴν
ὑγρότητα τῆς φύσιος καὶ τὴν μαλακίην. οὐ γὰρ δύνανται οὔτε τοῖς τόξοις ξυντείνειν οὔτε τῷ ἀκοντίῳ
ἐμπίπτειν τῷ ὤμῳ ὑπὸ ὑγρότητος καὶ ἀτονίης. ὁκόταν
δὲ καυθέωσιν, ἀναξηραίνεται ἐκ τῶν ἄρθρων τὸ πολὺ
τοῦ ὑγροῦ, καὶ ἐντονώτερα μᾶλλον γίνεται καὶ τροφι-
μώτερα καὶ ἠρθρωμένα τὰ σώματα μᾶλλον. ῥοικὰ δὲ
γίνεται καὶ πλατέα, πρῶτον μὲν ὅτι οὐ σπαργανοῦν-
ται ὥσπερ ἐν Αἰγύπτῳ—οὐδὲ νομίζουσι διὰ τὴν ἱπ-
πασίην, ὅκως ἂν εὔεδροι ἔωσιν—ἔπειτα δὲ διὰ τὴν
ἕδρην· τά τε γὰρ ἄρσενα, ἕως ἂν οὐχ οἷά τε ἐφ᾿ ἵππου
ὀχέεσθαι, τὸ πολὺ τοῦ χρόνου κάθηνται ἐν τῇ ἁμάξῃ
καὶ βραχὺ τῇ βαδίσει χρέονται διὰ τὰς μεταναστά-
σιας καὶ περιελάσιας· τὰ δὲ θήλεα θαυμαστὸν οἷον
ῥοικά ἐστι τὰ εἴδεα καὶ βραδέα. πυρρὸν δὲ τὸ γένος
ἐστὶ τὸ Σκυθικὸν διὰ τὸ ψῦχος, οὐκ ἐπιγινομένου
ὀξέος τοῦ ἡλίου. ὑπὸ δὲ τοῦ ψύχεος ἡ λευκότης ἐπι-
καίεται καὶ γίνεται πυρρή.

21. Πολύγονον δὲ οὐχ οἷόν τε εἶναι φύσιν τοιαύτην.
οὔτε γὰρ τῷ ἀνδρὶ ἡ ἐπιθυμίη τῆς μείξιος γίνεται
πολλὴ διὰ τὴν ὑγρότητα τῆς φύσιος καὶ τῆς κοιλίης
τὴν μαλακότητά τε καὶ τὴν ψυχρότητα, ἀφ᾿ ὅτων ἥκιστα εἰκὸς [εἶναι]⁸² ἄνδρα οἷόν τε λαγνεύειν· καὶ ἔτι ὑπὸ
76 | τῶν ἵππων ἀεὶ κοπτόμενοι ἀσθενέες γίνονται ἐς τὴν

128

20. I will put forth an important proof of their moistness. The majority of the Scythians, and all that are nomads, you will find cauterized on their shoulders, arms, wrists, chest, and loins, simply to counteract the moistness and softness of their constitution. For they are unable either to draw their bows or to put their weight behind their spear with their shoulder because of their moistness and flaccidness, but when they have been cauterized, most of the moisture is dried up out of their joints, and their bodies become more sinewy, better nourished, and better articulated. Their bodies have a slack, squat appearance, first because they are not swaddled, as the children in Egypt are—they reject this custom in order to have good riders—and then because of their sedentary lives; for the males, as long as they are not yet able to ride a horse, sit the greater part of the time in a wagon, and rarely employ walking during their migrations and tours; while the females are wonderfully slack and torpid in appearance. The Scythians race is ruddy because of the cold, never experiencing the piercing sun: from the cold the whiteness (sc. of their skin) is burned and becomes ruddy.

21. A nature of this kind prevents the Sythians from being prolific. The men have no great desire for intercourse because of the moistness of their nature and the softness and coldness of their cavity, from which conditions a man is least likely to be potent; furthermore, from being continually jolted by their horses, they lose any

---

[81] V: del. Jacoby (p. 550)
[82] V: om. Gad. (B)

μῖξιν. τοῖσι μὲν ἀνδράσιν αὗται αἱ προφάσιες γίνονται, τῇσι δὲ γυναιξὶν ἥ τε πιότης τῆς σαρκὸς καὶ
ὑγρότης· οὐ γὰρ δύνανται ἔτι ξυναρπάζειν αἱ μῆτραι
τὸν γόνον· οὔτε γὰρ ἐπιμήνιος κάθαρσις αὐτῇσι γίνεται ὡς χρεών ἐστιν, ἀλλ' ὀλίγον καὶ διὰ χρόνου, τό
τε στόμα τῶν μητρέων ὑπὸ πιμελῆς ξυγκλείεται καὶ
οὐχ ὑποδέχεται τὸν γόνον· αὐταί τε ἀταλαίπωροι καὶ
πίειραι[83] καὶ αἱ κοιλίαι ψυχραὶ καὶ μαλακαί. καὶ ὑπὸ
τούτων τῶν ἀναγκέων οὐ πολύγονόν ἐστι τὸ γένος τὸ
Σκυθικόν. μέγα δὲ τεκμήριον αἱ οἰκέτιδες ποιέουσιν·
οὐ γὰρ φθάνουσι παρὰ ἄνδρα ἀφικνεύμεναι καὶ ἐν
γαστρὶ ἴσχουσι διὰ τὴν ταλαιπωρίην καὶ ἰσχνότητα
τῆς σαρκός.

22. Ἔτι τε πρὸς τούτοισιν εὐνουχίαι γίνονται [οἱ][84]
πλεῖστοι ἐν Σκύθῃσι καὶ γυναικεῖα ἐργάζονται διαλέγονταί τε ὁμοίως καὶ αἱ γυναῖκες·[85] καλεῦνταί τε οἱ
τοιοῦτοι ⟨ἀν⟩ανδριεῖς.[86] οἱ μὲν οὖν ἐπιχώριοι τὴν αἰτίην προστιθέασι θεῷ καὶ σέβονταί τε τούτους τοὺς
ἀνθρώπους καὶ προσκυνέουσι, δεδοικότες περὶ γ' ἑωυτῶν ἕκαστοι. ἐμοὶ δὲ καὶ αὐτῷ δοκεῖ ταῦτα τὰ πάθεα
θεῖα εἶναι καὶ τἆλλα πάντα καὶ οὐδὲν ἕτερον ἑτέρου
78 θειότερον οὐδὲ ἀνθρωπινώτερον, ἀλλὰ πάντα | ὁμοῖα
καὶ πάντα θεῖα. ἕκαστον δὲ ἔχει φύσιν τῶν τοιούτων
καὶ οὐδὲν ἄνευ φύσιος γίνεται. καὶ τοῦτο τὸ πάθος ὥς
μοι δοκέει γίνεσθαι φράσω· ὑπὸ τῆς ἱππασίης αὐτοὺς

---

83 πίειραι Coray: πιεραὶ V
84 V: del. Coray

strength for intercourse. In the men these are the causes, but in the women it is the fatness and moistness of their flesh, which makes their uterus incapable of taking up the (sc. male) seed. For first their menstrual cleaning is not as it should be, but scanty and late, and furthermore the orifice of their uterus is blocked by fat and does not admit seed. Personally they are fat and idle, and their cavities are cold and soft. These are the reasons why the Scythian race is not prolific. A clear proof is afforded by their slave girls: these, because of their constant activity and their leanness of flesh, no sooner go to a man than they are pregnant.

22. Furthermore, there are many men among the Scythians who resemble eunuchs; they do women's work and talk like women: such men are called "unmanlies."[13] Now the natives attribute the cause to a god; they respect and worship these people, each one fearing on his own account. I too think that such diseases are divine, but all the other diseases are divine as well, none being more divine or more human than any other; all are alike in this regard, and all are divine. Each of these conditions has a nature of its own, and none arises without a natural cause. How, in my opinion, this disease arises I will explain. From

[13] Cornarius' change of ἀνδριεῖς (manlies) to ἀνανδριεῖς (un-manlies) does give a fitting sense to the Greek, but it seems more probable that ἀνδριεῖς is in fact the misreading of ἀναριεῖς (conjectured by Th. Gomperz), a Scythian word for these impotent men: cf. Herodotus 1, 105 ἐνάρεας, and 4, 67 ἐνάρεες.

85 διαλέγονταί τε ὁμοίως καὶ αἱ γυναῖκες Diller: καὶ αἱ γυναῖκες διαλέγονταί τε ὁμοίως V
86 ἀνανδριεῖς Froben: ἀνδριεῖς V

κέδματα λαμβάνει, ἅτε ἀεὶ κρεμαμένων ἀπὸ τῶν ἵππων τοῖς ποσίν· ἔπειτα ἀποχωλοῦνται καὶ ἑλκοῦνται
τὰ ἰσχία, οἳ ἂν σφόδρα νοσήσωσιν. ἰῶνται δὲ σφᾶς
αὐτοὺς τρόπῳ τοιῷδε. ὁκόταν ἄρχηται ἡ νοῦσος, ὄπισθεν τοῦ ὠτὸς ἑκατέρου φλέβα τάμνουσιν. ὁκόταν δὲ
ἀπορρυῇ τὸ αἷμα, ὕπνος ὑπολαμβάνει ὑπὸ ἀσθενείης
καὶ καθεύδουσιν. ἔπειτα ἀνεγείρονται, οἱ μέν τινες
ὑγιεῖς ἐόντες, οἱ δ᾽ οὔ. ἐμοὶ μὲν οὖν δοκεῖ ἐν ταύτῃ
τῇ ἰήσει διαφθείρεσθαι ὁ γόνος. εἰσὶ γὰρ παρὰ τὰ
ὦτα φλέβες, ἃς ἐάν τις ἐπιτάμῃ, ἄγονοι γίνονται οἱ
ἐπιτμηθέντες. ταύτας τοίνυν μοι δοκέουσι τὰς φλέβας
ἐπιτάμνειν. οἱ δὲ μετὰ ταῦτα ἐπειδὰν ἀφίκωνται παρὰ
γυναῖκας καὶ μὴ οἷοί τ᾽ ἔωσι χρῆσθαί σφισιν [αὐ-
ταῖς],[87] τὸ πρῶτον οὐκ ἐνθυμεῦνται, ἀλλ᾽ ἡσυχίην
ἔχουσιν. ὁκόταν δὲ δὶς καὶ τρὶς ⟨καὶ⟩[88] πλεονάκις
αὐτοῖσι πειρωμένοισι μηδὲν ἀλλοιότερον ἀποβαίνῃ,
νομίσαντές τι ἡμαρτηκέναι τῷ θεῷ, ὃν ἐπαιτιῶνται,
ἐνδύονται στολὴν γυναικείην καταγνόντες ἑωυτῶν
ἀνανδρείην· γυναικίζουσί τε καὶ ἐργάζονται μετὰ τῶν
γυναικῶν ἃ καὶ ἐκεῖναι.

Τοῦτο δὲ πάσχουσι Σκυθέων οἱ πλούσιοι, οὐχ οἱ
80 κάκιστοι | ἀλλ᾽ οἱ εὐγενέστατοι καὶ ἰσχὺν πλείστην
κεκτημένοι, διὰ τὴν ἱππασίην· οἱ δὲ πένητες ἧσσον·
οὐ γὰρ ἱππάζονται. καίτοι ἐχρῆν, εἴ γε[89] θειότερον
τοῦτο τὸ νόσευμα τῶν λοιπῶν ἐστιν, οὐ τοῖς γενναιο-
τάτοις τῶν Σκυθέων καὶ τοῖς πλουσιωτάτοις προσπί-

87 V: del. Coray   88 Add. Froben

riding on horseback, swellings arise at their joints, because their legs are always hanging down from their horses; then they become lame, and develop lesions at their hips when the disease becomes serious. They cure themselves in the following way. At the beginning of the disease they cut the vessel behind each ear. After the blood flows out, sleep comes over them from their weakness, and they go to bed. Later they wake up, some being cured and others not. Now, in my opinion, by this treatment their seed is destroyed; for by the side of the ear are vessels which, when someone cuts them, make the person cut sterile, and so I believe it is these vessels they are cutting.[14] After this treatment, when the Scythians approach women and find themselves to be impotent, at first they take no notice and ignore it. But when on making two, three or even more attempts the same thing happens to them, thinking that they have wronged the divinity to which they attribute the cause, they put on women's clothing, admit that they have lost their manhood, and so play the woman and do the same work with women that they do.

This is what the well-to-do among the Scythians suffer—not the common people, but the noblest and those with the most power—as a result of their riding; the poor suffer less, since they do not ride. And yet if this disease was more divine than the rest, it should befall not only the noblest and the richest of the Scythians, but all alike—and

[14] Cf. *Generation* 2; Loeb *Hippocrates*, vol. 10, pp. 9–11.

---

89 εἴ γε Jouanna: ἐπεὶ V

πτειν μούνοις, ἀλλὰ τοῖς ἅπασιν ὁμοίως, καὶ μᾶλλον
τοῖσιν ὀλίγα κεκτημένοισιν, οὐ <τοῖσι>[90] τιμωμένοισιν
ἤδη, εἰ χαίρουσιν οἱ θεοὶ καὶ θαυμαζόμενοι ὑπ' ἀνθρώπων καὶ ἀντὶ τούτων χάριτας ἀποδιδοῦσιν. εἰκὸς
γὰρ τοὺς μὲν πλουσίους θύειν πολλὰ τοῖς θεοῖς καὶ
ἀνατιθέναι ἀναθήματα ἐόντων χρημάτων καὶ τιμέων,[91]
τοὺς δὲ πένητας ἧσσον διὰ τὸ μὴ ἔχειν, ἔπειτα καὶ
ἐπιμεμφομένους ὅτι οὐ διδόασι χρήματα αὐτοῖσιν,
ὥστε τῶν τοιούτων ἁμαρτιῶν τὰς ζημίας τοὺς ὀλίγα
κεκτημένους φέρειν μᾶλλον ἢ τοὺς πλουσίους. ἀλλὰ
γάρ, ὥσπερ καὶ πρότερον ἔλεξα, θεῖα μὲν καὶ ταῦτά
ἐστιν ὁμοίως τοῖς ἄλλοις· γίνεται δὲ κατὰ φύσιν ἕκαστα. καὶ ἡ τοιαύτη νοῦσος ἀπὸ τοιαύτης προφάσιος
τοῖς Σκύθαις γίνεται οἵην εἴρηκα. ἔχει δὲ καὶ κατὰ
τοὺς λοιποὺς ἀνθρώπους ὁμοίως. ὅκου γὰρ ἱππάζονται μάλιστα καὶ πυκνότατα, ἐκεῖ πλεῖστοι ὑπὸ
κεδμάτων καὶ ἰσχιάδων καὶ ποδαγριῶν ἁλίσκονται
καὶ λαγνεύειν κάκιστοί εἰσι. ταῦτα δὲ τοῖσι Σκύθησι
πρόσεστι, καὶ εὐνουχοειδέστατοί εἰσιν ἀνθρώπων διὰ
82 τὰς <προειρημένας>[92] προφάσιας καὶ | ὅτι ἀναξυρίδας
ἔχουσιν ἀεὶ καί εἰσιν ἐπὶ τῶν ἵππων τὸ πλεῖστον τοῦ
χρόνου, ὥστε μήτε χειρὶ ἅπτεσθαι τοῦ αἰδοίου, ὑπό
τε τοῦ ψύχεος καὶ τοῦ κόπου ἐπιλαθέσθαι τοῦ ἱμέρου
καὶ τῆς μείξιος, καὶ μηδὲν παρακινεῖν πρότερον ἢ ἀν-
δρωθῆναι. περὶ μὲν οὖν τῶν Σκυθέων οὕτως ἔχει τοῦ
γένεος.

23. Τὸ δὲ λοιπὸν γένος τὸ ἐν τῇ Εὐρώπῃ διάφορον
αὐτὸ ἑωυτῷ ἐστι καὶ κατὰ τὸ μέγεθος καὶ κατὰ τὰς

especially those possessing little rather than the highly honored, if the gods take pleasure in being worshipped by human beings and repay this with favors. For naturally the rich, possessing great wealth and honors, offer great sacrifices to the gods and set up many votive offerings, whereas the poor can do less since they lack these means, and besides they even find fault with the gods for not giving them wealth, so that the penalties for such blasphemy should be sustained more by the poor than by the rich. But in truth, as I said above, these (sc. diseases) are equally divine with the others, and each of them arises in accordance with nature. Such a disease arises among the Scythians for the reason I have stated: it also happens to other people in the same way, for wherever men ride very much and very frequently, then the majority are attacked by swellings at the joints, sciatica, and gout, and they are sexually worst off. These things happen to the Scythians, making them the most eunuch-like of men, for the reasons I have indicated, and also because they always wear trousers and spend most of their time on their horses, so that they cannot touch their penis with their hand, but owing to cold and fatigue they forget about desire and intercourse, feeling no sexual impulse before they reach full manhood.[15] Such, then, is the condition of the Scythian race.

23. The remaining population in Europe differs within itself both in size and in bodily form, because of the

[15] That is, abnormally late, not at puberty as is natural.

---

[90] Add. Coray in note [91] τιμέων Jouanna: τιμᾶν V
[92] Add. Littré after Lat. predictas

μορφὰς διὰ τὰς μεταλλαγὰς τῶν ὡρέων, ὅτι μεγάλαι
γίνονται καὶ πυκναί, καὶ θάλπεά τε ἰσχυρὰ καὶ χειμῶνες καρτεροὶ καὶ ὄμβροι πολλοὶ καὶ αὖτις αὐχμοὶ
πολυχρόνιοι καὶ πνεύματα, ἐξ ὧν μεταβολαὶ πολλαὶ
84 καὶ παντοδαπαί. | ἀπὸ τούτων εἰκὸς αἰσθάνεσθαι καὶ
τὴν γένεσιν ἐν τῇ συμπήξει τοῦ γόνου ἄλλην, καὶ μὴ
τῷ αὐτῷ αὐτὴν γίνεσθαι ἔν τε τῷ θέρει καὶ τῷ χειμῶνι
μηδὲ ἐν ἐπομβρίῃ καὶ αὐχμῷ. διότι τὰ εἴδεα διηλ-
λάχθαι νομίζω τῶν Εὐρωπαίων μᾶλλον ἢ τῶν Ἀσιη-
νῶν καὶ τὰ μεγέθεα διαφορώτατα αὐτὰ ἑωυτοῖς εἶναι
κατὰ πόλιν ἑκάστην. αἱ γὰρ φθοραὶ πλείονες ἐγγίνονται τοῦ γόνου ἐν τῇ ξυμπήξει ἐν τῇσι μεταλλαγῇσι
τῶν ὡρέων πυκνῇσιν ἐούσῃσιν ἢ ἐν τῇσι παραπλη-
σίῃσι καὶ ὁμοίῃσι. περί τε τῶν ἠθέων ὁ αὐτὸς λόγος·
τό τε ἄγριον καὶ τὸ ἄμεικτον καὶ τὸ θυμοειδὲς ἐν τῇ
τοιαύτῃ φύσει ἐγγίνεται. αἱ γὰρ ἐκπλήξιες πυκναὶ γινόμεναι τῆς γνώμης τὴν ἀγριότητα ἐντιθέασι, τὸ δὲ
ἥμερόν τε καὶ ἤπιον ἀμαυροῦσι. διότι[93] εὐψυχοτέρους
νομίζω τοὺς τὴν Εὐρώπην οἰκέοντας εἶναι ἢ τοὺς τὴν
Ἀσίην. ἐν μὲν γὰρ τῷ αἰεὶ παραπλησίῳ αἱ ῥᾳθυμίαι
ἔνεισιν, ἐν δὲ τῷ μεταβαλλομένῳ αἱ ταλαιπωρίαι τῷ
σώματι καὶ τῇ ψυχῇ. καὶ ἀπὸ μὲν ἡσυχίης καὶ ῥᾳθυ-
μίης ἡ δειλίη αὔξεται, ἀπὸ δὲ τῆς ταλαιπωρίης καὶ
τῶν πόνων αἱ ἀνδρεῖαι. διὰ τοῦτό εἰσι μαχιμώτεροι οἱ
τὴν Εὐρώπην οἰκεῦντες καὶ διὰ τοὺς νόμους, ὅτι οὐ
βασιλεύονται ὥσπερ οἱ Ἀσιηνοί. ὅκου γὰρ βασιλεύονται, ἐκεῖ ἀνάγκη δειλοτάτους εἶναι· εἴρηται δέ μοι καὶ
πρότερον. αἱ γὰρ ψυχαὶ δεδούλωνται καὶ οὐ βού-

changes of the seasons, which are extreme and frequent; there are powerful heat waves, severe winters, copious rains followed by long droughts, and winds causing many changes of various kinds. From these changes it is natural for the population to be influenced, and for generation in the coagulation of the seed to vary, and not to be the same in the same person in summer as in winter, nor in rain as in drought. It is for this reason, I think, that the bodily form of Europeans varies more than that of Asiatics, and that their sizes vary greatly in each town. For there are more faults in the coagulation of the seed when the changes of the seasons are frequent than when the seasons are close and similar. With regard to character, the same principle applies. Wildness, unsociability and impetuosity are all engendered in such a natural setting: for the frequent shocks to the mind impart wildness, which displaces tameness and gentleness. Thus I think the inhabitants of Europe are also more courageous than Asiatics. For in a situation that is always the same resides complacency, while in one that is always changing there is activity of both body and mind; from comfort and complacency grows cowardice, from activity and effort bravery. For this reason the inhabitants of Europe are more warlike, as well as because of their laws, since they are not subject to kings as the Asiatics are. Thus, as I said above, where people are subject to kings, they must be most cowardly: for if men's minds are enslaved, they refuse to run risks readily and recklessly, in order to increase

---

93 διότι Froben: καὶ ὅτι V: διὸ καὶ Gad. (B)

86 λονται παρακινδυνεύειν | ἑκόντες εἰκῇ ὑπὲρ ἀλλοτρίης
δυνάμιος. ὅσοι δὲ αὐτόνομοι—ὑπὲρ ἑωυτῶν γὰρ τοὺς
κινδύνους αἱρεῦνται καὶ οὐκ ἄλλων—προθυμεῦνται
ἑκόντες καὶ ἐς τὸ δεινὸν ἔρχονται. τὰ γὰρ ἀριστεῖα
τῆς νίκης αὐτοὶ φέρονται. οὕτως οἱ νόμοι οὐχ ἥκιστα
τὴν εὐψυχίην ἐργάζονται. τὸ μὲν οὖν ὅλον καὶ τὸ
ἅπαν οὕτως ἔχει περί τε τῆς Εὐρώπης καὶ τῆς Ἀσίης.

24. Ἔνεισι δὲ καὶ ἐν τῇ Εὐρώπῃ φῦλα διάφορα
ἕτερα ἑτέροισι καὶ τὰ μεγέθεα καὶ τὰς μορφὰς καὶ τὰς
ἀνδρείας. τὰ δὲ διαλλάσσοντα ταῦτά[94] ἐστιν, ἃ καὶ ἐπὶ
τῶν πρότερον εἴρηται. ἔτι δὲ σαφέστερον φράσω. ὁκόσοι μὲν χώρην ὀρεινήν τε οἰκέουσι καὶ τρηχεῖαν καὶ
ὑψηλὴν καὶ εὔυδρον, καὶ αἱ μεταβολαὶ αὐτοῖσι γίνονται τῶν ὡρέων μεγάλα διάφοροι, ἐνταῦθα εἰκὸς εἴδεα
μεγάλα εἶναι καὶ πρὸς τὸ ταλαίπωρον καὶ τὸ ἀνδρεῖον
εὖ πεφυκότα, καὶ τό τε ἄγριον καὶ τὸ θηριῶδες αἱ
τοιαῦται φύσιες οὐχ ἥκιστα ἔχουσιν. ὁκόσοι δὲ κοῖλα
χωρία καὶ λειμακώδεα καὶ πνιγηρά, καὶ τῶν θερμῶν
πνευμάτων πλέον μέρος μετέχουσιν ἢ τῶν ψυχρῶν |
88 ὕδασί τε χρέονται θερμοῖσιν, οὗτοι δὲ μεγάλοι μὲν
οὐκ ἂν εἴησαν οὐδὲ κανονίαι, ἐς εὖρος δὲ πεφυκότες
καὶ σαρκώδεες καὶ μελανότριχες, καὶ αὐτοὶ μέλανες
μᾶλλον ἢ λευκότεροι, φλεγματίαι δὲ ἧσσον ἢ χολώ-
δεες· τό τε ἀνδρεῖον καὶ τὸ ταλαίπωρον ἐν τῇ ψυχῇ
φύσει μὲν οὐκ ἂν ὁμοίως ἐνείη, νόμος δὲ προσγενό-
μενος ἀπεργάσοιτο ἂν ὡσεὶ τοῦ εἴδεος οὐχ ὑπάρχοντος.[95] καὶ εἰ μὲν ποταμοὶ ἐνείησαν ἐν τῇ χώρῃ, οἵτινες
ἐκ τῆς χώρης ἐξοχετεύουσι τό τε στάσιμον καὶ τὸ

the power of someone else. But independent people—
taking risks on their own behalf and not on behalf of others—go willingly and eagerly into danger, since it is they
themselves that carry off the prize of victory. And thus
laws contribute not the least part to the generation of
courage. Such then, in outline and in general, is the situation of Europe and Asia.

24. In Europe too there are tribes differing from one
another in size, in bodily form, and in courage, and the
things that cause these differences are the same causes I
have discussed above: I will now explain in more detail.
The inhabitants of a region that is mountainous, rugged,
high, and well-watered, and for whom the changes of the
seasons exhibit sharp contrasts, are likely to be of large
form and well made for endurance and bravery, and such
natures possess not a little of wildness and ferocity. The
inhabitants of hollow regions, that are meadowy, stifling,
and visited more by hot winds than by cold winds, and who
employ waters that are warm, will be neither large nor
physically upright, but tend to be broad, fleshy, and dark-
haired; their bodies will incline to be dark rather than fair,
and less subject to phlegm than to bile. Equal bravery and
endurance are not present by nature in their character,
although if law reinforces these traits they can be attained,
even if they are not innate. If there are rivers in the region
that drain the ground of its standing water as well as the

---

94 ταὐτά Zwinger: ταῦτ' V
95 ὡσεὶ τοῦ . . . ὑπάρχοντος Gad. (B): om. V

ὄμβριον, οὗτοι ὑγιηρότεροι ἂν[96] εἴησαν καὶ λαπαροί.[97]
εἰ μέντοι ποταμοὶ μὴ εἴησαν, τὰ δὲ ὕδατα κρηναῖά τε
καὶ στάσιμα πίνοιεν καὶ ἑλώδεα, ἀνάγκη τὰ τοιαῦτα
90 εἴδεα προγαστρότερα εἶναι καὶ σπληνώδεα. | ὁκόσοι
δὲ ὑψηλήν τε οἰκέουσι χώρην καὶ λείην καὶ ἀνεμώδεα
καὶ εὔυδρον, εἴη ἂν ‹τὰ›[98] εἴδεα μεγάλα καὶ ἑωυτοῖσι
παραπλήσια· ἀνανδρότεραι δὲ καὶ ἡμερώτεραι αἱ
γνῶμαι. ὅσοι γὰρ ἐν εὐκρήτῳ καὶ ὕδασί τε πλείστοισι
καὶ ἀγαθοῖσι χρέονται, τούτοισι καὶ αἱ μορφαὶ καὶ τὰ
ἤθεα ἀγαθά, καὶ παχεῖς καὶ μεγαλόμορφοι καὶ ὅμοιοι
ἀλλήλοισιν.[99] ὁκόσοι δὲ λεπτά τε καὶ ἄνυδρα καὶ ψιλὰ
καὶ τῇσι δὲ μεταβολῇσι τῶν ὡρέων οὐκ εὔκρητα, ἐν
ταύτῃ τῇ χώρῃ τὰ εἴδεα εἰκὸς σκληρὰ εἶναι καὶ ἔντονα
καὶ ξανθότερα ἢ μελάντερα καὶ τὰ ἤθεα καὶ τὰς ὀργὰς αὐθάδεάς τε καὶ ἰδιογνώμονας. ὅκου γὰρ ‹αἱ›[100]
μεταβολαί εἰσι πυκνόταται τῶν ὡρέων καὶ πλεῖστον
διάφοροι αὐταὶ ἑωυτῆσιν, ἐκεῖ καὶ τὰ εἴδεα καὶ τὰ
ἤθεα καὶ τὰς φύσιας εὑρήσεις πλεῖστον διαφερούσας.

Μέγισται μὲν οὖν εἰσιν αὗται τῆς φύσιος αἱ διαλ-
λαγαί, ἔπειτα δὲ καὶ ἡ χώρη, ἐν ᾗ ἄν τις τρέφηται καὶ
τὰ ὕδατα. εὑρήσεις γὰρ ἐπὶ τὸ πλῆθος τῆς χώρης τῇ
φύσει ἀκολουθέοντα καὶ τὰ εἴδεα τῶν ἀνθρώπων καὶ
τοὺς τρόπους. ὅκου μὲν γὰρ ἡ γῆ πίειρα καὶ μαλθακὴ
καὶ εὔυδρος, καὶ τὰ ὕδατα κάρτα μετέωρα, ὥστε
θερμὰ εἶναι τοῦ θέρεος καὶ τοῦ χειμῶνος ψυχρά, καὶ
τῶν ὡρέων καλῶς κέεται, ἐνταῦθα καὶ οἱ ἄνθρωποι

---

[96] ὑγιηρότεροι ἂν Jouanna: ἂν ὑγιηροὶ τὲ V

water that falls as rain, the people will be of quite good health and free of constipation. If however no rivers are present, and the people drink spring water mixed with stagnant water and water from marshes, they will display protruding bellies and suffer in their spleens. People who dwell on a high plain that is windy and well-watered, will be of good size and similar to one another, but their characters will be somewhat unmanly and weak. Those again who live in a temperate climate and employ copious, good waters will have good physiques and good morals; they are thickly built and large shaped, and resemble one another. As to those who dwell on thin, dry, and bare soil, and where the climate is not temperate in the changes of the seasons, it is likely that in such a country the people's bodies will be hard and vigorous, blond rather than dark-haired, and in character and temper stubborn and independent. For where the changes of the seasons are most frequent and most sharply contrasted, you will find the greatest diversity in bodily form, in character, and in nature.

These then are the most important determinants of people's nature; next come the land in which a person is nourished, and the waters. For in general you will find both the bodily form and the character of inhabitants assimilated to the nature of their land. Where the land is rich, soft, and well-watered, and the water is held very near to the surface of the earth so that it is hot in summer and in winter cold, and if the situation is favorable as re-

---

97 λαπαροί P: λαμπροί V   98 Add. Ermerins
99 The sentence ὅσοι γὰρ ἐν εὐκρήτῳ . . . ἀλλήλοισιν is transmitted only by P   100 Add. Barb.

σαρκώδεες εἰσὶ καὶ ἄναρθροι καὶ ὑγροὶ καὶ ἀταλαί-
92 πωροι καὶ τὴν ψυχὴν κακοὶ | ὡς ἐπὶ τὸ πολύ. τό τε
ῥᾴθυμον καὶ τὸ ὑπηνηρὸν ἔνεστιν ἐν αὐτοῖσιν ἔς τε τὰς
τέχνας παχέες καὶ οὐ λεπτοὶ οὐδ' ὀξέες. ὅκου δ' ἐστὶν
ἡ χώρη ψιλή τε καὶ ἄνυδρος[101] καὶ τρηχεῖα καὶ ὑπὸ
τοῦ χειμῶνος πιεζομένη καὶ ὑπὸ τοῦ ἡλίου ἐκκεκαυ-
μένη,[102] ἐνταῦθα δὲ σκληρούς τε καὶ ἰσχνοὺς καὶ δι-
ηρθρωμένους καὶ ἐντόνους καὶ δασέας ἴδοις. τό τε
ἐργατικὸν ὀξὺ ἐνεὸν ἐν τῇ φύσει τῇ τοιαύτῃ καὶ τὸ
ἄγρυπνον, τά τε ἤθεα καὶ τὰς ὀργὰς αὐθάδεας καὶ
ἰδιογνώμονας, τοῦ τε ἀγρίου μᾶλλον μετέχοντας ἢ
τοῦ ἡμέρου, ἔς τε τὰς τέχνας ὀξυτέρους τε καὶ συν-
ετωτέρους καὶ τὰ πολέμια ἀμείνους εὑρήσεις· καὶ
τἆλλα τὰ ἐν τῇ γῇ φυόμενα πάντα ἀκόλουθα ἐόντα
τῇ γῇ. αἱ μὲν ἐναντιώταται φύσιές τε καὶ ἰδέαι ἔχουσιν οὕτως. ἀπὸ δὲ τούτων τεκμαιρόμενος τὰ λοιπὰ
ἐνθυμέεσθαι, καὶ οὐχ ἁμαρτήσῃ.

101 ἄνυδρος P: ἀνώχυρος V
102 ἐκκεκ. P: κεκ. V

gards the seasons, there the inhabitants are fleshy, showing no joints, moist, adverse to effort, and generally cowardly in character. Complacency and indolence dominate
them, and as far as the arts are concerned they are thick-
witted, and neither subtle nor sharp. But where the land
is bare, waterless, rough, oppressed by winter's storms and
burned by the sun, there you will discover people who are
hard, lean, showing obvious joints, vigorous, and hairy; in
such natures there are fierce energy and vigilance, as well
as stubborn, independent characters and tempers, wild
rather than tame; in the arts they have considerable sharp-
ness and intelligence, and in war exceptional courage. The
rest of the things that grow in the earth are also all adapted
to the earth. This is the situation with the most sharply
contrasted natures and forms. If you take these observations as your model in considering the rest, you will not go
wrong.

EPIDEMICS I/III

INTRODUCTION

Epidemics I and *III* appear to be based on actual clinical records kept around the year 409 BC on the island of Thasos,[1] but also in other towns in the northern Aegean. They consist of two kinds of accounts:

1. four annual community health reports (*catastasis* = state, condition, constitution) for particular places, running from one fall to the following summer
2. forty-two individual patient histories describing the course of acute diseases from their onset to recovery (seventeen) or a fatal outcome (twenty-five).

The overall structure of the two works (by chapters) is:

Epidemics I
    1–3   *catastasis* I (hot/dry)
    4–10 *catastasis* II (cold/wet)

---

[1] The patient Antiphon, son of Critobulus, named near the end of chapter 15 in *Epidemics I* is documented in a Thrasian inscription (IG XII,8,263,1.2), which names him as a magistrate in a legal case; a decree cited in the inscription dates it to the year 408 or 409 BC, more likely the latter (cf. Jouanna, pp. cxxii–cxxiv). The temporal relationship between Antiphon's disease and his magistracy, however, is unknown, leaving the dating of the medical text approximate.

The public health reports begin with a brief account of the dominant weather conditions during the year described, including its hotness or coldness, moistness or dryness, and how the Etesian winds blew. They then discuss in detail the symptoms, course, and prevalence through the year of the diseases most common in the populace (νόσημα ἐπίδημον). Of the four *catastases*, which vary considerably in their length, organization, and contents, the first three are based on information gathered in Thasos, while for the fourth no location is given.

The case histories follow an individual patient, identified by name, address, and/or familial relationship, in strict chronological sequence from a disease's onset to its conclusion. These accounts achieve a high degree of compression by making use of an established technical terminology, by employing a telegraphic reporting style, and by omitting all superfluous information, including entire days on which no significant signs appear. In the first two series of case histories, a4, a9, and b4 are assigned explicitly to Thasos, while none is set in any other location; patients in the third series are spread among five locations: Thasos (c1–c3, c11, and c15), Larissa (c5, c12), Abdera (c6–c10 and c13), Cyzicus (c14), Meliboea (c16).

Two of the twenty-six patients referred in the third *catastasis* are also present as case histories in the first series, showing that the same collection of clinical data served as the basis of both types of description.[2]

Epidemics I and *III* belong to a group of seven *Epidemics* (Ἐπιδημίαι ζ΄)[3] included by Erotian as "mixed" (ἐπίμικτα) works in the census of titles presented in the introduction to his Hippocratic *Glossary*. Among about a dozen glosses attributable to each of the books, he assigns A6 and Φ1 explicitly to *Epidemics I*. In his glosses A6, Σ2, and Φ5, he refers to explanations given by the glossator Bacchius of Tanagra (third century BC), providing evidence that *Epidemics I* and *III* were already the object of scholarly study in Hellenistic times.

Galen of Pergamum studied the *Epidemics*, including books *I* and *III*, intensively throughout his life, and left his views on their form, content, and transmission in:

1. a glossary of Hippocratic vocabulary containing articles on specific words,[4]

---

[2] Philiscus in chapter 14 and Hermippus of Clazomenae in chapter 20 of *Epidemics I* (*catastasis* III) reappear as case histories a1 and a10.

[3] For *Epidemics II, IV–VII*, see W. D. Smith, Loeb *Hippocrates*, vol. 7 (1994).

[4] See Perilli.

2. extensive commentaries on *Epidemics I* and *III* devoted to their exegesis,[5]
3. many explanatory references in several other of his works.[6]

He was well aware that *Epidemics I* and *III* belonged together and that their traditional numbering was erroneous:

> In my opinion earlier scholars are correct in saying that Thessalus added material to *Epidemics II*, as well as in asserting that only the first and the third books were written by Hippocrates and intended for publication. The most convincing proof of this is their diction, for we find the style in the first book and in the one that bears the title "third"—but not correctly so—is identical. I have indicated that this book is not correctly called the "third," because the book with this name should have been called the "second," and the one now called the "second" should have been called the "third."[7]

Evidence that Galen is correct in concluding that originally book *III* followed directly after book *I* is provided by the earliest extant Greek manuscript (A) of the works,

[5] Edited in Wenkebach (I), pp. 1–151 (= GalL) and *Testimonien* vol. II,1, pp. 172–99 (= GalT); Wenkebach (II) (= GalL); and *Testimonien* vol. II,1, pp. 228–41 (= GalT). Cf. Manetti/Roselli.

[6] Cf. *Testimonien* vol. II,2, pp. 122–52 and 178–99.

[7] See Wenkebach (I), p. 310, 27–38; Vagelpohl (II), p. 617, 8–17.

where at the end of the text of *Epidemics I* a mutilated
catchword—πυθιώνιος ᾤκει· παραγείσιρον ἤρξατο· τρόμος ἀπὸ χειρῶν—leads directly to the first sentence of the
text of *Epidemics III*.[8]

Epidemics I and *III* are often cited and quoted by ancient and medieval grammarians and medical writers, and
were well known in the Islamic world in the complete
Arabic translation, in the form of lemmata in Galen's *Commentaries* to the books, made by Ḥunayn ibn Isḥāq.[9]

The Greek text of *Epidemics I* is contained in three independent Greek manuscripts, A, V, and I, the last repre-
senting M, which has a lacuna at this point, while *Epidemics III* is available only in V and I (which again represents
M). Galen's *Commentaries* to the books, which contain the
complete Hippocratic texts as lemmata, also represent an
important witness to the text; these are available both in

8 This catchword stems from a time when such texts were
written on scrolls, rather than in codices, and served as an indica-
tor to later scribes showing how the text on the following scroll
should begin: in this case, however, the scribe failed to be guided
by the catchword, and chose as model for his next scroll a scroll
with the text of *Epidemics II*, which displaced the text introduced
by the catchword to a third scroll, resulting in its receiving the
title *Epidemics III*. Cf. Jouanna, pp. xxx f., who remarks that while
the ideal would be to make a clean sweep of all the arbitrary
subdivisions, it is impossible to ignore the partition in *Epid. I* and
III that has existed since the Hellenistic period.

9 Cf. *Testimonien* vol. III, pp. 175–87 and 195–98; Ullmann,
pp. 30 and 61f.; Vagelpohl (I), pp. 65–68

their Greek original[10] and in the Arabic translation of Ḥunayn.[11] Finally, two ancient Greek papyri make small but not insignificant contributions to the transmission:

Π₂₄	POxy 80.5222	(II/III c.)[12]
Π₅	PSI 2.116	(III c.)[13]

Epidemics I and *III* are present in all the collected Hippocratic editions and translations, including Adams, Kühlewein, and Chadwick, as well as in many special studies noted by Littré (vol. 2, pp. 593–97). Among more recent works are:

Baader, G., and R. Winau, eds. *Die hippokratischen Epidemien. Theorie—Praxis—Tradition.* Verhandlungen des Vᵉ Colloque International Hippocratique. Berlin, 1984. Sudhoffs Archiv Beiheft 27. Stuttgart, 1989.

Jouanna, J. *Hippocrate. Epidémies I et III.* Budé IV(1). Paris, 2016.

The present edition depends almost entirely on the work of Jouanna; Littré chapter numbers are supplied in parentheses.

---

[10] See n. 5 above.

[11] *Epidemics I* is edited and translated into English in Vagelpohl (I); *Epidemics III* is transmitted in the manuscript Scorialensis Arab. 804 (XIII c.), fols. 124ʳ–82ᵛ.

[12] D. Leith, *Oxyrhynchus Papiri* 80 (London, 2014), pp. 10f.

[13] L. Stephani, *Papiri greci e latini (Pubblicazioni della Società italiana . . . per la ricerca dei papyri . . . in Egitto),* vol. 2, (Florence, 1913), pp. 6f.

ΕΠΙΔΗΜΙΩΝ Α

II 598
Littré

1. Ἐν Θάσῳ, φθινοπώρου περὶ ἰσημερίην καὶ ὑπὸ πληϊάδα, ὕδατα πολλά, συνεχέα μαλθακῶς, ἐν νοτίοισι. χειμὼν νότιος, σμικρὰ βόρεια, αὐχμοί· τὸ σύνολον ἔς γε χειμῶνα οἷον ἔαρ γίνεται. ἔαρ δὲ νότιον ψυχεινόν, σμικρὰ ὕσματα. θέρος ὡς ἐπὶ τὸ πολὺ ἐπινέφελον· ἀνυδρίαι. ἐτησίαι ὀλίγα, σμικρά, διεσπαρμένως ἔπνευσαν.

Γενομένης δὲ τῆς ἀγωγῆς ὅλης ἐπὶ τὰ νότια καὶ μετ' αὐχμῶν, πρωὶ μὲν τοῦ ἦρος, ἐκ τῆς πρόσθεν καταστάσιος ὑπεναντίης καὶ βορείου γενομένης ὀλί-

600 γοισιν ἐγίνοντο | καῦσοι· καὶ τούτοισι πάνυ εὐσταθέα, καὶ ὀλίγοισιν ἡμορράγει· οὐδ' ἀπέθνησκον ἐκ τούτων. ἐπάρματα δὲ παρὰ τὰ ὦτα πολλοῖσιν ἑτερόρροπα καὶ ἐξ ἀμφοτέρων, τοῖσι πλείστοισιν ἀπύροισιν ὀρθοστάδην· ἔστι δὲ οἳ καὶ σμικρὰ ἐπεθερμαίνοντο. κατέσβη πᾶσιν ἀσινέως· οὐδ' ἐξεπύησεν οὐδενὶ ὥσπερ τὰ ἐξ

602 ἄλλων προφασίων. ἦν δὲ ὁ τρόπος αὐτῶν | χαῦνα, μεγάλα, κεχυμένα, οὐ μετὰ φλεγμονῆς, ἀνώδυνα· πᾶσιν ἀσήμως ἠφανίσθη. ἐγίνετο δὲ ταῦτα μειρακίοισι,

EPIDEMICS I

First Catastasis

1. In Thasos during autumn toward the equinox and at the time of (sc. the setting of) the Pleiades, there were abundant rains, continuous and falling gently, with southerly winds. Winter southerly, light north winds, droughts; on the whole, during the winter it was like a spring. Spring southerly and chilly, light showers. Summer for the most part cloudy; no rain; the Etesian winds, few and weak, blew sporadically.

The whole course (sc. of the year) tended to be southerly with droughts, but early in the spring, because the previous constitution had been the opposite and northerly, a few patients suffered from ardent fevers; during these the course was very regular, and hemorrhaging occurred in a few patients; none died. Many had swellings beside one ear, or both ears, in most cases unattended with fever, so that they did not have to be confined to bed. In some cases, however, there was slight heat. In all cases the swellings subsided without causing harm, and they never suppurated as swellings of other origin do. This is what the swellings were like: flabby, big, spreading, with neither inflammation nor pain; in every case they disappeared without a sign. They occurred in adolescents, young men,

153

νέοισιν, ἀκμάζουσι, καὶ τούτων τοῖσι περὶ παλαί-
στρην καὶ γυμνάσια πλείστοισι· γυναιξὶ δὲ ὀλίγῃσιν
ἐγίνετο. πολλοῖσι δὲ βῆχες ξηραὶ βήσσουσι καὶ οὐδὲν ἀνάγουσιν· φωναὶ βραγχώδεες. οὐ μετὰ πολύ,
τοῖσι δὲ καὶ μετὰ χρόνον, φλεγμοναὶ μετ᾽ ὀδύνης ἐς
ὄρχιν ἑτερόρροπα, τοῖσι δὲ ἐς ἀμφοτέρους· πυρετοὶ |
604 τοῖσι μέν, τοῖσι δ᾽ οὔ· ἐπιπόνως ταῦτα τοῖσι πλείστοισι. τὰ δ᾽ ἄλλα, ὅσα κατ᾽ ἰητρεῖον, ἀνόσως διῆγον.

2. Πρωῒ δὲ τοῦ θέρεος ἀρξάμενοι, διὰ θέρεος καὶ
κατὰ χειμῶνα, πολλοὶ τῶν ἤδη πολὺν χρόνον ὑποφε-
ρομένων φθινώδεες κατεκλίνησαν, ἐπεὶ καὶ τοῖσιν ἐν-
δοιαστῶς ἔχουσι, πολλοῖσιν ἐβεβαίωσε τότε. ἔστι δ᾽
οἷσιν ἤρξατο πρῶτον τότε, οἷσιν ἔρρεπεν ἡ φύσις ἐπὶ
606 τὸ φθινῶδες. | ἀπέθανον δὲ πολλοὶ καὶ πλεῖστοι τού-
των, καὶ τῶν κατακλινέντων οὐκ οἶδα εἴ τις καὶ[1]
μέτριον χρόνον περιεγένετο. ἀπέθνησκον δὲ ὀξυτέρως
ἢ ὡς εἴθισται διάγειν τὰ τοιαῦτα. ὡς τά γε ἄλλα καὶ
μακρότερα ἐν πυρετοῖσιν ἐόντα εὐφόρως ἤνεγκαν καὶ
οὐκ ἀπέθνησκον, περὶ ὧν γεγράψεται. μοῦνον γὰρ καὶ
μέγιστον τῶν γενομένων νοσημάτων τοὺς πολλοὺς τὸ
φθινῶδες ἔκτεινεν.

Ἦν δὲ τοῖσι πλείστοισιν αὐτῶν τὰ παθήματα·[2]
608 φρικώδεες πυρετοί, ξυνεχέες, ὀξέες, τὸ μὲν ὅλον | οὐ
διαλείποντες, ὁ δὲ τρόπος ἡμιτριταῖος· μίαν κουφότε-
ροι, τῇ ἑτέρῃ παροξυνόμενοι, καὶ τὸ ὅλον ἐπὶ τὸ

---

[1] καὶ VI: οὐδ᾽ εἰ A   [2] Add. τοιάδε VI

and men in their prime, and for the most part in those who frequented the wrestling school and gymnasia; in women there were few occurrences. In many patients there were dry coughs, but when they coughed they brought nothing up; their voices were hoarse. Soon after, though in some cases after some time, painful inflammations occurred either in one testicle or in both, sometimes accompanied with fever, in other cases not. In most instances these caused much suffering. From the other disorders that are treated in the physician's office, the people were spared.

2. Beginning early in the summer, throughout the summer, and in winter many of those who had been ailing a long time took to their beds in a state of consumption, while at the same time in many who had hitherto been doubtful sufferers the disease was confirmed. In others the disease began now for the first time, these being inclined in their nature toward a consumptive state. Many, in fact most of these, died; of those who took to their beds I know of none who survived for even a middling time. They died more rapidly than is usual in consumptive conditions. As for the other longer complaints attended with fever which I will describe below, these patients made out well and were not endangered. For consumption was the worst of the diseases that occurred, and it alone killed the greatest numbers.

In the majority of cases the symptoms were these: Fever with shivering, continuous, acute, not intermitting completely, but of the semitertian type; remitting for one day and being exacerbated on the next, they became on

ὀξύτερον ἐπιδιδόντες. ἱδρῶτες αἰεί, οὐ δι᾽ ὅλου· ψῦξις
ἀκρέων πολλή, καὶ μόγις ἀναθερμαινόμενα. κοιλίαι
ταραχώδεες χολώδεσιν, ὀλίγοις, ἀκρήτοισι, λεπτοῖσι,
δακνώδεσι· πυκνὰ ἀνίσταντο. οὖρα ἢ λεπτὰ καὶ
ἄχροα καὶ ἄπεπτα[3] καὶ ὀλίγα, ἢ πάχος ἔχοντα καὶ
σμικρὴν ὑπόστασιν, οὐ καλῶς καθιστάμενα, ἀλλ᾽
ὠμῇ τινι καὶ ἀκαίρῳ ὑποστάσει. ἔβησσον δὲ σμικρά,
πυκνά, πέπονα, κατ᾽ ὀλίγα μόγις ἀνάγοντες. οἶσι δὲ
τὰ βιαιότατα ξυμπίπτοι, οὐδ᾽ ἐς ὀλίγον πεπασμὸν
ἤει, ἀλλὰ διετέλεον ὠμὰ πτύοντες. φάρυγγες δὲ τοῖσι
πλείστοισι τούτων ἐξ ἀρχῆς καὶ διὰ τέλεος ἐπώδυνοι·
εἶχον ἔρευθος μετὰ φλεγμονῆς· ῥεύματα σμικρά, λεπτά, δριμέα. ταχὺ τηκόμενοι καὶ κακούμενοι, ἀπόσι-
610 τοι πάντων γευμάτων διὰ τέλεος, ἄδιψοι· καὶ | παράλη-
ροι πολλοὶ περὶ θάνατον. περὶ μὲν τὰ φθινώδεα
ταῦτα.

3. Κατὰ δὲ θέρος ἤδη καὶ φθινόπωρον πυρετοὶ
πολλοὶ ξυνεχέες οὐ βιαίως, μακρὰ δὲ νοσέουσιν οὐδὲ
περὶ τὰ ἄλλα δυσφόρως διάγουσιν ἐγένοντο· κοιλίαι
τε ταραχώδεες[4] τοῖσι πλείστοισι πάνυ εὐφόρως καὶ
οὐδὲν ἄξιον λόγου προσέβλαπτον. οὖρά τε τοῖσι
πλείστοισιν εὔχροα μὲν καὶ καθαρά, λεπτὰ δέ, καὶ
μετὰ χρόνον περὶ κρίσιν πεπαινόμενα. βηχώδεες οὐ
λίην, οὐδὲ τὰ βησσόμενα δυσκόλως· οὐδ᾽ ἀπόσιτοι,
ἀλλὰ καὶ διδόναι πάνυ ἐνεδέχετο. τὸ μὲν ὅλον ὑπενό-
612 σεον οἱ φθίνοντες, οὐ τὸν φθινώδεα | τρόπον, πυρε-

---

3 καὶ ἄπεπτα VI: om. A   4 ταραχώδεες VI: γὰρ A

the whole more acute. Sweats were continual, but not over the whole body. Severe chilling in the extremities, which could hardly be warmed back up. Bowels disordered with bilious, scanty, unmixed, thin, smarting stools, causing the patient to get up often. Urines either thin, colorless, unconcocted and scanty, or thick and with a slight deposit, not settling properly, but with a crude and unfavorable deposit. These patients coughed up small, compact, concocted sputa, brought up little by little with difficulty. Those suffering the symptoms in their most violent form showed no concoction at all, but went on expectorating crude sputa. In the majority of these cases, the throat was painful throughout right from the beginning, being red and inflamed. Fluxes slight, thin, pungent. Patients quickly wasted away and declined, being throughout averse to all food and experiencing no thirst. Delirium in many cases as death approached. Such were the symptoms of this consumption.

3. Already during the summer and autumn, many continuous but moderate fevers arose in persons who were sick for a long time, but did not suffer any other particular distress. Disturbances of the bowels that occurred were in most cases quite easy to bear, and caused no appreciable harm. Urines were in most cases of good color and clear, but thin, and after a time they became concocted around the crisis. Coughing was light, and what was coughed up caused no distress. No lack of appetite; in fact it was quite possible to give even food. In general such consumptives who lacked the characteristics of the true consumptive

τοῖσι φρικώδεσι, σμικρὰ ὑφιδροῦντες, ἄλλοτε ἀλλοίως παροξυνόμενοι πεπλανημένως, τὸ μὲν ὅλον οὐκ
ἐκλείποντες, παροξυνόμενοι δὲ τριταιοφύεα τρόπον.
ἔκρινε τούτων οἷσι τὰ βραχύτατα γίνοιτο περὶ εἰκοστήν, τοῖσι πλείστοισι δὲ⁵ περὶ τεσσαρακοστήν, πολλοῖσι δὲ περὶ τὰς ὀγδοήκοντα. ἔστι δ' οἷσιν οὐδ'
οὕτως, ἀλλὰ πεπλανημένως καὶ ἀκρίτως ἐξέλιπον.
τούτων δὲ τοῖσι πλείστοισιν οὐ πολὺν διαλείποντες
χρόνον ὑπέστρεψαν οἱ πυρετοὶ πάλιν· ἐκ δὲ τῶν ὑποστροφέων ἐν τῇσιν αὐτῇσι περιόδοισιν ἐκρίνοντο·
614 πολλοῖσι δὲ αὐτῶν | ἀνήγαγον, ὥστε καὶ ὑπὸ χειμῶνα
νοσέειν·

Ἐκ πάντων δὲ τῶν ὑπογεγραμμένων ἐν τῇ καταστάσει ταύτῃ μόνοισι τοῖσι φθινώδεσι θανατώδεα
συνέπεσεν· ἐπεὶ τοῖσί γε ἄλλοισι πᾶσιν εὐφόρως· καὶ
θανατώδεες ἐν τοῖσιν ἄλλοισι πυρετοῖσιν οὐκ ἐγένοντο.

Κατάστασις δευτέρη

4. Ἐν Θάσῳ πρωὶ τοῦ φθινοπώρου, χειμῶνες οὐ κατὰ
καιρόν, ἀλλ' ἐξαίφνης ἐν βορείοισι καὶ νοτίοισι πολλοῖς, ὑγροὶ καὶ προεκρηγνύμενοι· ταῦτα δὴ ἐγένετο
τοιαῦτα μέχρι πληϊάδος⁶ καὶ ὑπὸ πληϊάδα. χειμὼν δὲ
βόρειος· ὕδατα πολλά, λάβρα, μεγάλα, χιόνες· μει-
ξαίθρια τὰ πλεῖστα. ταῦτα δὲ γίνεται μὲν πάντα, οὐ

---

⁵ πλείστοισι δὲ A: δὲ πλείστοισι VI   ⁶ Add. δύσεως VI

state were only mildly ill with shivering fevers, and their sweats were slight, their paroxysms being variable and irregular; on the whole they were without remissions, having instead paroxysms like those of tertian fevers. The earliest crisis was about the twentieth day; in most cases the crisis was about the fortieth day, though in many it was about the eightieth. In some cases the illness did not end in this way, but left off in an irregular manner without a crisis. In the majority of these cases the fevers relapsed after a brief remission, and after the relapse a crisis occurred at the end of the same periods as before. The disease in many of these instances was so protracted that it even lasted through the winter.

Out of all those described in this constitution, only the consumptives suffered deaths, while all the other patients with fevers bore their diseases well without fatalities.

Second Catastasis

4. In Thasos early in the autumn there were unseasonable storms, occurring suddenly with many north and south winds, and rains bursting out before expected. These conditions continued during the Pleiades and until their setting. Winter was northerly; frequent violent torrential rains; snows; in general the weather was mixed. With all this, however, the cold weather was not very unseasonable.

λίην δὲ ἀκαίρως τὰ τῶν ψυχέων. ἤδη δὲ μεθ' ἡλίου
τροπὰς χειμερινάς, καὶ ἡνίκα ζέφυρος πνεέιν ἄρχεται,
ὀπισθοχειμῶνες μεγάλοι, βόρεια πολλά, χιὼν καὶ
616 ὕδατα πολλὰ συνεχέως· | οὐρανὸς λαιλαπώδης καὶ
ἐπινέφελος. ταῦτα δὲ συνέτεινε καὶ οὐκ ἀνίει μέχρι
ἰσημερίης. ἔαρ δὲ ψυχρόν, βόρειον, ὑδατῶδες, ἐπινέφελον. θέρος οὐ λίην καυματῶδες ἐγίνετο· ἐτησίαι
ξυνεχέες ἔπνευσαν. ταχὺ δὲ περὶ ἀρκτοῦρον ἐν βορείοισι πολλὰ πάλιν ὕδατα.

5. Γενομένου δὲ τοῦ ἔτεος ὅλου ὑγροῦ καὶ ψυχροῦ
καὶ βορείου κατὰ χειμῶνα μὲν ὑγιηρῶς εἶχον τὰ
πλεῖστα· πρῶι δὲ τοῦ ἦρος, πολλοί τινες καὶ οἱ πλεῖστοι διῆγον ἐπινόσως. ἤρξαντο μὲν οὖν τὸ πρῶτον
ὀφθαλμίαι ῥοώδεες, ὀδυνώδεες, ὑγραὶ ἀπέπτως· σμικρὰ λημία δυσκόλως πολλοῖσιν ἐκρηγνύμενα· τοῖσι
πλείστοισιν ὑπέστρεφον· ἀπέλιπον ὀψὲ πρὸς τὸ φθινόπωρον. κατὰ δὲ θέρος καὶ φθινόπωρον δυσεντεριώδεες καὶ τεινεσμοὶ καὶ λειεντεριώδεες. καὶ διάρροιαι
χολώδεσι, πολλοῖσι, λεπτοῖσιν,[7] ὠμοῖσι, καὶ δακνώδε-
σιν· ἔστι δ' οἷσι καὶ ὑδατώδεες. πολλοῖσι δὲ καὶ
περίρροιαι μετὰ πόνου χολώδεες,[8] ξυσματώδεες, πυώ-
618 δεες· στραγγουριώδεες, | οὐ νεφριτικά· ἀλλὰ τούτοι-
σιν ἀντ' ἄλλων ἄλλα. ἔμετοι φλεγματώδεες, χολώδεες
καὶ σιτίων ἀπέπτων ἀναγωγαί. ἱδρῶτες πᾶσι πάντοθεν, πολὺς πλάδος. ἐγίνετο δὲ ταῦτα πολλοῖσιν ὀρθοστάδην ἀπύροισι, πολλοῖσι δὲ πυρετοί, περὶ ὧν
γεγράψεται. ἐνίοισι δὲ ὑπεφαίνετο πάντα τὰ ὑπογε-

But immediately after the winter solstice, when the west wind begins to blow, there was a severe bout of late-winter weather, much north wind, snow and continuous heavy rains; sky stormy and clouded. These conditions persisted and did not relent before the equinox. Spring cold, northerly, wet, cloudy. Summer did not turn out excessively hot, the Etesian winds blew continuously. But soon after, near (sc. the rising of) Arcturus, there was heavy rain again, with northerly winds.

5. The whole year having been wet, cold and northerly, in the winter there was good health in most respects, but in early spring many, in fact most, suffered illnesses. Now ophthalmias first began with fluxes, pain, and unconcocted discharges: small eyesores that caused distress to most patients when they broke out. These relapsed in the great majority, and went away only after a long time with the approach of autumn. In the summer and autumn dysenteric diseases, tenesmus and lientery; diarrheas with copious bilious, thin, crude, smarting stools; in some cases they were also watery. Many also had painful (sc. alvine) discharges that were bilious, purulent, and contained shreds (sc. of flesh). Strangury, no kidney disease. In these cases the various symptoms succeeded one another in varying orders. Vomitings containing phlegm and bile, and bringing up undigested foods. Sweating in all patients and in every part of the body; copious moisture. These things happened in many afebrile patients who were not confined to bed; but in many others there was fever, as I shall describe. In some patients all the symptoms mentioned above appeared

---

[7] πολλοῖσι, λεπτοῖσι A: λεπτοῖσιν, πολλοῖσιν VI
[8] ὑδατώδεες add. VI

γραμμένα· μετὰ πόνου φθινώδεες. ἤδη δὲ φθινοπώρου
καὶ ὑπὸ χειμῶνα πυρετοὶ ξυνεχέες—καί τισιν αὐτῶν
ὀλίγοισι καυσώδεες—ἡμερινοί, νυκτερινοί, ἡμιτριταῖοι, τριταῖοι ἀκριβέες, τεταρταῖοι, πλάνητες. ἕκαστοι δὲ τῶν ὑπογεγραμμένων πυρετῶν πολλοῖσιν ἐγίνοντο.

6. Οἱ μὲν οὖν καῦσοι ἐλαχίστοισί τε ἐγίνοντο, καὶ
620 ἥκιστα | τῶν καμνόντων οὗτοι ἐπόνησαν· οὔτε γὰρ
ἡμορράγει, εἰ μὴ πάνυ σμικρὰ καὶ ὀλίγοισιν, οὔτε οἱ
παράληροι. τά τε ἄλλα πάντ' εὐφόρως. ἔκρινε τούτοισι
πάνυ εὐτάκτως, τοῖσι πλείστοισι σὺν τῇσι διαλειπού-
σῃσιν ἐν ἑπτακαίδεκα ἡμέρῃσιν· οὐδὲ ἀποθανόντα
οὐδένα οἶδα τότε καύσῳ, οὐδὲ φρενιτικὰ τότε γενόμενα. οἱ δὲ τριταῖοι, πλείους μὲν τῶν καύσων καὶ ἐπι-
πονώτεροι· εὐτάκτως δὲ τούτοισι πᾶσιν ἀπὸ τῆς πρώ-
της λήψιος τέσσαρας περιόδους· ἐν ἑπτὰ δὲ τελέως
ἔκριναν, οὐδ' ὑπέστρεψαν οὐδενὶ τούτων. οἱ δὲ τεταρταῖοι πολλοῖσι μὲν ἐξ ἀρχῆς ἐν τάξει τεταρταίου ἤρ-
ξαντο· ἔστι δ' οἷς οὐκ ὀλίγοισιν ἐξ ἄλλων πυρετῶν
καὶ νοσημάτων ἀποστάσιες[9] ἐς τεταρταίους[10] ἐγέ-
622 νοντο· | μακρὰ δὲ καὶ ὡς εἴθισται, τούτοισι καὶ ἔτι
μακρότερα συνέπιπτεν. ἀμφημερινοὶ δὲ καὶ νυκτερι-
νοὶ καὶ πλάνητες πολλοῖσι πολλοὶ καὶ πολὺν χρόνον
παρέμενον, ὀρθοστάδην τε καὶ κατακειμένοισι. τοῖσι
πλείστοισι τούτων ὑπὸ πληϊάδα καὶ μέχρι χειμῶνος
οἱ πυρετοὶ παρείποντο. σπασμοὶ δὲ πολλοῖσι, μᾶλλον
δὲ παιδίοισιν, ἐξ ἀρχῆς καὶ ὑπεπύρεσσον· καὶ ἐπὶ

in a mild form; there was wasting accompanied by pain. As soon as autumn came, and during winter, continuous fevers—in some few cases ardent—day fevers, night fevers, semitertians, exact tertians, quartans, and irregular fevers: each of the fevers recorded was present in many patients.

6. Now the ardent fevers attacked the fewest persons, and these were distressed less than any of the others. There was no bleeding (sc. from the nose), except very slight discharges in a few instances, and no delirium. All the other symptoms were easily born; the crises in these diseases were very regular, generally on the seventeenth day, counting the days of intermission, and I know of no ardent fever proving fatal at that time, nor any that then developed into phrenitis. The tertians were more numerous than the ardent fevers, and more troublesome, but all the ones that then developed went regularly through four periods from the first onset, had complete crises in seven periods, and in no case relapsed. But the quartans, while in many instances they began from their onset with quartan periodicity, in not a few cases apostases from other fevers and illnesses turned into quartan fevers. They were protracted, the way quartans usually are, and in these cases even more protracted. Quotidians, night fevers, and irregular fevers befell many persons, and persisted for a long time, with the patients either walking about or in bed. In most of these cases the fevers continued through the Pleiades and even until winter; in many of them, especially with children, there were convulsions and slight feverishness from the beginning; sometimes, too, convulsions

---

9 ἀποστάσιες I: ἀποστάσει A: ἀποστάσηες V
10 ἐς τεταρταίους VI: τεταρταῖοι A

πυρετοῖσιν ἐγίνοντο σπασμοί· χρόνια μὲν τοῖσι πλείστοισι τούτων, ἀβλαβέα δέ, εἰ μὴ τοῖσι καὶ ἐκ τῶν
ἄλλων πάντων ὀλεθρίως ἔχουσιν.

7. Οἱ δὲ δὴ ξυνεχέες μὲν τὸ ὅλον καὶ οὐδὲν ἐκλείποντες, παροξυνόμενοι δὲ πᾶσι τριταιοφυέα τρόπον,
μίαν ὑποκουφίζοντες, καὶ μίαν παροξυνόμενοι, πάντων βιαιότατοι τῶν τότε γενομένων, καὶ μακρότατοι,
καὶ μετὰ πόνων μεγίστων γενόμενοι· πρηέως ἀρχόμενοι, τὸ δ' ὅλον ἐπιδιδόντες αἰεὶ καὶ παροξυνόμενοι καὶ
ἀνάγοντες ἐπὶ τὸ κάκιον· σμικρὰ διακουφίζοντες καὶ
624 ταχὺ | πάλιν ἐξ ἐπισχέσιος βιαιοτέρως παροξυνόμε-
νοι· ἐν κρισίμοις ὡς ἐπὶ τὸ πολὺ κακούμενοι. ῥίγεα δὲ
πᾶσι μὲν ἀτάκτως καὶ πεπλανημένως ἐγίνετο, ἐλάχιστα δὲ καὶ ἥκιστα τούτοισιν, ἀλλ' ἐπὶ τῶν ἄλλων
πυρετῶν μέζω. ἱδρῶτες πολλοί, τούτοισιν ἐλάχιστοι,
κουφίζοντες οὐδέν, ἀλλ' ὑπεναντίον βλάβας φέροντες.
ψῦξις δὲ πολλὴ τούτοισιν ἀκρέων, καὶ μόγις ἀναθερμαινόμενα. οὐδ' ἄγρυπνοι τὸ σύνολον καὶ μάλιστα
οὗτοι, καὶ πάλιν κωματώδεες. κοιλίαι δὲ πᾶσι μὲν ταραχώδεες καὶ κακαί, πολὺ δὲ τούτοισι κάκισται. οὖρα
δὲ τοῖσι πλείστοισι τούτων, ἢ λεπτὰ καὶ ὠμὰ καὶ
ἄχρω καὶ μετὰ χρόνον σμικρὰ πεπαινόμενα κρισί-
μως, ἢ πάχος μὲν ἔχοντα, θολερὰ δὲ καὶ οὐδὲν |
626 καθιστάμενα, οὐδ' ὑφιστάμενα, ἢ σμικρὰ καὶ κακὰ
καὶ ὠμὰ τὰ ὑφιστάμενα· κάκιστα δὲ ταῦτα πάντων.[11]
βῆχες δὲ παρείποντο μὲν τοῖς πυρετοῖσι, γράψαι δὲ
οὐκ ἔχω βλάβην οὐδ' ὠφελείην γενομένην διὰ βηχὸς
τότε.

supervened upon the fevers. Generally these illnesses were protracted, but not dangerous, except in cases where, on the basis of all the other circumstances, patients could be expected to die.

7. Those fevers, however, that were perfectly continuous and never intermitted at all, but in all cases grew worse after the manner of semitertians, with a remission on one day followed by an exacerbation on the next, were the most severe of all the fevers which occurred at the time, the longest and the most troublesome. They began mildly, increased continually in every feature, became more acute, and changed for the worse; they had slight remissions followed quickly after the abatement by more violent exacerbations, generally becoming worse on the critical days. Rigors occurring in all the cases were irregular and erratic, appearing most rarely and least in the semitertians, but more forcefully in the other fevers. Many patients had sweats—again least those with semitertians—that brought no relief, but on the contrary caused harm. Such patients suffered great chilling in their extremities, which were difficult to warm back up; these in particular were not always sleepless, and sometimes became somnolent. In all cases the bowels were disturbed and in a bad state, but in the semitertians they were by far the worst. In many cases the urines were either thin, crude and colorless, after a time becoming slightly concocted during the crisis, or thick enough but turbid, in no way settling or forming a sediment, or they contained small, malignant, crude sediments, which were the worst of all. Coughs attended the fevers, but I cannot state that either harm or good resulted from the coughing on this occasion.

---

11 πάντων A: πάντα VI

8. Χρόνια μὲν οὖν καὶ δυσχερέα καὶ πάνυ ἀτάκτως καὶ πεπλανημένως καὶ ἀκρίτως τὰ πλεῖστα τούτων διετέλει γινόμενα, καὶ τοῖσι πάνυ ὀλεθρίως ἔχουσι καὶ τοῖσι μή. εἰ γάρ τινα[ς][12] αὐτῶν καὶ διαλίποι σμικρά, ταχὺ πάλιν ὑπέστρεφεν. ἔστι δ' οἷσιν ἔκρινεν αὐτῶν ὀλίγοισιν· οἷσι τὰ βραχύτατα γένοιτο, περὶ ὀγδοη-κοστὴν ἐοῦσι· καὶ τούτων ἐνίοις ὑπέστρεφεν, ὥστε κατὰ χειμῶνα τοὺς πλείστους αὐτῶν ἔτι νοσέειν. τοῖσι δὲ πλείστοισιν ἀκρίτως ἐξέλειπεν. ὁμοίως δὲ ταῦτα συνέπιπτεν τοῖς περιγινομένοισιν καὶ τοῖσιν οὔ. πολλῆς δέ τινος γινομένης ἀκρισίης καὶ ποικιλίης ἐπὶ τῶν νοσημάτων, καὶ μεγίστου μὲν σημείου καὶ κακίστου διὰ τέλεος παρεπομένου τοῖσι πλείστοισιν |

628 ἀποσίτοις εἶναι πάντων γευμάτων, μάλιστα δὲ τούτων, οἷσι καὶ τἆλλα ὀλεθρίως ἔχοι, διψώδεες οὐ λίην ἀκαί-ρως ἦσαν ἐπὶ τοῖσι πυρετοῖσι τούτοισι. γενομένων δὲ χρόνων μακρῶν καὶ πόνων πολλῶν καὶ κακῆς συν-τήξιος ἐπὶ τούτοισιν, ἀποστάσιες ἐγίνοντο, ἢ μέζους ὥστε ὑποφέρειν μὴ δύνασθαι, ἢ μείους, ὥστε μηδὲν ὠφελέειν, ἀλλὰ ταχὺ παλινδρομέειν καὶ ξυνεπείγειν ἐπὶ τὸ κάκιον.

9. Ἦν δὲ τούτοισι τὰ γινόμενα δυσεντεριώδεα, καὶ τεινεσμοί, καὶ λειεντερικοί, καὶ ῥοώδεες. ἔστι δ' οἷσι καὶ ὕδρωπες, μετὰ τούτων καὶ ἄνευ τούτων. ὅ τι δὲ παραγένοιτο τούτων βιαίως, ταχὺ συνῄρει, ἢ πάλιν, ἐπὶ τὸ μηδὲν ὠφελέειν. ἐξανθήματα σμικρὰ καὶ οὐκ ἀξίως τῆς περιβολῆς τῶν νοσημάτων, καὶ ταχὺ πάλιν ἀφανιζόμενα· ἢ παρὰ τὰ ὦτα οἰδήματα μωλυόμενα

8. Now the greatest number of these diseases continued to be protracted, difficult, very irregular, very erratic, and without any critical signs, both in those who were mortally ill and in those who were not. For even if they remitted a little in some patients, they quickly relapsed. In a few of them there was a crisis, the earliest being about the eightieth day, some of these having a relapse, so that in the winter most of them were still ill. The greatest number had no crisis before the disease disappeared. The following occurred in those who recovered just as much as in those who did not: There was a notable absence of any crisis and a great variation in the diseases; the most striking and the worst symptom, which attended the great majority throughout, was a complete loss of appetite for all foods, especially in those who were generally in a dangerous condition; they did not, however, suffer much from unseasonable thirst in these fevers. After long intervals with many pains and pernicious dissolution during the fevers, apostases developed which were either too severe to be endurable, or too slight to be beneficial, so that there was a speedy return of the original symptoms, and an aggravation of the mischief.

9. The symptoms from which these patients suffered were dysenteries, tenesmus, lienteries and fluxes; some had dropsies too, either with or without the other symptoms. Whenever any of these pressed violently they were quickly fatal, or, if the opposite, they provided no relief. Eruptions which were slight and did not match the extent of the diseases, and quickly disappeared again, or swellings beside the ears which subsided and never amounted

---

¹² τινα Jouanna: τινας AVI

630   καὶ οὐδὲν ἀποσημαίνοντα· ἔστι δ' οἷς | ἐς ἄρθρα,
μάλιστα δὲ κατὰ ἰσχίον, ὀλίγοισι κρισίμως ἀπολεί-
ποντα καὶ ταχὺ πάλιν ἐπικρατεύμενα ἐπὶ τὴν ἐξ ἀρχῆς ἕξιν.

    10. Ἔθνῃσκον δὲ[13] πάντων μέν, πλεῖστοι δὲ τούτων,
καὶ τούτων παιδία, ὅσα ἀπὸ γάλακτος ἤδη, καὶ πρε-
σβύτερα, ὀκταέτεα καὶ δεκαέτεα, καὶ ὅσα πρὸ ἥβης.
ἐγίνετο δὲ τούτοισι ταῦτα οὐκ ἄνευ τῶν πρώτων γεγραμμένων, τὰ δὲ πρῶτα πολλοῖσιν ἄνευ τούτων.
μοῦνον δὲ χρηστὸν καὶ μέγιστον τῶν γενομένων σημείων, καὶ πλείστους ἐρρύσατο τῶν ἐόντων ἐπὶ τοῖσι
μεγίστοισι κινδύνοισι, οἷσιν ἐπὶ τὸ στραγγουριῶδες
632   ἐτράπετο καὶ ἐς τοῦτο | ἀποστάσιες ἐγίνοντο. συνέ-
πιπτε δὲ καὶ τὸ στραγγουριῶδες τῇσιν ἡλικίῃσιν
ταύτῃσιν γίνεσθαι μάλιστα. ἐγίνετο δὲ καὶ τῶν ἄλλων πολλοῖσιν ὀρθοστάδην καὶ ἐπὶ τῶν νοσημάτων.
ταχὺ δὲ καὶ μεγάλη τις ἡ μεταβολὴ τούτοισι πάντων
ἐγίνετο. κοιλίαι τε γάρ, καὶ εἰ τύχοιεν ἐφυγραινόμεναι
κακοήθεα τρόπον, ταχὺ συνίσταντο, γεύμασίν τε πᾶσιν ἡδέως εἶχον, οἱ πυρετοὶ πρηέες μετὰ ταῦτα.
χρόνια δὲ καὶ τούτοισι τὰ περὶ τὴν στραγγουρίην καὶ
ἐπιπόνως. οὖρα δὲ τούτοισιν ᾔει πολλὰ παχέα[14] ποικίλα καὶ ἐρυθρά, μιξόπυα[15] μετ' ὀδύνης. περιεγένοντο
δὲ πάντες οὗτοι, καὶ οὐδένα τούτων οἶδα ἀποθανόντα.

    11. (5 L.) Ὁκόσα δὲ[16] διὰ κινδύνων, πεπασμοὺς τῶν
634   ἀπιόντων πάντας | πάντοθεν ἐπικαίρους ἢ καλὰς καὶ

---

[13] Add. ἐκ VI   [14] Add. καὶ VI

to anything: in some patients the swellings migrated to the joints, especially the hip joint; in a few instances they went away with a crisis, or in other instances they quickly re-established themselves in their original site.

10. From all the diseases some patients died, but the greatest number from these fevers, especially children—those just weaned, older children of eight or ten years, and those approaching puberty. These latter symptoms did not appear in the absence of the first ones I have described above, but the first ones occurred in many patients without the latter ones. The unique good sign and the most signifi-cant of those that occurred, which saved very many pa-tients who were in the greatest danger, was when there was a turn to strangury and apostases developed in this way. The strangury, too, developed mostly in patients of the ages mentioned, although it did happen in many of the others, either when they were up or when they were ill. A great change rapidly took place in every aspect of their disease, since the bowels, even if they happened to be charged with pernicious fluids, quickly recovered, the pa-tients' appetite for everything returned, and after that the fevers were mild. But the symptoms of the strangury, even in these cases, endured for a long time and were trouble-some. Their urine passed in great amounts, thick, varied, red, mixed with pus, and painful. But they all survived, and I know of none of these that died.

11. In all dangerous cases look for all favorable con-coctions in the evacuations from all sources, or for fair

---

κρισίμους ἀποστάσιας σκοπεῖσθαι. πεπασμοὶ ταχυ
τῆτα κρίσιος, ἀσφάλειαν ὑγιέα[17] σημαίνουσιν. ὠμὰ
δὲ καὶ ἄπεπτα καὶ ἐς κακὰς ἀποστάσιας τρεπόμενα
ἀκρισίας ἢ πόνους ἢ χρόνους ἢ θανάτους ἢ τῶν
αὐτῶν ὑποστροφάς. ὅ τι δὲ τούτων ἔσται μάλιστα,
σκεπτέον ἐξ ἄλλων. λέγειν τὰ προγενόμενα, γινώ
σκειν τὰ παρεόντα, προλέγειν τὰ ἐσόμενα· μελετᾶν
636   ταῦτα. ἀσκεῖν περὶ δύο τὰ νοσήματα·[18] ὠφελέειν | ἢ
μὴ βλάπτειν. ἡ τέχνη διὰ τριῶν, τὸ νόσημα καὶ ὁ
νοσέων καὶ ὁ ἰητρός· ὁ ἰητρὸς ὑπηρέτης τῆς τέχνης·
ὑπεναντιοῦσθαι τῷ νοσήματι τὸν νοσέοντα μετὰ τοῦ
ἰητροῦ.

12. (6 L.) Τὰ περὶ κεφαλὴν καὶ τράχηλον ἀλγήματα
καὶ βάρεα μετ' ὀδύνης ἄνευ πυρετῶν καὶ ἐν πυρετοῖσι·
φρενιτικοῖσι μὲν σπασμοί, καὶ ἰώδεα ἐπανεμεῦσιν,
ἔνιοι ταχυθάνατοι τούτων. ἐν καύσοισι δὲ καὶ τοῖς
ἄλλοισι πυρετοῖσιν, οἷσι μὲν τραχήλου πόνος καὶ
κροτάφων βάρος καὶ σκοτώδεα τὰ[19] περὶ τὰς ὄψιας
καὶ ὑποχονδρίου ξύντασις μετ' ὀδύνης γίνεται, τούτοι
638   σιν αἱμορραγέει | διὰ ῥινῶν· οἷσι δὲ βάρεα μὲν ὅλης
τῆς κεφαλῆς, καρδιωγμοὶ δὲ καὶ ἀσώδεές εἰσιν, ἐπα
νεμέουσιν χολώδεα καὶ φλεγματώδεα. τὸ πολὺ δὲ
παιδίοισιν ἐν τοῖσι τοιούτοισιν οἱ σπασμοὶ μάλιστα,
γυναιξὶ δὲ καὶ ταῦτα καὶ ἀπὸ ὑστερέων πόνοι, πρε
σβυτέροισι δὲ καὶ ὅσοις ἤδη τὸ θερμὸν κρατέεται,
παραπληγικὰ ἢ μανικὰ ἢ στερήσις ὀφθαλμῶν.

170

and critical apostases. Concoctions signify proximity of the crisis and sure recovery of health; but crude and unconcocted humors turning into unfavorable apostases portend the failure of crisis to appear, and pain, prolonged illness, and death, or relapse of the same symptoms: but which of these is most likely to occur must be learned from different signs. Declare the past, recognize the present, foretell the future: attend to these things. As to diseases, make a habit of two things—to help, or at least to do no harm. The art has three factors, the disease, the patient, the physician. The physician is the servant of the art. The patient must cooperate with the physician in combating the disease.

12. Pains about the head and neck, and heaviness combined with pain, occur both when fever is absent and when it is present. Sufferers from phrenitis have convulsions, and afterward vomit material like verdigris; some of these die very quickly. But in ardent and the other fevers, those with pain in the neck, heaviness of the temples, obscurity of vision, and painful tension of the hypochondrium, bleed from the nose; those with heaviness of the whole head, heartburn, and nausea, later vomit material with bile and phlegm. Generally children in this state are most likely to suffer from convulsions; women have both these symptoms and pains in the uterus; older people and those whose natural heat is already failing have paralysis, raving, or loss of vision.

---

[17] ὑγιέα Potter: ὑγιεῖν A: ὑγείην VI
[18] δύο τὰ νοσήματα A: τὰ νοσήματα δύο VI
[19] τὰ om. VI

Κατάστασις τρίτη

13. (7 L.) Ἐν Θάσῳ, πρὸ ἀρκτούρου ὀλίγον καὶ ἐπ᾽ ἀρκτούρου, ὕδατα πολλὰ μεγάλα ἐν βορείοισι. περὶ δὲ ἰσημερίην καὶ μέχρι πληϊάδος, νότια ὕσματα ὀλίγα. χειμὼν βόρειος, αὐχμοί, ψύχεα, πνεύματα με-
640 γάλα, χιόνες. περὶ δὲ ἰσημερίην | χειμῶνες μέγιστοι. ἔαρ βόρειον· αὐχμοί, ὕσματα ὀλίγα, ψύχεα. περὶ δὲ ἡλίου τροπὰς θερινάς, ὕδατα ὀλίγα, μεγάλα ψύχεα μέχρι κυνὸς ἐπλησίασε. μετὰ δὲ κύνα μέχρι ἀρκτού-ρου, θέρος θερμόν· καύματα μεγάλα καὶ οὐκ ἐκ προσ-αγωγῆς, ἀλλὰ συνεχέα καὶ βίαια· ὕδωρ οὐκ ἐγένετο· ἐτησίαι ἔπνευσαν. περὶ ἀρκτοῦρον, ὕσματα νότια μέ-χρι ἰσημερίης.

14. (8 L.) Ἐν τῇ καταστάσει ταύτῃ, κατὰ χειμῶνα μὲν ἤρξαντο παραπληγίαι καὶ πολλοῖσιν ἐγίνοντο· καί τινες αὐτῶν ἔθνησκον διὰ ταχέων· καὶ γὰρ ἄλλως τὸ νόσημα ἐπίδημον ἦν· τὰ δὲ ἄλλα διετέλεον ἄνοσοι. πρωῒ δὲ τοῦ ἦρος ἤρξαντο καῦσοι καὶ διετέλεον μέχρι
642 ἰσημερίης καὶ πρὸς τὸ θέρος. ὅσοι μὲν | οὖν ἦρος καὶ θέρεος ἀρξαμένου[20] αὐτίκα νοσέειν ἤρξαντο, οἱ πλεῖ-στοι διεσώζοντο· ὀλίγοι δέ τινες ἔθνησκον. ἤδη δὲ τοῦ φθινοπώρου καὶ τῶν ὑσμάτων γενομένων, θανατώδεες ἦσαν καὶ πλείους ἀπώλλυντο.

Ἦν δὲ τὰ παθήματα τῶν καύσων· οἶσι μὲν καλῶς καὶ δαψιλέως ἐκ ῥινῶν αἱμορραγῆσαι, διὰ τούτου μάλιστα σῴζεσθαι· καὶ οὐδένα οἶδα, εἰ καλῶς αἱμορ-

Third Catastasis

13. In Thasos a little before and at the time of (sc. the rising of) Arcturus frequent violent rains with northerly winds. About the equinox until (sc. the cosmic setting of) the Pleiades infrequent southerly rains. Winter northerly, droughts, cold periods, strong winds, snow. About the equinox very severe storms. Spring northerly, droughts, infrequent rains, periods of cold. About the summer solstice light showers, periods of great cold until the approach of Sirius. After (sc. the rising of) Sirius, until (the rising of) Arcturus, hot summer, great heat with no variation, but continuous and intense. No rain fell. The Etesian winds blew. About Arcturus southerly rains until the equinox.

14. In this catastasis, during winter paralyses began, which attacked many, some of whom quickly died. In fact, the disease was also present through the whole populace, but otherwise the populace was spared from disease. Early in spring ardent fevers began which continued until the equinox and on into summer. Now those who in spring or at the beginning of summer suddenly began to be ill, in most cases survived, although a few died. But when autumn and the rains came, the cases were dangerous, and more died.

The symptoms of the ardent fevers were as follows: Patients in whom there was a proper and copious bleeding from the nose were generally saved by this; in fact I do not know of a single case in this catastasis that proved fatal

---

[20] ἀρξαμένου Froben: -μενοι AV: -μενος I

ραγῆσαι, ἐν τῇ καταστάσει ταύτῃ ἀποθανόντα. Φιλί-
σκῳ γὰρ καὶ Ἐπαμίνονι καὶ Σιληνῷ τεταρταίῳ καὶ
πεμπταίῳ σμικρὸν ἀπὸ ῥινῶν ἔσταξεν· ἀπέθανον. οἱ
μὲν οὖν πλεῖστοι τῶν νοσησάντων περὶ κρίσιν ἐπερρίγουν, καὶ μάλιστα οἷσι μὴ αἱμορραγῆσαι· ἐπερρίγουν δὲ καὶ οὗτοι καὶ ἐφίδρουν.

644   15. Ἔστι δὲ οἷσιν ἴκτεροι ἑκταῖοις, | ἀλλὰ τούτοις
ἢ κατὰ κύστιν κάθαρσις ἢ κοιλίη ἐκταραχθεῖσα
ὠφέλει ἢ δαψιλὴς αἱμορραγίη, οἷον Ἡρακλείδῃ, ὃς
κατέκειτο παρὰ Ἀριστοκύδει. καίτοι τούτῳ καὶ ἐκ ῥινῶν ἡμορράγησε, καὶ ἡ κοιλίη ἐπεταράχθη, καὶ κατὰ
κύστιν ἐκαθήρατο· ἐκρίθη εἰκοσταῖος· οὐχ οἷον ὁ Φα-
ναγόρεω οἰκέτης, ᾧ οὐδὲν τούτων ἐγένετο· ἀπέθανεν.
ἡμορράγει δὲ τοῖσι πλείστοισι, μάλιστα δὲ μειρα-
κίοισι καὶ ἀκμάζουσι· καὶ ἔθνησκον πλεῖστοι τούτων,
οἷσι μὴ αἱμορραγῆσαι. πρεσβυτέροισι δὲ ἐς ἰκτέρους
ἢ κοιλίαι ταραχώδεες ἢ δυσεντεριώδεες,[21] οἷον Βίωνι
τῷ παρὰ Σιληνὸν κατακειμένῳ. ἐπεδήμησαν δὲ καὶ
δυσεντερίαι κατὰ θέρος· καί τισι καὶ τῶν διανοσησάντων, οἷσι καὶ αἱμορραγίαι ἐγένοντο, ἐς δυσεντεριώδεα
ἐτελεύτησεν, οἷον τῷ Ἐράτωνος παιδὶ καὶ Μύλλῳ·
πολλῆς αἱμορραγίης γενομένης ἐς δυσεντεριώδεα κα-
646   τέστη· | περιεγένοντο.

Πολὺ[22] μὲν οὖν μάλιστα οὗτος ὁ χυμὸς ἐπεπόλα-
σεν· ἐπεὶ καὶ οἷσι περὶ κρίσιν οὐχ ἡμορράγησεν,
ἀλλὰ παρὰ τὰ ὦτα ἐπαναστάντα ἠφανίσθη—τούτων
δὲ ἀφανισθέντων παρὰ τὸν κενεῶνα βάρος τὸν ἀρι-

when a proper bleeding occurred. For it was Philiscus, Epaminon and Silenus, who had only a slight epistaxis on the fourth and fifth days, that died. Now the majority of the patients had rigors near the crisis, especially patients that had no epistaxis; these had sweats, too, as well as the rigors.

15. Some had jaundice on the sixth day, but these were benefited by either a cleaning through the bladder or a disturbance of the bowels, or by a copious hemorrhage, as with Heraclides, who lay sick at the house of Aristocydes. This patient, in fact, not only bled from the nose, but also experienced a disturbance of the bowels and a cleaning through the bladder; he had a crisis on the twentieth day. Not so with the servant of Phanagoras, who had none of these symptoms, and died. But the majority had hemorrhages, especially the adolescents and those in the prime of life, and most of those who died had had no hemorrhage. Older people had jaundice or disordered bowels, or dysentery, as for example Bion, who lay sick at the house of Silenus. Dysenteries were also present through the populace during the summer. And some patients who had had hemorrhages ended up with dysentery, as for example happened to the slave of Erato and to Myllus, who after a copious hemorrhage lapsed into dysentery: they recovered.

This humor (i.e., blood) then, in particular, was in great abundance, since even those who had no hemorrhage near the crisis, but swellings beside the ears which later disappeared—and after their disappearance were followed by

---

21 ἢ δυσεντεριώδεες VI: om. A

22 Πολὺ A: πολὺς V: πολλοι (sic) I: πολλοῖς recc.

στερὸν καὶ ἄκρον ἰσχίον—ἀλγήματος μετὰ κρίσιν
γενομένου καὶ οὔρων λεπτῶν διεξιόντων, αἱμορραγέ-
ειν σμικρὰ ἤρξατο περὶ τετάρτην καὶ εἰκοστήν, καὶ
ἐγένοντο ἐς αἱμορραγίην ἀποστάσιες· Ἀντιφῶντι
Κριτοβούλου ἀπεπαύσατο καὶ ἐκρίθη τελέως περὶ
τεσσαρακοστήν.

16. Γυναῖκες δὲ ἐνόσησαν μὲν πολλαί, ἐλάσσους
δὲ ἢ ἄνδρες, καὶ ἔθνησκον ἧσσον. ἐδυστόκεον δὲ αἱ
πλεῖσται, καὶ μετὰ τοὺς τόκους ἐπενόσεον, καὶ ἔθνησκον αὗται μάλιστα, οἷον ἡ Τελεβούλου θυγάτηρ
ἀπέθανεν ἑκταίη ἐκ τόκου. τῇσι μὲν οὖν πλείστῃσιν
ἐν τοῖσι πυρετοῖσι γυναικεῖα ἐπεφαίνετο, ἔστι δ᾽ ἧσιν
648   ἡμορράγησεν ἐκ ῥινῶν· καὶ παρθένοισι | πολλῇσι
τότε πρῶτον ἐγένετο· ἔστι δ᾽ ὅτε καὶ ἐκ ῥινῶν καὶ τὰ
γυναικεῖα τῇσιν αὐτῇσιν ἐπεφαίνετο, οἷον τῇ Δαϊθάρσεος θυγατρὶ παρθένῳ ἐπεφάνη τότε πρῶτον καὶ ἐκ
ῥινῶν λάβρον ἐρρύη· καὶ οὐδεμίαν οἶδα ἀποθανοῦ-
σαν, ᾗσι τούτων τι καλῶς γένοιτο. ᾗσι δὲ συνεκύρη-
σεν ἐν γαστρὶ ἐχούσῃσι νοσῆσαι, πᾶσαι ἀπέφθειραν,
ἃς καὶ ἐγὼ οἶδα.

17. Οὖρα δὲ τοῖσι πλείστοισιν εὔχροα μέν, λεπτὰ
δὲ καὶ ὑποστάσιας ὀλίγας ἔχοντα· κοιλίαι δὲ ταραχώδεες τοῖσι πλείστοισι, διαχωρήμασι λεπτοῖσι καὶ χολώδεσι. πολλοῖσι δέ, τῶν ἄλλων κεκριμένων, ἐς δυσεντερίας ἐτελεύτα, οἷον Ξενοφάνει καὶ Κριτίᾳ. οὖρα δὲ
ὑδατώδεα πολλὰ[23] καὶ λεπτὰ καὶ μετὰ κρίσιν, καὶ
ὑποστάσιος καλῆς γενομένης καὶ τῶν ἄλλων καλῶς
650   κεκριμένων· ἀναμνήσομαι | οἷσιν ἐγένετο Βίωνι, ὃς

heaviness in the left flank and the extremity of the hip—after the crisis had pain, passed thin urine, and then about the twenty-fourth day began to hemorrhage a little, with apostases emptying into the hemorrhage. In the case of Antiphon, the son of Critobulus, the illness ceased and came to a complete crisis about the fortieth day.

16. Though many women fell ill, they were fewer than the men, and they died in smaller numbers. Most however had difficult childbirths, and after giving birth they would fall ill, and these in particular died, as did the daughter of Telebulus on the sixth day after her delivery. Now the menses appeared during these fevers in most cases, and some women bled from the nose; in many girls the menses occurred then for the first time. Sometimes both epistaxis and menstruation appeared in the same women; for example, the young daughter of Daïtharses had her first menstruation then and also a violent discharge from her nose. I know of no woman who died if any of these symptoms showed themselves properly; those who chanced to fall ill while they were pregnant all had abortions, as far as I can tell.

17. Urines in most cases were of a good color, but thin and with a few sediments, and the bowels of most were disordered with thin, bilious excretions. After a crisis of the other symptoms, many ended up with dysentery, as did Xenophanes and Critias. I will mention cases in which copious, watery, clear and thin urines were passed, even after a crisis in other respects favorable, and with the urines containing a favorable sediment: Bion who lay sick

---

23 Add. καθαρὰ VI

κατέκειτο παρὰ Σιληνόν, Κράτιδι τῇ παρὰ Ξενοφά-
νεος, Ἀρέτωνος παιδί, Μνησιστράτου γυναικί. μετὰ
δὲ δυσεντεριώδεες ἐγένοντο οὗτοι πάντες. ἦρα ὅτι οὔ-
ρησαν ὑδατώδεα, σκεπτέον.

Περὶ δὲ ἀρκτοῦρον ἑνδεκαταίοισι πολλοῖσιν ἔκρινε,
καὶ τούτοισιν οὐδ' αἱ κατὰ λόγον γινόμεναι ὑποστροφαὶ ὑπέστρεφον· ἦσαν δὲ καὶ κωματώδεες περὶ τὸν
χρόνον τοῦτον, πλείω δὲ παιδία, καὶ ἔθνησκον ἥκιστα
οὗτοι πάντων.

18. (9 L.) Περὶ δὲ ἰσημερίην καὶ μέχρι πληϊάδος
καὶ ὑπὸ χειμῶνα παρείποντο μὲν οἱ καῦσοι· ἀτὰρ
καὶ οἱ φρενιτικοὶ τηνικαῦτα πλεῖστοι ἐγένοντο, καὶ
ἔθνησκον τούτων οἱ πλεῖστοι. ἐγένοντο δὲ καὶ κατὰ
θέρος ὀλίγοι. τοῖσι μὲν οὖν καυσώδεσιν ἀρχομένοισιν
ἐπεσήμαινεν, οἷσι τὰ ὀλέθρια ἐνέπιπτεν·[24] αὐτίκα
652 γὰρ ἀρχομένοισι πυρετὸς ὀξύς· | σμικρὰ ἐπερρίγουν,
ἄγρυπνοι, διψώδεες, ἀσώδεες· σμικρὰ ἐφίδρουν περὶ
μέτωπον καὶ κληῗδας, οὐδεὶς δι' ὅλου· πολλὰ παρ-
έλεγον, φόβοι, δυσθυμίαι, ἄκρεα περίψυχρα, πόδες
ἄκροι, μᾶλλον δὲ τὰ περὶ χεῖρας. οἱ παροξυσμοὶ ἐν
ἀρτίῃσι· τοῖσι δὲ πλείστοισι τεταρταίοισιν οἱ πόνοι
μέγιστοι καὶ ἱδρὼς ἐπὶ πλεῖστον ὑπόψυχρος καὶ
ἄκρεα οὐκέτι ἀνεθερμαίνοντο, ἀλλὰ πελιδνὰ καὶ ψυ-
χρά· οὐδ' ἐδίψων ἔτι· ἐπὶ τούτοισιν οὖρα τούτοις
ὀλίγα, μέλανα, λεπτά· καὶ κοιλίαι τούτοις[25] ἐφίσταντο·
οὐδ' ἡμορράγησεν ἐκ ῥινῶν οὐδενί, οἷσι ταῦτα συμ-
654 πίπτοι, ἀλλὰ | σμικρὰ ἔσταξεν· οὐδ' ἐς ὑποστροφὴν
οὐδενὶ τούτων ἦλθεν, ἀλλ' ἑκταῖοι ἀπέθνησκον σὺν

178

at the house of Silenus, Cratia who lodged with Xenophanes, the slave of Areto, and the wife of Mnesistratus: afterward all these suffered from dysentery. Is it that they
passed watery urines? This must be investigated.

Toward (sc. the rising of) Arcturus many had a crisis on
the eleventh day, and these did not suffer the normal
relapses. There were also somnolent fevers about this
time, usually in children, and these died the least of all.

18. About the equinox and until (sc. the cosmic setting
of) the Pleiades, as well as during winter, although the
ardent fevers continued, yet cases of phrenitis became the
most frequent ailment at this time, and most of these patients died—in the summer, too, a few cases of phrenitis
had occurred. Now the sufferers from ardent fever who
displayed fatal symptoms showed these signs at the beginning: for immediately from the time they became ill there
was acute fever accompanied by slight rigors, sleeplessness, thirst, nausea, slight sweats about the forehead and
collarbones but in no case over the whole body, much
delirium, fears, depression, and very cold extremities—
toes and hands, especially the latter. The exacerbations
occurred on even days; but in most cases the pains were
greatest on the fourth day, with sweat generally tending to
be cold, and the extremities could no longer be warmed
back up, but remained livid and cold; the thirst disappeared. Besides, their urine was scanty, dark, and thin, and
their bowels were constipated. Nor were there hemorrhages from the nose in any case when these symptoms
occurred, or at most a slight epistaxis. None of these cases
suffered a relapse, but they died on the sixth day, with

---

²⁴ ἐνέπιπτεν A: ξυνέπ- VI   ²⁵ τούτοις om. VI

ἱδρῶτι. τοῖσι δὲ φρενιτικοῖσι συνέπιπτε μὲν καὶ τὰ
ὑπογεγραμμένα πάντα, ἔκρινε δὲ τούτοισιν ὡς τὸ
πολὺ ἑνδεκαταίοισιν· ἔστι δ' οἷσι καὶ εἰκοσταίοισι. |

656   19. Πλῆθος μὲν οὖν τῶν νοσημάτων ἐγένετο, ἐκ δὲ
τῶν καμνόντων ἀπέθνησκον μάλιστα μειράκια, νέοι,
ἀκμάζοντες, λεῖοι, ὑπολευκόχρωτες, ἰθύτριχες, μελανότριχες, μελανόφθαλμοι, οἱ εἰκῇ καὶ ἐπὶ τὸ ῥάθυμον
βεβιωκότες, ἰσχνόφωνοι, τρηχύφωνοι, τραυλοί, ὀρ-
γίλοι. καὶ γυναῖκες πλεῖσται ἐκ τούτου τοῦ εἴδεος
ἀπέθνησκον. ἐν δὲ τῇ καταστάσει ταύτῃ ἐπὶ σημείων
μάλιστα τεσσάρων διεσῴζοντο· οἷσι γὰρ ἢ διὰ ῥινῶν
καλῶς αἱμορραγῆσαι ἢ κατὰ κύστιν οὖρα πολλὰ καὶ
658 πολλὴν καὶ καλὴν | ὑπόστασιν ἔχοντα ἔλθοι, ἢ κατὰ
κοιλίην ταραχώδεα χολώδεσιν ἐπικαίρως, ἢ δυσεντε-
ρικοὶ γενοίατο. πολλοῖσι δὲ συνέπιπτε μὴ ἐφ' ἑνὸς
κρίνεσθαι τῶν ὑπογεγραμμένων σημείων, ἀλλὰ διεξι-
έναι διὰ πάντων τοῖσι πλείστοισι καὶ δοκέειν μὲν
ἔχειν ὀχληροτέρως· διεσῴζοντο δὲ πάντες, οἷσι ταῦτα
συμπίπτοι. γυναιξὶ δὲ καὶ παρθένοισι συνέπιπτε μὲν
καὶ τὰ ὑπογεγραμμένα σημεῖα πάντα· ᾗσι δὲ ἢ τούτων τι καλῶς γένοιτο ἢ τὰ γυναικεῖα δαψιλέως ἐπιφα-
νείη, διὰ τούτων ἐσῴζοντο καὶ ἔκρινε· καὶ οὐδεμίαν
οἶδα ἀπολομένην, ᾗσι τούτων τι καλῶς γένοιτο.
Φίλωνος γὰρ θυγατρὶ ἐκ ῥινῶν λάβρον ἐρρύη, ἑβδομαίη ἐοῦσα ἐδείπνησεν ἀκαιροτέρως· ἀπέθανεν.

    Οἷσιν ἐν πυρετοῖσιν ὀξέσι, μᾶλλον δὲ καυσώδεσιν,
ἀέκουσιν δάκρυα παραρρεῖ, τούτοισιν ἀπὸ ῥινῶν αἱμορραγίην προσδέχεσθαι, ἢν καὶ τἆλλα ὀλεθρίως μὴ

sweating. The cases of phrenitis had all the symptoms described, but their crisis generally occurred on the eleventh day, although in some it was on the twentieth day.

19. Now a number of illnesses occurred, and of their sufferers', death took chiefly adolescents, young people, people in their prime, the smooth, the slightly fair-skinned, the straight-haired, the black-haired, the black-eyed, those who had lived thoughtlessly and carelessly, the thin-voiced, the rough-voiced, those who lisped, and the irascible. Most women who died were of the same type. In this catastasis there were four symptoms in particular that determined recovery: a proper hemorrhage through the nostrils; copious discharges via the bladder of urine with much sediment of a favorable character; the bowels being disordered with bilious evacuations at the right time; the appearance of dysenteric symptoms. The crisis in many cases did not come with only one of the symptoms described, but in most cases with all of them, and the patients appeared to be in a very distressed state: nevertheless all who had these symptoms were saved. Women and girls experienced all the symptoms described, and those in whom either one symptom was particularly favorable or a copious menstruation supervened, were saved by these and had a crisis. In fact I know of no woman who died when any of these symptoms developed favorably, for the daughter of Philo, though she had a violent epistaxis, dined rather unseasonably on the seventh day, and so died.

In acute fevers, but more particularly in ardent fevers, when there is a flow of tears, epistaxis is to be expected if the patients have no other fatal symptoms—when

660 ἔχωσιν—ἐπεὶ τοῖσί γε φλαύρως ἔχουσιν οὐχ | αἱμορ-
ραγίην, ἀλλὰ θάνατον σημαίνει.

20. Τὰ παρὰ τὰ ὦτα ἐν πυρετοῖσιν ἐπαιρόμενα μετ᾽
ὀδύνης ἔστιν οἷσιν ἐκλείποντος τοῦ πυρετοῦ κρισίμως
οὔτε καθίστατο, οὔτε ἐξεπύει· τούτοισι διάρροιαι χολωδέων, ἢ δυσεντερίη, ἢ παχέων οὔρων ὑπόστασις
γενομένη, ἔλυσεν, οἷον Ἑρμίππῳ τῷ Κλαζομενίῳ. τὰ
δὲ περὶ τὰς κρίσιας ἐξ ὧν καὶ διεγινώσκομεν, ἢ ὅμοια
ἢ ἀνόμοια· οἷον οἱ δύο ἀδελφεοί, οἳ ἤρξαντο ὁμοῦ τὴν
αὐτὴν ὥρην· κατέκειντο παρὰ τὸ θέρετρον Ἐπιγένεος
ἀδελφεοί· τούτων τῷ πρεσβυτέρῳ ἔκρινεν ἑκταίῳ, τῷ
δὲ νεωτέρῳ ἑβδομαίῳ. ὑπέστρεψεν ἀμφοτέροισιν ὁμοῦ
662 τὴν αὐτὴν ὥρην, | καὶ διέλιπεν ἡμέρας ⟨ἐξ τῷ μὲν
ἑτέρῳ, τῷ δ᾽ ἑτέρῳ⟩²⁶ πέντε. ἐκ δὲ τῆς ὑποστροφῆς
ἐκρίθη ἀμφοτέροισιν ὁμοῦ τὸ σύμπαν ἑπτακαιδεκα-
ταίοισιν. ἔκρινε δὲ τοῖσι πλείστοισιν ἑκταίοις. διέλειπεν ἕξ· ἐκ δὲ τῶν ὑποστροφέων ἔκρινε πεμπταίοις.
οἷσι δ᾽ ἔκρινεν ἑβδομαίοισι, διέλιπεν ἑπτά, ἐκ δὲ τῆς
ὑποστροφῆς ἔκρινε τριταίοις. οἷσι δ᾽ ἔκρινεν ἑβδομαίοισι, διαλείποντα τρεῖς, ἔκρινεν ἑβδομαίοις. οἷσι
664 δ᾽ ἔκρινεν ἑκταίοισι, διαλείποντα | ἕξ, ἐλάμβανε τρι-
σίν· διέλιπε μίαν, μίαν ἐλάμβανεν, ἔκρινεν, οἷον
Εὐάγοντι τῷ Δαϊθάρσεος. οἷσι δ᾽ ἔκρινεν ἑκταίοισι,

---

²⁶ ἐξ τῷ μὲν ἑτέρῳ, τῷ δ᾽ ἑτέρῳ propos. GalT

1 As the following tabulation shows, the author considers all
the cases to be "similar" because they share a total course of

however they are in a bad way, such weeping portends not a hemorrhage but death.

20. Swellings beside the ears with pain during fevers, in some cases when the fever ceased after a crisis, neither subsided nor suppurated. These were resolved by bilious diarrhea, or dysentery, or the formation of a sediment in thick urine, as in Hermippus of Clazomenae. The features of crises from which we derive our judgments are their similarities or dissimilarities.[1] For example, the two brothers who fell ill at the same time (the brothers of Epigenes) lay ill near the summer place. The elder of these had a crisis on the sixth day, the younger on the seventh. Both suffered a relapse at the same time; then there was an intermission of ‹six days in the one, and in the other of› five days. After the relapse both had a complete crisis together on the seventeenth day. But the great majority had a crisis on the sixth day, with an intermission of six days: after the relapse a crisis followed on the fifth day. Those who had a crisis on the seventh day had an intermission of seven days, with a crisis on the third day after the relapse. Others with a crisis on the seventh day had an intermission of three days, with a crisis on the seventh day. Some who had a crisis on the sixth day had an intermission of six days and a return of the disease for three; then an intermission of one day and a relapse of one, followed by a crisis, as happened to Euagon the son of Daïtharses. Others with a crisis on the sixth day had an

seventeen days, although "dissimilar" in the specific components of the course: $6 + 6 + 5$; $7 + 5 + 5$; $6 + 6 + 5$; $7 + 7 + 3$; $7 + 3 + 7$; $6 + 6 + 3 + 1 + 1$; $6 + 7 + 4$.

διέλειπεν ἑπτά, ἐκ δὲ τῆς ὑποστροφῆς ἔκρινε τεταρ-
ταίοις, οἷον τῇ Ἀγλαΐδου θυγατρί. οἱ μὲν οὖν πλεῖστοι τῶν νοσησάντων ἐν τῇ καταστάσει ταύτῃ, τούτῳ
τῷ τρόπῳ διενόσησαν· καὶ οὐδένα οἶδα τῶν περιγενο-
μένων, ᾧτινι οὐχ ὑπέστρεψαν αἱ κατὰ λόγον ὑποστροφαὶ γενόμεναι· καὶ διεσῴζοντο πάντες, οὓς κἀγὼ
οἶδα, οἷσιν αἱ ὑποστροφαὶ διὰ τοῦ εἴδεος τούτου γενοίατο. οὐδὲ τῶν διανοσησάντων διὰ τούτου τοῦ τρόπου οὐδενὶ οἶδα ὑποστροφὴν γενομένην πάλιν.

21. Ἔθνησκον δὲ τοῖσι νοσήμασι τούτοις οἱ πλεῖστοι ἑκταῖοι, οἷον Ἐπαμίνων[δας][27] καὶ Σιληνὸς καὶ
666 Φιλίσκος ὁ Ἀνταγόρεω. οἷσι δὲ | τὰ παρὰ τὰ ὦτα
γενοίατο, ἔκρινε μὲν εἰκοσταίοισι, κατέσβη δὲ πᾶσι
καὶ οὐκ ἐξεπύησεν, ἀλλ᾽ ἐπὶ κύστιν ἐτράπετο. Κρατι-
στώνακτι, ὃς παρ᾽ Ἡρακλεῖ ᾤκει, καὶ Σκύμνου τοῦ
γναφέως θεραπαίνῃ ἐξεπύησεν· ἀπέθανον οἷσι δ᾽
ἔκρινεν ἑβδομαίοισι, διέλειπεν ἐννέα, ὑπέστρεφεν,
ἔκρινεν ἐκ τῆς ὑποστροφῆς τεταρταίοισι, Πάντακλεῖ,
ὃς ᾤκει παρὰ Διονύσιον. οἷσι δ᾽ ἔκρινεν ἑβδομαίοισι,
διέλειπεν ἕξ· ὑποστροφή, ἐκ δὲ τῆς ὑποστροφῆς ἔκρινεν ἑβδομαίοισι, Φανοκρίτῳ, ὃς κατέκειτο παρὰ Γνά-
θωνι τῷ γναφεῖ.

22. Ὑπὸ δὲ χειμῶνα περὶ ἡλίου τροπὰς χειμερινὰς
καὶ μέχρι ἰσημερίης παρέμενον μὲν καὶ οἱ καῦσοι καὶ
τὰ φρενιτικά, καὶ ἔθνησκον πολλοί, αἱ μέντοι κρίσιες
μετέπεσον· καὶ ἔκρινε τοῖσι πλείστοισιν ἐξ ἀρχῆς

[27] Ἐπαμίνων Jouanna: Ἐπαμε(ι)νώνδας AVI

intermission of seven days, and after the return of the disease a crisis on the fourth day, as happened to the daughter of Aglaïdas. Now most of those who fell ill in this constitution went through their illness in this order; and none of those who recovered, so far as I know, failed to suffer relapses according to the pattern of these cases; all, so far as I know, recovered if their relapses took place after this fashion. Further, I know of none who suffered a fresh relapse after going through the illness in the way described.

21. In these diseases most died on the sixth day, as did Epaminon, Silenus and Philiscus the son of Antagoras. Those who had the swellings beside the ears had a crisis on the twentieth day; the swellings all subsided without suppuration, being diverted to the bladder. In Cratistonax, who lived near (sc. the temple of) Heracles, and the serving maid of the fuller Scymnus the swellings suppurated: they died. Those in whom there was a crisis on the seventh day, had an intermission of nine days followed by a relapse, and there was a second crisis on the fourth day of the relapse, as in the case of Pantacles, who lived by the temple of Dionysus. When there was a crisis on the seventh day, with an intermission of six days followed by a relapse, there was a second crisis on the seventh day after the relapse—in the case of Phanocritus, for example, who lay at the house of Gnathon the fuller.

22. During winter near the time of the winter solstice, and continuing until the equinox, the ardent fevers and phrenitis continued with many deaths, but their crises were different. Most cases had a crisis on the fifth day

668 πεμπταίοισι, | διέλειπεν τέσσαρας, ὑπέστρεφεν· ἐκ δὲ
τῆς ὑποστροφῆς, ἔκρινε πεμπταίοισι, τὸ σύμπαν τεσ-
σαρεσκαιδεκαταίοις. ἔκρινε δὲ παιδίοισιν οὕτω τοῖσι
πλείστοισιν, ἀτὰρ καὶ πρεσβυτέροισιν. ἔστι δὲ οἷσιν
ἔκρινεν ἑνδεκαταίοισιν· ὑποστροφὴ τεσσαρεσκαιδεκα-
ταίοις· ἔκρινε τελέως εἰκοσταίοισι. εἰ δέ τινες ἐπερρίγουν περὶ τὴν εἰκοστήν, τούτοισιν ἔκρινε τεσσαρα-
κοσταίοις. ἐπερρίγουν δ' οἱ πλεῖστοι περὶ κρίσιν τὴν
ἐξ ἀρχῆς· οἱ δ' ἐπιρριγώσαντες ἐξ ἀρχῆς περὶ κρίσιν,
καὶ ἐν τῇσιν ὑποστροφῇσιν ἅμα κρίσει. ἐρρίγουν δ'
ἐλάχιστοι μὲν τοῦ ἦρος, θέρεος πλείους, φθινοπώρου
ἔτι πλείους, ὑπὸ δὲ χειμῶνος πολὺ πλεῖστοι· αἱ δὲ
αἱμορραγίαι ὑπέληγον.

23. (10 L.) Τὰ δὲ περὶ τὰ νοσήματα, ἐξ ὧν διεγινώ-
670 σκομεν, μαθόντες | ἐκ τῆς κοινῆς φύσιος ἁπάντων καὶ
τῆς ἰδίης ἑκάστου· ἐκ τοῦ νοσήματος, ἐκ τοῦ νοσέοντος, ἐκ τῶν προσφερομένων, ἐκ τοῦ προσφέροντος—
ἐπὶ τὸ ῥᾷστον γὰρ καὶ χαλεπώτατον[28] ἐκ τούτου[29]—ἐκ
τῆς καταστάσιος ὅλης καὶ κατὰ μέρεα τῶν οὐρανίων
καὶ χώρης ἑκάστης, ἐκ τοῦ ἔθεος, ἐκ τῆς διαίτης, ἐκ
τῶν ἐπιτηδευμάτων, ἐκ τῆς ἡλικίης ἑκάστου· λόγοισι,
τρόποισι, σιγῇ, διανοήμασιν, ὕπνοισιν, οὐχ ὕπνοισιν,
ἐνυπνίοισι οἵοισι καὶ ὅτε, τιλμοῖσι, κνησμοῖσι, δάκρυ-
σιν· ἐκ τῶν παροξυσμῶν διαχωρήμασιν, οὔροισιν,
πτυάλοισιν, ἐμέτοισι· καὶ ὅσαι ἐξ οἵων ἐς οἷα διαδο-
χαὶ νοσημάτων καὶ ἀποστάσιες ἐπὶ τὸ ὀλέθριον καὶ

---

[28] ῥᾷστον . . . χαλεπώτατον A: ῥᾷον . . . χαλεπώτερον VI

from the onset, then intermitted four days, relapsed, had a crisis on the fifth day of the relapse, that is, on the fourteenth day in total. In children the crises were for the most part thus, but also in older people. In some there was a crisis on the eleventh day, a relapse on the fourteenth, and a definitive crisis on the twentieth. But if some patients had a rigor about the twentieth day, their crisis came on the fortieth day. Most had rigors around the first crisis from the disease's onset, and in those who did so (sc. there were) also rigors again in the relapses at the time of the crisis. Fewest experienced rigors in the spring, more in summer, more still in autumn, but by far the most during winter. Hemorrhages, however, gradually stopped.

23. The following were the circumstances in the diseases, from which I framed my judgments, learning from the common nature of all and the particular nature of the individual; from the disease, the ill person, the regimen prescribed and the prescriber—for a case may tend toward being very easy or very difficult on this account—from the constitution, both as a whole and with respect to specific factors of the weather and of each region; from the custom, regimen, inclinations, and age of each patient; from words, behaviors, silence, thoughts, sleep or its absence, the nature and time of dreams, pluckings, scratchings, and tears shed; from exacerbations, stools, urine, sputa, vomitus; from which disease came first and which followed which in the successions of diseases, and from apostases leading to a fatal issue or to a crisis; from sweat,

---

29 τούτου A²: τοῦτο A: τούτων VI

κρίσιμον· ἱδρώς, ῥῖγος, ψύξις, βήξ, πταρμοί, λυγμοί,
πνεύματα, ἐρεύξιες, φῦσαι σιγώδεες ψοφώδεες, αἱμορραγίαι, αἱμορροΐδες. ἐκ τούτων καὶ ὅσα διὰ τούτων
σκεπτέον.

24. (11 L.) Πυρετοὶ οἱ μὲν συνεχέες, οἱ δ᾽ ἡμέρην
672 ἔχουσι, | νύκτα διαλείπουσι, νύκτα ἔχουσιν, ἡμέρην
διαλείπουσιν· ἡμιτριταῖοι, τριταῖοι, τεταρταῖοι, πεμ-
πταῖοι, ἑβδομαῖοι, ἐναταῖοι. εἰσὶ δὲ ὀξύταται μὲν καὶ
μέγισται καὶ χαλεπώταται νοῦσοι καὶ θανατωδέστα-
674 ται ἐν τῷ ξυνεχεῖ | πυρετῷ. ἀσφαλέστατος δὲ πάντων
καὶ ῥήϊστος καὶ μακρότατος πάντων ὁ τεταρταῖος· οὐ
γὰρ μόνος αὐτὸς ἐφ᾽ ἑωυτοῦ τοιοῦτός ἐστιν, ἀλλὰ καὶ
νοσημάτων ἑτέρων μεγάλων ῥύεται. ἐν δὲ τῷ ἡμιτρι-
ταίῳ καλεομένῳ ξυμπίπτει μὲν καὶ ὀξέα νοσήματα
γίνεσθαι, καὶ ἔστι τῶν λοιπῶν οὗτος θανατωδέστα-
τος· ἀτὰρ καὶ φθινώδεες καὶ ὅσοι ἄλλα μακρότερα
νοσήματα νοσέουσιν, ἐπὶ τούου μάλιστα νοσέουσι.
νυκτερινὸς οὐ λίην θανατώδης, μακρὸς δέ. ἡμερινὸς
μακρότερος· ἔστι δ᾽ οἷσι ῥέπει καὶ ἐπὶ τὸ φθινῶδες.
ἑβδομαῖος μακρός, οὐ θανατώδης. ἐναταῖος ἔτι μα-
κρότερος, οὐ θανατώδης. τριταῖος ἀκριβὴς ταχυκρί-
σιμος καὶ οὐ θανατώδης. ὁ δὲ πεμπταῖος πάντων μὲν
κάκιστος· καὶ γὰρ πρὸ φθίσιος καὶ ἤδη φθίνουσιν
ἐπιγινόμενος κτείνει.

25. Εἰσὶ δὲ τρόποι καὶ καταστάσιες καὶ παροξυ-
676 σμοὶ | τούτων ἑκάστου τῶν πυρετῶν. αὐτίκα γὰρ συν-
εχής· ἔστιν οἷσιν ἀρχόμενος ἀνθεῖ καὶ ἀκμάζει μάλιστα καὶ ἀνάγει ἐπὶ τὸ χαλεπώτατον, περὶ δὲ κρίσιν

rigor, chill, cough, sneezes, hiccoughs, breathing, belchings, flatulence silent and noisy, hemorrhages, and hemorrhoids. Of these things and their effects consideration must also be taken.

24. Fevers: Some are continuous, some have an access during the day and an intermission during the night, others have an access during the night and an intermission during the day; semitertians, tertians, quartans, quintans, septans, nonans. The most acute, severe, difficult and fatal diseases are those with a continuous fever. The safest of all fevers and the easiest and longest of them is the quartan; not only is it such alone in itself, but it also checks other serious diseases. In the fever called semitertian, acute diseases are also likely to occur, and this fever is the deadliest of the rest; furthermore consumptives and those who suffer from other, longer diseases are befallen mainly by this fever. The nocturnal fever is not very fatal, but it is long. The diurnal fever is even longer, and in some cases there is a tendency toward consumption. The septan is long but not fatal. The nonan is even longer but not fatal. The exact tertian has a speedy crisis and is not fatal. But the quintan is the worst of all, for if it comes on before consumption or during consumption, it is fatal.

25. Each of these fevers has its behaviors, its peculiar catastases, and its exacerbations. For example, a continuous fever: in some cases, from its very beginning it is fully developed and at its high point, raised to its severest state,

καὶ ἅμα κρίσει λεπτύνεται·[30] ἔστι δ᾽ οἷσιν ἄρχεται
μαλακῶς καὶ ὑποβρύχια, ἐπαναδιδοῖ δὲ καὶ παροξύνεται καθ᾽ ἡμέρην ἑκάστην, περὶ δὲ κρίσιν ἅλις
ἐξέλαμψεν· ἔστι δ᾽ οἷσιν ἀρχόμενος πρηέως ἐπιδιδοῖ
καὶ παροξύνεται, καὶ μέχρι τινὸς ἀκμάσας, πάλιν
ὑφίησι μέχρι κρίσιος καὶ περὶ κρίσιν. συμπίπτει δὲ
ταῦτα γίνεσθαι ἐπὶ παντὸς πυρετοῦ καὶ νοσήματος.
δεῖ δὲ τὰ διαιτήματα σκοπεύμενον ἐκ τούτων προσφέρειν. πολλὰ δὲ καὶ ἄλλα ἐπίκαιρα σημεῖα τούτοις
ἐστὶν ἠδελφισμένα, περὶ ὧν τὰ μέν πού τι γέγραπται,
678 τὰ δὲ καὶ γεγράψεται. | πρὸς ἃ[31] διαλογιζόμενον δοκι-
μάζειν καὶ σκοπεῖσθαι, τίνι τούτων ὀξὺ καὶ θανατῶδες ἢ περιεστικόν, καὶ τίνι μακρὸν καὶ θανατῶδες ἢ
περιεστικὸν καὶ τίνι[32] προσαρτέον ἢ οὔ, καὶ πότε, καὶ
πόσον, καὶ τί τὸ προσφερόμενον ἔσται.

26. (12 L.) Τὰ δὲ παροξυνόμενα ἐν ἀρτίῃσι κρίνεται
ἐν ἀρτίῃσιν· ὧν δὲ οἱ παροξυσμοὶ ἐν περισσῇσι, κρίνεται ἐν περισσῇσιν.[33] ἔστι δὲ πρώτη περίοδος τῶν
ἐν τῇσιν ἀρτίῃσι κρινόντων τετάρτη· ἕκτῃ, ὀγδόῃ,
δεκάτῃ, τεσσαρεσκαιδεκάτῃ, εἰκοστῇ, τετάρτῃ καὶ
εἰκοστῇ, τριακοστῇ, τεσσαρακοστῇ, ἑξηκοστῇ, ὀγδο-
680 ηκοστῇ, | εἰκοστῇ καὶ ἑκατοστῇ.[34] τῶν δ᾽ ἐν τῇσι
περισσῇσι κρινόντων περίοδος πρώτη, τρίτῃ· πέμπτῃ,
ἑβδόμῃ, ἐνάτῃ, ἑνδεκάτῃ, ἑπτακαιδεκάτῃ, εἰκοστῇ
πρώτῃ, εἰκοστῇ ἑβδόμῃ, τριακοστῇ πρώτῃ.[35] εἰδέναι

---

[30] λεπτύνεται A: ἀπολεπτύνεται VI   [31] Add. δεῖ VI
[32] μακρὸν καὶ θανατῶδες ἢ περιεστικὸν καὶ τίνι om. VI

190

but around and just at its crisis it moderates; in other cases it begins gently and in a subdued manner, but increases and becomes more acute each day, and bursts out violently near the crisis. In yet other cases it begins mildly, strengthens and becomes more acute, but then after reaching some particular point it declines again until the crisis or near the crisis. These things can happen in any fever and in any disease. It is necessary to take these factors into account when prescribing a patient's regimen. Many other important signs are also related to these, some of which I have described, and others I shall describe later. These must be weighed when deciding and considering in which patients the disease will be acute and fatal, or survivable, and in whom it will be chronic and fatal, or survivable; and in whom food must be prescribed or not, and when, and in what amount, and what it should be.

26. When the exacerbations of diseases are on even days, the crises are on even days, while diseases exacerbated on odd days have their crises on odd days. The first period of diseases with crises on the even days lasts until the fourth day: then the sixth, eighth, tenth, fourteenth, twentieth, twenty-fourth, thirtieth, fortieth, sixtieth, eightieth, hundred and twentieth. Of those with a crisis on the odd days the first period lasts until the third day: then the fifth, seventh, ninth, eleventh, seventeenth, twenty-first, twenty-seventh, thirty-first. Further, one must know

---

33 ἐν περισσῆσιν AV: om. I

34 τετάρτῃ, ἕκτῃ . . . εἰκοστῇ καὶ ἑκατοστῇ Jouanna: δ΄, ϛ΄ . . . κ΄ καὶ ρ΄ A: τέταρτη, ἕκτη . . . ἑκατοστὴ εἰκοστή VI

35 τρίτῃ, πέμπτῃ . . . τριακοστῇ πρώτῃ Jouanna: γ΄, ε΄ . . . λα΄ AI: τρίτη, πέμπτη . . . τριακοστὴ πρώτη V

δὲ χρὴ ὅτι ἢν ἄλλως κριθῇ ἔξω τῶν ὑπογεγραμμένων,
ἐσομένας ὑποστροφάς· γένοιτο δὲ ἂν καὶ ὀλέθρια. δεῖ
δὴ προσέχειν τὸν νόον καὶ εἰδέναι ἐν τοῖσι χρόνοισι
τούτοισι τὰς κρίσιας ἐσομένας ἐπὶ σωτηρίην ἢ τὸ
ὄλεθρον, ἢ ῥοπὰς ἐπὶ τὸ ἄμεινον ἢ τὸ χεῖρον. πλάνη-
682 τες δὲ πυρετοὶ καὶ τεταρταῖοι | καὶ πεμπταῖοι καὶ
ἑβδομαῖοι καὶ ἐναταῖοι, ἐν ᾗσι περιόδοισι κρίνονται,
σκεπτέον.

Περὶ ἀρρώστων[36]

27. (13 L.) α΄. Φιλίσκος ᾤκει παρὰ τὸ τεῖχος· κατεκλίνη.

Τῇ πρώτῃ πυρετὸς ὀξύς, ἵδρωσεν, ἐς νύκτα ἐπιπόνως.

Δευτέρῃ πάντα παρωξύνθη· ὀψὲ δὲ ἀπὸ κλυσματίου
καλῶς διῆλθε· νύκτα δι᾽ ἡσυχίης.

Τρίτῃ πρωῒ καὶ μέχρι μέσου ἡμέρης ἔδοξε γενέσθαι ἄπυρος· πρὸς δείλην δὲ πυρετὸς ὀξὺς μετὰ ἱδρῶ-
τος, διψώδης, γλῶσσα ἐπεξηραίνετο, μέλανα οὔρησε·
νύκτα δυσφόρως, οὐκ ἐκοιμήθη, πάντα παρέκρουσε.

Τετάρτῃ πάντα παρωξύνθη, οὖρα μέλανα· νύκτα
εὐφορωτέρην, οὖρα εὐχροώτερα.

Πέμπτῃ περὶ μέσον ἡμέρης σμικρὸν ἀπὸ ῥινῶν
ἔσταξεν ἄκρητον· οὖρα δὲ ποικίλα, ἔχοντα ἐναιω-
684 ρήματα στρογγύλα, | γονοειδέα, διεσπασμένα, οὐχ
ἱδρύετο· προσθεμένῳ δὲ βάλανον φυσώδεα σμικρὰ
διῆλθε· νύκτα ἐπιπόνως, ὕπνοι σμικροί, λόγοι, λῆρος,

that if crises occur on days other than the ones named there will be relapses, and maybe even fatalities. One must pay attention and know that at these times crises will supervene leading to recovery or death, or to changes for the better or the worse. As for irregular fevers, quartans, quintans, septans and nonans, the periods in which they have their crises must be considered.

Cases a1–a14

27. (1) Philiscus lived by the wall; he took to his bed.

First day. Acute fever and sweating; night uncomfortable.

Second day. General exacerbation; later a small clyster moved the bowels well. Restful night.

Third day. Early and until midday he appeared to be without fever; but toward evening acute fever with sweating; thirst; dry tongue; he passed dark urine; uncomfortable night, without sleep; completely out of his mind.

Fourth day. All symptoms exacerbated; urines dark; more comfortable night; urines of a better color.

Fifth day. About midday slight epistaxis of unmixed blood; urines varied, with scattered, round particles suspended in them, resembling semen; they did not settle. On the application of a suppository the patient passed, with flatulence, a little excreta. A distressing night, spells of sleep, talking; wandering of the mind; extremities ev-

---

36 Περὶ Ἀρρώστων I: ἄρρωστος α᾽ V: om. A

ἄκρεα πάντοθεν ψυχρὰ καὶ οὐκέτι ἀναθερμαινόμενα,
οὔρησε μέλανα, ἐκοιμήθη σμικρὰ πρὸς ἡμέρην· ἄφωνος, ἵδρωσε ψυχρῷ, ἄκρεα πελιδνά.

Περὶ δὲ μέσον ἡμέρης ἑκταῖος ἀπέθανεν.

Τούτῳ πνεῦμα διὰ τέλεος, ὥσπερ ἀνακαλεομένῳ,
ἀραιὸν γὰρ μέγα·[37] σπλὴν ἐπήρθη περιφερεῖ κυρ-
τώματι, ἱδρῶτες ψυχροὶ διὰ τέλεος, οἱ παροξυσμοὶ ἐν
ἀρτίῃσιν.

β΄. Σιληνὸς ᾤκει ἐπὶ τοῦ πλαταμῶνος πλησίον τῶν
Εὐαλκίδεω. ἐκ κόπων καὶ πότων καὶ γυμνασίων
ἀκαίρων πῦρ ἔλαβεν. ἤρξατο δὲ πονέειν κατ᾽ ὀσφῦν·
καὶ κεφαλῆς βάρος καὶ τραχήλου ξύντασις.

Ἀπὸ δὲ κοιλίης τῇ πρώτῃ χολώδεα, ἄκρητα, ἔπαφρα, κατακορέα πολλὰ διῆλθεν· οὖρα μέλανα, μέλαι-
ναν ὑπόστασιν ἔχοντα, διψώδης, γλῶσσα ἐπίξηρος·
νυκτὸς οὐδὲν | ἐκοιμήθη.

Δευτέρῃ πυρετὸς ὀξύς, διαχωρήματα πλείω, λεπτό-
τερα, ἔπαφρα, οὖρα μέλανα· νύκτα δυσφόρως, σμικρὰ
παρέκρουσε.

Τρίτῃ πάντα παρωξύνθη· ὑποχονδρίου ξύντασις ἐξ
ἀμφοτέρων παραμήκης πρὸς ὀμφαλόν, ὑπολάπαρος·
διαχωρήματα λεπτά, ὑπομέλανα, οὖρα θολερά, ὑπο-
μέλανα· νυκτὸς οὐδὲν ἐκοιμήθη, λόγοι πολλοί, γέλως,
ᾠδή, κατέχειν οὐκ ἠδύνατο.

Τετάρτῃ διὰ τῶν αὐτῶν.

Πέμπτῃ διαχωρήματα ἄκρητα, χολώδεα, λεῖα, λιπαρά, οὖρα λεπτά, διαφανέα· σμικρὰ κατενόει.

erywhere cold, and difficult to warm back up; he passed dark urine; spells of sleep toward dawn; speechless; cold sweat; extremities livid.

Sixth day, around noon the patient died.

The breathing throughout was as though the patient was recollecting to do it: rare and large. Spleen raised in a round swelling; cold sweats all the time. The exacerbations on even days.

(2) Silenus lived on the flat rock near the place of Eualcides. After overexertion, drinking, and exercises at the wrong time, fever seized him. He began by having pains in the loins, with heaviness in the head and tightness of the neck.

First day. From the bowels discharges of bilious, unmixed, frothy, and highly colored material passed in copious amounts; urines dark, with a dark sediment; thirst; tongue dry; no sleep at night.

Second day. Acute fever, stools more copious, thinner, frothy; urines dark; uncomfortable night; slightly out of his mind.

Third day. General exacerbation; tightness of the hypochondrium on both sides lengthwise toward the navel, slightly concave; stools thin, darkish; urines turbid, darkish; no sleep at night; much talk, laughter, singing; could not keep himself still.

Fourth day. About the same.

Fifth day. Stools unmixed, bilious, smooth, greasy; urines thin, transparent; lucid intervals.

---

37 ἀνακαλουμένῳ . . . μέγα A: ἀνακαλουμέγα V: ἀνακαλουμένῳ ἀραιὸν μέγα I

Ἕκτῃ περὶ κεφαλὴν σμικρὰ ἐφίδρωσεν, ἄκρεα ψυχρά, πελιδνά, πολὺς βληστρισμός, ἀπὸ κοιλίης οὐδὲν διῆλθεν, οὖρα ἐπέστη, πυρετὸς ὀξύς.

Ἑβδόμῃ ἄφωνος, ἄκρεα οὐκέτι ἀνεθερμαίνετο, οὔρησεν οὐδέν.

Ὀγδόῃ ἵδρωσε δι᾽ ὅλου ψυχρῷ· ἐξανθήματα μετὰ ἱδρῶτος ἐρυθρά, στρογγύλα, σμικρὰ οἷον ἴονθοι, παρέμενεν, οὐ καθίστατο· ἀπὸ δὲ κοιλίης ἐρεθισμῷ σμικρῷ κόπρανα λεπτά, οἷα ἄπεπτα, πολλὰ διῄει μετὰ πόνου· οὔρει μετ᾽ ὀδύνης | δακνώδεα· ἄκρεα σμικρὰ ἀνεθερμαίνετο, ὕπνοι λεπτοί, κωματώδης, ἄφωνος, οὖρα λεπτὰ διαφανέα.

Ἐνάτῃ διὰ τῶν αὐτῶν.

Δεκάτῃ ποτὰ κατεδέχετο,[38] κωματώδης, ὕπνοι λεπτοί· ἀπὸ δὲ κοιλίης ὅμοια, οὔρησεν ἀθρόον ὑπόπαχυ· κειμένῳ ὑπόστασις κριμνώδης λευκή, ἄκρεα πάλιν ψυχρά.

Ἑνδεκάτῃ ἀπέθανεν.

Ἐξ ἀρχῆς τούτῳ καὶ διὰ τέλεος πνεῦμα ἀραιόν, μέγα· ὑποχονδρίου παλμὸς συνεχής, ἡλικίη ὡς περὶ ἔτεα εἴκοσιν.

γ΄. Ἡροφῶντι πυρετὸς ὀξύς, ἀπὸ κοιλίης ὀλίγα, τεινεσμώδεα κατ᾽ ἀρχάς, μετὰ δὲ λεπτὰ διῄει χολώδεα, ὑπόσυχνα· ὕπνοι οὐκ ἐνῆσαν, οὖρα μέλανα λεπτά.

Πέμπτῃ πρωῒ κώφωσις, παρωξύνθη πάντα, σπλὴν ἐπήρθη, ὑποχονδρίου ξύντασις, ἀπὸ κοιλίης ὀλίγα διῆλθε μέλανα, παρεφρόνησεν.

Sixth day. Slight sweats about the head; extremities cold and livid; much tossing; nothing passed from the bowels; urines suppressed; acute fever.

Seventh day. Speechless; extremities could no longer be warmed up; passed no urine.

Eighth day. Cold sweat over the whole body; red skin eruptions with sweat—round, small like acne—which persisted and did not subside. From the bowels with slight stimulus a discharge of stools, thin and seemingly unconcocted, passed in a large amount with straining. He passed irritating urines accompanied by pain. Extremities warmed up a little; fitful sleep; somnolent; speechless; thin, transparent urines.

Ninth day. About the same.

Tenth day. He accepted no drinks; somnolent; fitful sleep. Discharges from the bowels as before; had a copious discharge of thickish urine, which on standing left a farinaceous, white deposit; extremities again cold.

Eleventh day. He died.

From the beginning and through to the end in this case, the breathing was rare and large. Continuous quivering of the hypochondrium; age about twenty years.

(3) Herophon had acute fever; scanty stools with tenesmus at the beginning, afterward becoming thin, bilious and fairly frequent. No sleep; urines dark and thin.

Fifth day. Early in the morning deafness; general exacerbation; spleen swollen; tension of the hypochondrium; from the bowels a little dark material passed; he lost his reason.

---

³⁸ κατεδέχετο A: οὐκ ἐδέχετο VI

Ἕκτῃ ἐλήρει, ἐς νύκτα ἱδρώς, ψύξις, παράληρος
παρέμενεν.

690 Ἑβδόμῃ περιέψυκτο, διψώδης, | παρέκρουσε· νύκτα
κατενόει, κατεκοιμήθη.

Ὀγδόῃ ἐπύρεσσε, σπλὴν ἐμειοῦτο, κατενόει πάντα,
ἤλγησεν τὸ πρῶτον κατὰ βουβῶνα, σπληνὸς κατ᾽
ἴξιν, ἔπειτα δὲ πόνοι ἐς ἀμφοτέρας κνήμας· νύκτα εὐφόρως, οὖρα εὐχροώτερα, ὑπόστασιν εἶχε σμικρήν.

Ἐνάτῃ ἵδρωσεν, ἐκρίθη, διέλειπεν.

Πέμπτῃ ὑπέστρεψεν· αὐτίκα δὲ σπλὴν ἐπήρθη, πυρετὸς ὀξύς, κώφωσις πάλιν.

Μετὰ δὲ τὴν ὑποστροφὴν τρίτῃ σπλὴν ἐμειοῦτο,
κώφωσις ἦσσον, σκέλεα ἐπωδύνως· νύκτα ἵδρωσεν.
ἐκρίθη περὶ ἑπτακαιδεκάτην· οὐδὲ παρέκρουσεν ἐν τῇ
ὑποστροφῇ.

δ΄. Ἐν Θάσῳ Φιλίνου γυναῖκα, θυγατέρα τεκοῦσαν,
κατὰ φύσιν καθάρσιος γενομένης καὶ τὰ ἄλλα κού-
φως διάγουσαν, τεσσαρεσκαιδεκαταίην ἐοῦσαν μετὰ
τὸν τόκον, πῦρ ἔλαβε μετὰ ῥίγεος· ἤλγεε δὲ ἀρχομένη
καρδίην καὶ ὑποχόνδριον δεξιόν· γυναικείων πόνοι·
κάθαρσις ἐπαύσατο. προσθεμένῃ δὲ ταῦτα μὲν ἐκου-
692 φίσθη, κεφαλῆς | δὲ καὶ τραχήλου καὶ ὀσφύος πόνοι
παρέμενον· ὕπνοι οὐκ ἐνῆσαν, ἄκρεα ψυχρά, διψώδης,
κοιλίη συνεκαύθη, σμικρὰ διῄει, οὖρα λεπτά, ἄχρω
κατ᾽ ἀρχάς.

Ἑκταίη ἐς νύκτα παρέκρουσε πολλά, καὶ πάλιν
κατενόει.

Sixth day. Delirious; at night sweat and chill; the delirium persisted.

Seventh Day. Chill all over; thirst; out of his mind. During the night he became rational, and slept.

Eighth day. Fever; spleen shrank; completely rational; pain at first in the groin, on the side of the spleen; then the pains extended to both lower legs. Night comfortable; urines of a better color, with a little deposit.

Ninth day. Sweat, crisis, intermission.

Fifth day (sc. after the crisis) the patient relapsed. Immediately the spleen swelled; acute fever; return of deafness.

Third day after the relapse the spleen shrank and the deafness diminished, but legs painful. During the night he sweated. The crisis was about the seventeenth day. There was no delirium during the relapse.

(4) In Thasos the wife of Philinus gave birth to a daughter; her (sc. lochial) cleaning occurred according to nature, and the mother was doing well when on the fourteenth day after her delivery she was seized with fever accompanied by rigor. At first she suffered discomfort in the cardia and the right hypochondrium, and then pains in her genitalia. The cleaning ended. By the application (sc. of a pessary) her troubles were lightened, but pains persisted in her head, neck and loins. No sleep; extremities cold; thirst; bowels burned; passed a few stools; urines thin, and at first colorless.

Sixth day. At night frequent delirium, followed by recovery of reason.

Ἑβδόμῃ διψώδης, διαχωρήματα ὀλίγα χολώδεα κατακορέα.

Ὀγδόῃ ἐπερρίγωσε, πυρετὸς ὀξύς, σπασμοὶ πολλοὶ μετὰ πόνου· πολλὰ παρέλεγεν, ἐξανίστατο· βάλανον προσθεμένη πολλὰ διῆλθε μετὰ περιρρόου χολώδεος· ὕπνοι οὐκ ἐνῆσαν.

Ἐνάτῃ σπασμοί.

Δεκάτῃ σμικρὰ κατενόει.

Ἑνδεκάτῃ ἐκοιμήθη, πάντων ἀνεμνήσθη, ταχὺ δὲ πάλιν παρέκρουσεν· οὔρει δὲ μετὰ σπασμῶν ἀθρόον πολὺ ὀλιγάκις ἀναμιμνησκόντων παχὺ λευκόν, οἷον γίνεται ἐκ τῶν καθισταμένων, ὅταν ἀναταραχθῇ· κείμενον πολὺν χρόνον οὐ καθίστατο· χρῶμα καὶ πάχος ἴκελον οἷον γίνεται ὑποζυγίου. τοιαῦτα οὔρει, οἷα κἀγὼ εἶδον.

Περὶ δὲ τεσσαρεσκαιδεκάτην ἐούσῃ παλμὸς δι' ὅλου τοῦ σώματος, λόγοι πολλοί, σμικρὰ κατενόει· διὰ ταχέων δὲ πάλιν παρέκρουσεν.

694   Περὶ δὲ | ἑπτακαιδεκάτην ἐοῦσα ἄφωνος.

Εἰκοστῇ ἀπέθανεν.

ε΄. Ἐπικράτεος γυναῖκα, ἣ κατέκειτο παρὰ ἀρχηγέτην, περὶ τόκον[39] ἐοῦσαν ῥῖγος ἔλαβεν ἰσχυρῶς· οὐκ ἐθερμάνθη, ὡς ἔλεγον. καὶ τῇ ὑστεραίῃ τὰ αὐτά. τρίτῃ δ' ἔτεκεν θυγατέρα καὶ τἆλλα πάντα κατὰ λόγον ἦλθε.

Δευτεραίην μετὰ τὸν τόκον ἔλαβε πυρετὸς ὀξύς· καρδίης πόνος καὶ γυναικείων. προσθεμένη δὲ ταῦτα μὲν ἐκουφίσθη· κεφαλῆς δὲ καὶ τραχήλου καὶ ὀσφύος

Seventh day. Thirst; stools few, bilious, highly colored.

Eighth day. Rigor; acute fever; many painful convulsions; much delirium. She got up, and with the application of a suppository copious stools passed with a bilious coating on them. No sleep.

Ninth day. Convulsions.

Tenth day. She regained her reason a little.

Eleventh day. Slept; she remembered everything, but then quickly lost her reason again. With spasms, she passed urine in large amounts all at once, but this only rarely when her attendants reminded her: it was white and thick, like urine with a sediment that has been shaken; it stood for a long time without forming a sediment; its color and consistency were like those of the urine of cattle. Such was the urine she passed, as I myself saw.

About the fourteenth day there was quivering over the whole body; much talking, slight return to reason, but she quickly became delirious again.

About the seventeenth day she became speechless.

Twentieth day. She died.

(5) The wife of Epicrates, who lay at (sc. the statue of) the founder,[2] when near her delivery was seized with severe rigor without, it was said, becoming warm. Next day the same. On the third day she gave birth to a daughter, and the delivery went completely according to rule.

On the second day after the delivery she was seized with acute fever and pain in the cardia and genitalia. A pessary relieved these symptoms, but there was pain in the

---

[2] Scholars are divided on whether the founder of the city or a person named Archegetes is meant.

πόνος· ὕπνοι οὐκ ἐνῆσαν· ἀπὸ δὲ κοιλίης ὀλίγα χολώδεα λεπτὰ διῄει ἄκρητα· οὖρα λεπτὰ ὑπομέλανα.

Ἀφ' ἧς δὲ ἔλαβε τὸ πῦρ, ἐς νύκτα ἑκταίη παρέκρουσεν.

Ἑβδομαίη πάντα παρωξύνθη, ἄγρυπνος, παρέκρουσεν, διψώδης, διαχωρήματα χολώδεα κατακορέα.

Ὀγδόη ἐπερρίγωσεν καὶ[40] ἐκοιμήθη πλείω.

Ἐνάτη διὰ τῶν αὐτῶν.

Δεκάτη σκέλεα ἐπιπόνως ἤλγει, καρδίης πάλιν
ὀδύνη, καρηβαρίη, οὐ παρέκρουεν, | ἐκοιμᾶτο μᾶλλον,
κοιλίη ἐπέστη.

Ἑνδεκάτη οὔρησεν εὐχροώτερα, ἡσυχῇ ὑπόστασιν
ἔχοντα· διῆγε κουφότερον.

Τεσσαρεσκαιδεκάτη ἐπερρίγωσεν, πυρετὸς ὀξύς.

Πεντεκαιδεκάτη ἤμεσε χολώδεα ξανθὰ ὑπόσυχνα,
ἵδρωσεν ἄπυρος· ἐς νύκτα δὲ πυρετὸς ὀξύς, οὖρα πάχος ἔχοντα, ὑπόστασις λευκή.

Ἑξκαιδεκάτη παρωξύνθη· νύκτα δυσφόρως, οὐχ
ὕπνωσεν, παρέκρουσεν.

Ὀκτωκαιδεκάτη διψώδης, γλῶσσα ἐπεκαύθη, οὐχ
ὕπνωσεν, παρέκρουσε πολλά, σκέλεα ἐπωδύνως εἶχεν.

Περὶ δὲ εἰκοστὴν πρωὶ σμικρὰ ἐπερρίγωσεν, κωματώδης, δι' ἡσυχίης ὕπνωσεν· ἤμεσε χολώδεα ὀλίγα
μέλανα· ἐς νύκτα κώφωσις.

Περὶ δὲ πρώτην καὶ εἰκοστὴν πλευροῦ ἀριστεροῦ
βάρος δι' ὅλου μετ' ὀδύνης, σμικρὰ ὑπέβησσεν. οὖρα
δὲ πάχος ἔχοντα, θολερά, ὑπέρυθρα· κείμενα οὐ καθίστατο· τὰ δ' ἄλλα κουφοτέρως· οὐκ ἄπυρος.

head, neck and loins. No sleep. From the bowels passed a few stools, bilious, thin and unmixed. Urines thin and darkish.

Delirium on the *night of the sixth day* from the day the fever began.

Seventh day. All symptoms exacerbated: sleeplessness, delirium, thirst; bilious, highly-colored stools.

Eighth day. Rigor; more sleep.

Ninth day. The same.

Tenth day. Severe pains in the legs; pain again in the cardia; heaviness in the head; no delirium; more sleep; bowels blocked.

Eleventh day. Urine of better color, with a deposit that gradually settled; she passed the time more easily.

Fourteenth day. Rigor; acute fever.

Fifteenth day. Vomited bilious, yellow material many times; sweated without fever; at night, however, acute fever; urines thick, a white sediment.

Sixteenth day. Exacerbation; an uncomfortable night; no sleep; delirium.

Eighteenth day. Thirst; tongue parched; no sleep; much delirium; pain in the legs.

About the twentieth day. Early in the morning slight rigors; somnolence; quiet sleep; scanty, bilious, dark vomitus; deafness at night.

About the twenty-first day. Heaviness all through the left side, with pain; slight coughing; urines thick, turbid, reddish, no sediment on standing. In other respects easier; but still had fever.

---

40 καὶ A: om. VI

Αὕτη ἐξ ἀρχῆς φάρυγγα ἐπώδυνος· ἔρευθος· κίων ἀνεσπασμένος· ῥεῦμα δριμύ, δακνῶδες, ἁλμυρῶδες | διὰ τέλεος παρέμενεν.

Περὶ δὲ εἰκοστὴν ἑβδόμην ἄπυρος, οὔροισιν ὑπόστασις, πλευρὸν ὑπήλγει.

Περὶ δὲ πρώτην καὶ τριακοστὴν πῦρ ἔλαβεν, κοιλίη χολώδεσιν ἐπεταράχθη.

Τεσσαρακοστῇ ἤμεσεν ὀλίγα χολώδεα.

Ἐκρίθη τελέως ἄπυρος ὀγδοηκοστῇ.

ϛʹ. Κλεανακτίδην, ὃς κατέκειτο ἐπάνω τοῦ Ἡρακλείου, πῦρ ἔλαβε πεπλανημένως· ἤλγει δὲ κεφαλὴν ἐξ ἀρχῆς καὶ πλευρὸν ἀριστερόν, καὶ τῶν ἄλλων πόνοι κοπιώδεα τρόπον· οἱ πυρετοὶ παροξυνόμενοι ἄλλοτ᾽ ἀλλοίως, ἀτάκτως· ἱδρῶτες ὁτὲ μέν, ὁτὲ δ᾽ οὔ· τὰ μὲν πλεῖστα ἐπεσήμαινον οἱ παροξυσμοὶ ἐν κρισίμοισι μάλιστα.

Περὶ δὲ εἰκοστὴν τετάρτην, χεῖρας ἄκρας ἐπόνησεν,[41] ἤμεσε χολώδεα ξανθά, ὑπόσυχνα, μετ᾽ ὀλίγον δὲ ἰώδεα· πάντων ἐκουφίσθη.

Περὶ δὲ τριακοστὴν ἐόντι ἤρξατο ἀπὸ ῥινῶν αἱμορραγέειν ἐξ ἀμφοτέρων, καὶ ταῦτα πεπλανημένως κατ᾽ ὀλίγον μέχρι κρίσιος· οὐκ ἀπόσιτος οὐδὲ διψώδης παρὰ πάντα τὸν χρόνον | οὐδ᾽ ἄγρυπνος· οὖρα δὲ λεπτά, οὐκ ἄχρω.

Περὶ δὲ τεσσαρακοστὴν ἐών, οὔρησεν ὑπέρυθρα ὑπόστασιν πολλὴν ἐρυθρὴν ἔχοντα· ἐκουφίσθη. μετὰ δὲ ποικίλως τὰ τῶν οὔρων· ὁτὲ μὲν ὑπόστασιν εἶχεν, ὁτὲ δ᾽ οὔ.

From the beginning she had pain in the throat with redness; uvula drawn back; an acrid irritating, salty flux persisted to the end.

About the twenty-seventh day. No fever; sediment in the urine; some pain in the side.

About the thirty-first day. Attacked by fever; bowels disordered by a bilious flux.

Fortieth day. A little bilious vomitus.

Eightieth day. Definitive crisis with cessation of fever.

(6) Cleanactides, who lay above the temple of Heracles, was seized by an irregular fever. He had at the beginning pains in the head and the left side, and in the other parts pains like those caused by fatigue. The fevers became acute at different times in different ways in an irregular pattern; sometimes there were sweats, at other times none. Generally the exacerbations manifested themselves mainly on the critical days.

About the twenty-fourth day. Pain at the tips of the hands; he passed bilious, yellow vomitus quite frequently, which after a short time resembled verdigris; general relief.

About the thirtieth day. Epistaxis from both nostrils began, and continued irregularly and in slight amounts until the crisis. He neither lost his appetite for food, nor suffered thirst, during the whole time, nor was he sleepless. Urine thin, but not colorless.

About the fortieth day. He passed reddish urine with an abundant, red deposit, which brought relief. Afterward the urine varied in color, and sometimes it had a sediment, while at other times none.

---

41 ἐπόνησεν A: ἐψύχετο VI

Ἑξηκοστῇ οὔροισιν ὑπόστασις πολλὴ καὶ λευκὴ καὶ λείη· ξυνέδωκε πάντα, πυρετοὶ διέλιπον, οὖρα δὲ πάλιν λεπτὰ μέν, εὔχρω δέ.

Ἑβδομηκοστῇ πυρετὸς διέλειπεν ἡμέρας δέκα.

Ὀγδοηκοστῇ ἐρρίγωσε, πυρετὸς ὀξὺς ἔλαβεν, ἵδρωσεν πολλῷ· οὔροισιν ὑπόστασις ἐρυθρή, λείη. τελέως ἐκρίθη.

ζ'. Μέτωνα πῦρ ἔλαβεν, ὀσφύος βάρος ἐπώδυνον.

Δευτέρῃ ὕδωρ πιόντι ὑπόσυχνον ἀπὸ κοιλίης καλῶς διῆλθε.

Τρίτῃ κεφαλῆς βάρος, διαχωρήματα λεπτά, χολώδεα, ὑπέρυθρα.

702 Τετάρτῃ[42] παρωξύνθη, | ἐρρύη ἀπὸ δεξιοῦ δὶς κατ' ὀλίγον. νύκτα δυσφόρως, διαχωρήματα ὅμοια τῇ τρίτῃ, οὖρα ὑπομέλανα· εἶχεν ἐναιώρημα ὑπομέλαν ἐόν, διεσπασμένον· οὐχ ἱδρύετο.

Πέμπτῃ ἐρρύη λάβρον ἐξ ἀριστεροῦ ἄκρητον, ἵδρωσεν, ἐκρίθη. μετὰ κρίσιν ἄγρυπνος, παρέλεγεν, οὖρα λεπτὰ ὑπομέλανα. λουτροῖσιν ἐχρήσατο κατὰ κεφαλῆς, ἐκοιμήθη, κατενόει.

Τούτῳ οὐχ ὑπέστρεψεν, ἀλλ' ἡμορράγει πολλάκις μετὰ κρίσιν.

η'. Ἐρασῖνον, ὃς[43] ᾤκει παρὰ Βοώτεω χαράδρην, πῦρ ἔλαβεν μετὰ δεῖπνον· νύκτα ταραχώδης.

Ἡμέρην τὴν πρώτην δι' ἡσυχίης· νύκτα ἐπιπόνως.

Δευτέρῃ πάντα παρωξύνθη· ἐς νύκτα παρέκρουσε.

Τρίτῃ ἐπιπόνως, πολλὰ παρέκρουσε.

Τετάρτῃ δυσφορώτατα· ἐς δὲ τὴν νύκτα οὐδὲν ἐκοι-

Sixtieth day. Urines had an abundant sediment, white and smooth; general improvement; fever intermitted; urine again thin but of good color.

Seventieth day. Fever, which intermitted for ten days.

Eightieth day. Rigor; attacked by acute fever; much sweat; in the urines a red, smooth sediment. Definitive crisis.

(7) Meton was seized with fever, and painful heaviness in the loins.

Second day. After drinking a good deal of water, his bowels were well moved.

Third day. Heaviness in the head; stools thin, bilious, reddish.

Fourth day. Exacerbation; slight epistaxis twice from the right nostril. An uncomfortable night; stools as on the third day; urines darkish, with a darkish cloud floating over them that dispersed; no deposit.

Fifth day. Violent epistaxis of unmixed blood from the left nostril; sweat; crisis. After the crisis sleeplessness; delirium; urines thin and darkish. His head was bathed; sleep; reason restored.

The patient suffered no relapse, but had frequent hemorrhages after the crisis.

(8) Erasinus, who lived by the gully of Boötes, was seized with fever after supper; a troubled night.

First day. Quiet, but the night was again troubled.

Second day. General exacerbation; delirium at night.

Third day. Distress and much delirium.

Fourth day. Very uncomfortable; at night no sleep;

---

[42] πάντα add. VI   [43] Ἐρασῖνον, ὃς Jouanna: Ἐράσινον, ὃς V: Ἐρασίνος A: ἐρασινὸς ὃς I

μήθη· ἐνύπνια καὶ λογισμοί· ἔπειτα χείρω, μεγάλα καὶ ἐπίκαιρα· φόβος, δυσφορίη.

Πέμπτη πρωῒ κατήρτητο, κατενόει πάντα· πολὺ δὲ πρὸ μέσου ἡμέρης ἐξεμάνη, κατέχειν οὐκ ἠδύνατο· ἄκρεα ψυχρὰ ὑποπέλια, οὖρα ἐπέστη· ἀπέθανε περὶ ἡλίου δυσμάς.

704   Τούτῳ | πυρετοὶ διὰ τέλεος σὺν ἱδρῶτι, ὑποχόνδρια μετέωρα, σύντασις μετ᾽ ὀδύνης· οὖρα⁴⁴ μέλανα ἔχοντα ἐναιωρήματα στρογγύλα· οὐχ ἱδρύετο· ἀπὸ δὲ κοιλίης κόπρανα διῄει· δίψα διὰ τέλεος, οὐ λίην· σπασμοὶ πολλοὶ ξὺν ἱδρῶτι περὶ θάνατον.

θ΄. Κρίτωνι ἐν Θάσῳ ποδὸς ὀδύνη ἤρξατο ἰσχυρὴ ἀπὸ δακτύλου τοῦ μεγάλου ὀρθοστάδην περιόντι. κατεκλίνη αὐθημερόν, φρικώδης, ἀσώδης, σμικρὰ ὑποθερμαινόμενος· ἐς νύκτα παρεφρόνησεν.

Δευτέρη οἴδημα δι᾽ ὅλου τοῦ ποδὸς καὶ περὶ σφυρὸν ὑπέρυθρον μετὰ ξυντάσιος· φλυκταινίδια μέλανα, πυρετὸς ὀξύς, ἐξεμάνη· ἀπὸ δὲ κοιλίης ἄκρητα, χολώδεα, ὑπόσυχνα διῆλθεν·⁴⁵ ἀπέθανεν ἀπὸ τῆς ἀρχῆς δευτεραῖος.

ι΄. Τὸν Κλαζομένιον, ὃς κατέκειτο παρὰ τὸ Φρυνιχίδεω φρέαρ, πῦρ ἔλαβε. ἤλγει δὲ κεφαλήν, τράχηλον, ὀσφῦν ἐξ ἀρχῆς· αὐτίκα δὲ κώφωσις, ὕπνοι οὐκ
706   ἐνῆσαν, πυρετὸς ὀξὺς ἔλαβεν, | ὑποχόνδριον ἐπῆρτο μετ᾽ ὄγκου οὐ λίην, ξύντασις, γλῶσσα ξηρή.

Τετάρτη ἐς νύκτα παρεφρόνει.

Πέμπτη ἐπιπόνως.

Ἕκτη πάντα παρωξύνθη.

dreams and talking. Then worse symptoms, of a striking and significant character; fear and discomfort.

Fifth day. Early in the morning he was composed and in complete possession of his senses. But quite a while before midday he became madly delirious; he could not keep himself still; extremities cold and rather livid; urine suppressed; died about sunset.

In this patient the fever was throughout accompanied by sweat; the hypochondria were raised and distended with pain. Urine dark, with round, suspended particles which did not settle. Stools were discharged from the bowels. Thirst throughout not very great. Many convulsions with sweating about the time of death.

(9) Crito, in Thasos, while up and walking about, was seized with a violent pain in the foot beginning from the great toe. He took to bed the same day; shivering, nausea, a little fever; at night he was delirious.

Second day. Swelling of the whole foot, and some redness around the ankle, with distension; dark blisters, acute fever, mad delirium. Alvine discharges passed unmixed, bilious and rather frequent. He died on the second day from the commencement.

(10) The man from Clazomenae, who lay by the well of Phrynichides, was seized with fever. He suffered pain at the beginning in his head, neck and loins, followed immediately by deafness. No sleep; seized with acute fever; hypochondrium raised, but not very much enlargement; distension; tongue dry.

Fourth day. Delirium at night.

Fifth day. Troubled.

Sixth day. All symptoms exacerbated.

---

44 Add. δὲ VI   45 διῆλθεν om. A

Περὶ δὲ ἑνδεκάτην σμικρὰ συνέδωκεν.

Ἀπὸ δὲ κοιλίης ἀπ᾽ ἀρχῆς καὶ μέχρι τεσσαρεσκαι-
δεκάτης λεπτά, πολλά, ὑδατόχολα διῄει· εὐφόρως τὰ
περὶ διαχώρησιν διῆγεν· ἔπειτα κοιλίη ἐπέστη. οὖρα
διὰ τέλεος λεπτὰ μέν, εὔχροα δὲ καὶ πολλά· εἶχεν
ἐναιώρημα ὑποδιεσπασμένον· οὐχ ἱδρύετο.

Περὶ δὲ ἕκτην καὶ δεκάτην οὔρησεν ὀλίγῳ πα-
χύτερα· εἶχε σμικρὴν ὑπόστασιν· ἐκούφισεν ὀλίγῳ·
κατενόει μᾶλλον.

Ἑπτακαιδεκάτη πάλιν λεπτά· παρὰ δὲ τὰ ὦτα ἀμ-
φότερα ἐπήρθη ξὺν ὀδύνῃ· ὕπνοι οὐκ ἐνῆσαν, παρελήρει, σκέλεα ἐπωδύνως εἶχεν.

Εἰκοστῇ ἄπυρος ἐκρίθη, οὐχ ἵδρωσε, πάντα κατενόει.

Περὶ δὲ εἰκοστὴν ἑβδόμην ἰσχίου ὀδύνη δεξιοῦ
ἰσχυρῶς· διὰ ταχέων ἐπαύσατο. τὰ δὲ παρὰ τὰ ὦτα
708 οὔτε καθίστατο οὔτε ἐξεπύει, | ἤλγει δέ.

Περὶ πρώτην καὶ τριηκοστὴν διάρροια πολλοῖσιν
ὑδατώδεσι μετὰ δυσεντεριωδέων· οὖρα παχέα οὔρει·
κατέστη τὰ παρὰ τὰ ὦτα.

Τεσσαρακοστῇ ὀφθαλμὸν δεξιὸν ἤλγει, ἀμβλύτε-
ρον ἑώρα· κατέστη.

ιαʹ. Τὴν Δρομεάδεω γυναῖκα θυγατέρα τεκοῦσαν
καὶ τῶν ἄλλων πάντων γενομένων κατὰ λόγον, δευτεραίην ἐοῦσαν ῥῖγος ἔλαβεν· πυρετὸς ὀξύς.

Ἤρξατο δὲ πονεῖν τῇ πρώτῃ περὶ ὑποχόνδριον·
ἀσώδης, φρικώδης, ἀλύουσα—καὶ τὰς ἐχομένας—οὐχ
ὕπνωσε· πνεῦμα ἀραιόν, μέγα, αὐτίκα ἀνεσπασμένον.

About the eleventh day a slight improvement.

From the beginning to *the fourteenth day* copious, thin stools of a watery biliousness were discharged from the bowels; this was well tolerated by the patient. Then the bowels were blocked. Urines through to the end were thin, but of a good color and copious; they had a cloud that spread somewhat through them, but did not settle.

About the sixteenth day the urines were somewhat thicker, and had a little sediment. The patient became a little easier, and regained more of his reason.

Seventeenth day. Urines thin again; swellings beside both ears, with pain. No sleep; delirium; pain in the legs.

Twentieth day. Fever remitted and the crisis occurred; no sweating; complete recovery of reason.

About the twenty-seventh day violent pain in the right hip, which quickly ceased. The swellings by the ears neither subsided nor suppurated, but continued painful.

About the thirty-first day diarrhea with copious, watery discharges and signs of dysentery. He passed thick urines; the swellings beside his ears went down.

Fortieth day. Pain in the right eye; vision rather impaired. Recovery.

(11) The wife of Dromeades gave birth to a daughter, and, the delivery going otherwise completely according to rule, was seized on the second day with rigor; acute fever.

On the first day she began to feel pain in the region of the hypochondrium; nausea; shivering; restlessness—also on the following days—she did not sleep. Respiration rare, large, interrupted by a sudden inspiration.

Δευτέρῃ ἀφ᾿ ἧς ἐρρίγωσεν, ἀπὸ κοιλίης καλῶς κόπρανα διῆλθεν· οὖρα παχέα, λευκά, θολερά, οἷα γίνεται ἐκ τῶν καθισταμένων, ὅταν ἀναταραχθῇ· κείμενα χρόνον πολὺν οὐ καθίστατο. νύκτα οὐκ ἐκοιμήθη.

Τρίτῃ περὶ μέσον ἡμέρης ἐπερρίγωσε· πυρετὸς ὀξύς, οὖρα ὅμοια, ὑποχονδρίου πόνος, ἀσώδης· νύκτα 710 δυσφόρως, οὐκ ἐκοιμήθη· ἵδρωσε | δι᾿ ὅλου ψυχρῷ, ταχὺ δὲ πάλιν ἀνεθερμάνθη.

Τετάρτῃ περὶ ὑποχόνδριον σμικρὸν ἐκουφίσθη· κεφαλῆς δὲ βάρος μετ᾿ ὀδύνης· ὑπεκαρώθη· σμικρὰ ἀπὸ ῥινῶν ἔσταξε· γλῶσσα ἐπίξηρος, διψώδης, οὖρα σμικρὰ λεπτὰ ἐλαιώδεα· σμικρὰ ἐκοιμήθη.

Πέμπτῃ διψώδης, ἀσώδης, οὖρα ὅμοια, ἀπὸ κοιλίης οὐδέν· περὶ δὲ μέσον ἡμέρης πολλὰ παρέκρουσε καὶ πάλιν ταχὺ σμικρὰ κατενόει· ἀνισταμένη ὑπεκαρώθη, ψύξις· σμικρὰ νυκτὸς ἐκοιμήθη· παρέκρουσεν.

Ἕκτῃ πρωῒ ἐπερρίγωσεν, ταχὺ διεθερμάνθη, ἵδρωσε δι᾿ ὅλου· ἄκρεα ψυχρά, παρέκρουσεν, πνεῦμα μέγα, ἀραιόν· μετ᾿ ὀλίγον σπασμοὶ ἀπὸ κεφαλῆς ἤρξαντο· ταχὺ ἀπέθανεν.

ιβʹ. Ἄνθρωπος θερμαινόμενος ἐδείπνησεν καὶ ἔπιε πλέον. ἤμεσε πάντα νυκτός· πυρετὸς ὀξύς, ὑποχονδρίου δεξιοῦ πόνος, φλεγμονὴ ὑπολάπαρος ἐκ τοῦ ἔσω μέρεος· νύκτα δυσφόρως· οὖρα δὲ κατ᾿ ἀρχὰς 712 πάχος ἔχοντα, ἐρυθρά· | κείμενα οὐ καθίστατο· γλῶσσα ἐπίξηρος, οὐ λίην διψώδης.

Τετάρτῃ πυρετὸς ὀξύς, πόνοι πάντων.

Second day from rigor. Healthy discharge of stools from the bowels. Urines thick, white, and turbid like urine with a sediment that has been shaken; it stood for a long time without forming a sediment. No sleep at night.

Third day. At about midday she had a rigor; acute fever; urines as before; pain in the hypochondrium; nausea; an uncomfortable night without sleep; a cold sweat over the whole body, but the patient was quickly warmed up again.

Fourth day. About the hypochondrium slight relief; heaviness of the head with pain; somewhat somnolent; slight epistaxis; tongue dry; thirst; scanty urines, thin and oily; a little sleep.

Fifth day. Thirst; nausea; urines similar; no movement of the bowels; about midday much delirium, followed quickly by a partial recovery of reason; rose, but became somewhat somnolent; chilliness; slept a little at night; was delirious.

Sixth day. Early in the morning she had a rigor; quickly recovered heat; sweated over the whole body; extremities cold; was delirious; respiration large and rare. After a short while convulsions began from the head; she soon died.

(12) A man dined when heated and drank too much. During the night he vomited everything; acute fever; pain in the right hypochondrium; inflammation, with some slackness coming from the inside part; an uncomfortable night; urines at first thick and red; on standing they did not form a sediment; tongue dry; not too thirsty.

Fourth day. Acute fever; pains all over.

Πέμπτῃ οὔρησε λεῖον ἐλαιῶδες πολύ· πυρετὸς ὀξύς.
Ἕκτῃ δείλης πολλὰ παρέκρουσεν. οὐδὲ νύκτα ἐκοιμήθη.

Ἑβδόμῃ πάντα παρωξύνθη· οὖρα ὅμοια, λόγοι πολλοί, κατέχειν οὐκ ἠδύνατο· ἀπὸ δὲ κοιλίης ἐρεθισμῷ ὑγρὰ ταραχώδεα διῆλθεν μετὰ ἐλμίγγων. νύκτα ἐπιπόνως. πρωῒ δ᾽ ἐρρίγωσε· πυρετὸς ὀξύς, ἵδρωσε θερμῷ, ἄπυρος ἔδοξε γενέσθαι· οὐ πολὺ ἐκοιμήθη, ἐξ ὕπνου ψύξις· πτυαλισμός. δείλης πολλὰ παρέκρουσε· μετ᾽ ὀλίγον δὲ ἤμεσε μέλανα ὀλίγα χολώδεα.

Ἐνάτῃ ψύξις, παρελήρει πολλά, οὐχ ὕπνωσεν.

Δεκάτῃ σκέλεα ἐπωδύνως, πάντα[46] παρελήρει.

Ἑνδεκάτῃ ἀπέθανεν.

ιγ′. Γυναῖκα ἣ κατέκειτο ἐν ἀκτῇ, τρίμηνον πρὸς ἑωυτῇ ἔχουσαν, πῦρ ἔλαβεν· αὐτίκα δὲ ἤρξατο πονέειν ὀσφῦν.

Τρίτῃ πόνος τραχήλου καὶ κεφαλῆς κατὰ κληῖδα, 714 χεῖρα δεξιήν· | διὰ ταχέων δὲ γλῶσσα ἠφώνει, δεξιὴ χεὶρ παρελύθη μετὰ σπασμοῦ παραπληγικὸν τρόπον· παρελήρει πάντα. νύκτα δυσφόρως, οὐκ ἐκοιμήθη, κοιλίη ἐπεταράχθη χολώδεσιν ἀκρήτοισιν ὀλίγοισιν.

Τετάρτῃ γλῶσσα ἀσαφὴς ἦν· ἐλύθη· σπασμοί, πόνοι τῶν αὐτῶν παρέμενον· κατὰ ὑποχόνδριον ἔπαρμα σὺν ὀδύνῃ· οὐκ ἐκοιμᾶτο, παρέκρουσε πάντα, κοιλίη ταραχώδης, οὖρα λεπτά, οὐκ εὔχρω.

Πέμπτῃ πυρετὸς ὀξύς, ὑποχονδρίου πόνος, παρ-

---

[46] Add. παρωξύνθη VI

Fifth day. Passed much smooth, oily urine; acute fever.

Sixth day. In the evening much delirium. No sleep at night.

Seventh day. General exacerbation; urines similar; much talking; could not keep himself still; on stimulation the bowels passed watery, disturbed discharges, with worms. An uncomfortable night. *Early in the morning* he had a rigor; acute fever. Then he experienced a hot sweat, and seemed to lose his fever. A little sleep, followed by chilliness; expectoration. In the evening much delirium, and shortly afterward he vomited a little dark, bilious material.

Ninth day. Chill; much delirium; no sleep.

Tenth day. Legs painful; talked complete nonsense.

Eleventh day. Death.

(13) A woman lying by the promontory, who was three months pregnant, was seized with fever, and immediately began to feel pains in her loins.

Third day. Pain in the neck and head, and also in the region of the collarbone and the right arm. Quickly her tongue lost the power of speech; her right arm was paralyzed, in conjunction with a convulsion, after the manner of a stroke; completely delirious. An uncomfortable night; she did not sleep; bowels disordered with bilious, unmixed, stools in a small amount.

Fourth day. Her tongue was indistinct; it (i.e., the paralysis) was relieved, but the convulsions and the pains of the same parts persisted; swelling in the hypochondrium with pain; no sleep; utter delirium; bowels disordered; urines thin, and not of a good color.

Fifth day. Acute fever; pain in the hypochondrium;

ἔκρουε πάντα, διαχωρήματα χολώδεα. ἐς νύκτα ἵδρωσεν, ἄπυρος.

Ἕκτῃ κατενόει, πάντων⁴⁷ ἐκουφίσθη, περὶ δὲ κληῗδα ἀριστερὴν πόνος παρέμενε· διψώδης, οὖρα λεπτά,
οὐκ ἐκοιμήθη.

Ἑβδόμῃ τρόμος, ὑπεκαρώθη, σμικρὰ παρέκρου-
σεν· ἀλγήματα κατὰ κληῗδα καὶ βραχίονα ἀριστερὸν
παρέμενε, τὰ δ' ἄλλα διεκούφισεν· πάντα κατενόει.

716 Τρεῖς | διέλειπεν, ἄπυρος.

Ἑνδεκάτῃ ὑπέστρεψεν, ἐπερρίγωσεν, πῦρ ἔλαβεν.

Περὶ δὲ τεσσαρεσκαιδεκάτην ἤμεσε χολώδεα
ξανθὰ ὑπόσυχνα· ἵδρωσεν, ἄπυρος ἐκρίθη.

ιδ'. Μελιδίῃ, ἣ κατέκειτο παρὰ Ἥρης ἱρόν, ἤρξατο
κεφαλῆς καὶ τραχήλου καὶ στήθεος πόνος ἰσχυρῶς·
αὐτίκα δὲ πυρετὸς ὀξὺς ἔλαβεν· γυναικεῖα δὲ σμικρὰ
ἐπεφαίνετο· πόνοι τούτων πάντων ξυνεχέες.

Ἕκτῃ κωματώδης, ἀσώδης, φρικώδης, ἐρύθημα ἐπὶ
γνάθων, σμικρὰ παρέκρουσεν.

Ἑβδόμῃ ἵδρωσε, πυρετὸς⁴⁸ διέλιπεν, οἱ πόνοι παρέμενον.

Ὑπέστρεψεν· ὕπνοι σμικροί, οὖρα διὰ τέλεος εὔχρω-
μέν, λεπτὰ δέ· διαχωρήματα λεπτά, χολώδεα, δακνώ-
δεα, κατ' ὀλίγα,⁴⁹ μέλανα, δυσώδεα διῆλθεν· οὔροις
ὑπόστασις λευκή, λείη.

Ἵδρωσεν· ἐκρίθη τελέως ἑνδεκάτῃ.

---

⁴⁷ πάντων AI: πάντα· V
⁴⁸ πυρετὸς GalL: ἄπυρος AVI

utter delirium; bilious stools. At night she sweated; without fever.

Sixth day. She regained her reason; general relief, but pain remained around the left collarbone; thirst; urines thin; no sleep.

Seventh day. Trembling; some coma; slight delirium; pains continued in the region of the collarbone and the left upper arm; other symptoms relieved; in complete possession of her reason.

For three days there was an intermission; the patient was afebrile.

Eleventh day. Relapse; rigor; attack of fever.

But *about the fourteenth day* the patient vomited bilious, yellow matter fairly frequently; sweated; fever disappeared; it was the crisis.

(14) Melidia, who lay by the temple of Hera, was befallen by a violent pain beginning in her head, neck and chest. Immediately acute fever set in, and a slight menstrual flow appeared. Continuous pains in all these parts.

Sixth day. Somnolence; nausea; shivering; flushed cheeks; slight delirium.

Seventh day. Sweat; no fever; intermission; the pains persisted.

Relapse; a little sleep; urines throughout of good color but thin; stools thin, bilious, irritating, very scanty, dark in color and foul-smelling; sediment in the urine white and smooth.

She sweat; definitive crisis on the eleventh day.

---

[49] κατ᾽ ὀλίγα Jouanna in *app. crit.*: ὀλίγα A: κάρτα ὀλίγα VI

ΕΠΙΔΗΜΙΩΝ Γ

III 24
Littré

1. α΄. Πυθίωνι, ὃς ᾤκει παρὰ Γῆς ἱρόν, ἤρξατο τρόμος ἀπὸ χειρῶν.

Τῇ πρώτῃ πυρετὸς ὀξύς· λῆρος.

Δευτέρῃ πάντα παρωξύνθη.

Τρίτῃ τὰ αὐτά.

26 Τετάρτῃ ἀπὸ κοιλίης ὀλίγα, | ἄκρητα, χολώδεα διῆλθε.

Πέμπτῃ πάντα παρωξύνθη· ὕπνοι λεπτοί· κοιλίη ἔστη.

Ἕκτῃ πτύαλα ποικίλα, ὑπέρυθρα.

Ἑβδόμῃ στόμα παρειρύσθη.

28 Ὀγδόῃ | πάντα παρωξύνθη, τρόμοι παρέμενον· οὖρα δὲ κατ᾽ ἀρχὰς μὲν καὶ μέχρι τῆς ὀγδόης λεπτά, ἄχροα· ἐναιώρημα εἶχον ἐπινέφελον.

Ἑνδεκάτῃ ἵδρωσε, πτύαλα ὑποπέπονα, ἐκρίθη· οὖρα ὑπόλεπτα περὶ κρίσιν.

Μετὰ δὲ κρίσιν, τεσσαράκοντα ἡμέρῃσιν ὕστερον, ἐμπύημα περὶ ἕδρην, καὶ στραγγουριώδης ἐγένετο ἀπόστασις. |

EPIDEMICS III

Cases b1– b12

1. (1) Pythion, who lived by the temple of the Earth, began to tremble from his hands.

First day. Acute fever; delirium.

Second day. General exacerbation.

Third day. Same.

Fourth day. Stools passed scanty, uncompounded and bilious.

Fifth day. General exacerbation; fitful sleep; bowels blocked.

Sixth day. Varied, reddish sputa.

Seventh day. Mouth drawn awry.

Eighth day. General exacerbation; tremors persisted; urines from the beginning to the eighth day thin, colorless, with cloudy material floating in them.

Eleventh day. He sweat; sputa somewhat concocted; crisis occurred; urine thinnish around the time of the crisis.

After the crisis, forty days subsequent to it, abscess in the seat, and an apostasis by strangury.

32   β΄. Ἑρμοκράτην, ὃς κατέκειτο παρὰ τὸ καινὸν τεῖ-
χος, πῦρ ἔλαβεν. ἤρξατο δὲ ἀλγέειν κεφαλήν, ὀσφῦν·
34 ὑποχονδρίου | ἔντασις λαπαρῶς· γλῶσσα δὲ ἀρχο-
μένῳ ἐπεκαύθη· κώφωσις αὐτίκα· ὕπνοι οὐκ ἐνῆσαν·
διψώδης οὐ λίην· οὔρει παχέα, ἐρυθρά, κείμενα οὐ
καθίστατο· ἀπὸ δὲ κοιλίης ξυγκεκαυμένα διῄει.

Πέμπτῃ οὔρησε λεπτά, εἶχεν ἐναιώρημα, οὐχ
ἵδρυτο, ἐς νύκτα παρέκρουσεν.

Ἕκτῃ ἰκτεριώδης, πάντα παρωξύνθη, οὐ κατενόει.

Ἑβδόμῃ δυσφόρως, οὖρα λεπτά, ὅμοια.

Τὰς ἑπομένας παραπλησίως.

Περὶ δὲ ἑνδεκάτην ἐόντι πάντα ἔδοξε κουφισθῆναι·
κῶμα ἤρξατο, οὔρει παχύτερα, ὑπέρυθρα, κάτω λε-
36 πτά· [οὐ]¹ καθίστατο ἡσυχῇ· | κατενόει.

Τεσσαρεσκαιδεκάτῃ ἄπυρος, οὐχ ἵδρωσεν, ἐκοιμήθη, κατενόει πάντα, οὖρα παραπλήσια.

Περὶ δὲ ἑπτακαιδεκάτην ἐόντι ὑπέστρεψεν· ἐθερμάνθη.

Τὰς ἑπομένας πυρετὸς ὀξύς, οὖρα λεπτά, παρέκρουσε.

Πάλιν δὲ εἰκοστῇ ἐκρίθη, ἄπυρος, οὐχ ἵδρωσεν·
ἀπόσιτος παρὰ πάντα τὸν χρόνον, κατενόει, διαλέγε-
σθαι οὐκ ἠδύνατο, γλῶσσα ἐπίξηρος· οὐκ ἐδύψη· κατ-
εκοιμᾶτο σμικρά, κωματώδης. |

38   Περὶ πρώτην καὶ εἰκοστὴν ἐπεθερμάνθη, κοιλίη
ὑγρὴ πολλοῖσι λεπτοῖσι—καὶ τὰς ἑπομένας—πυρετὸς
ὀξύς, γλῶσσα συνεκαύθη.

Ἑβδόμῃ καὶ εἰκοστῇ ἀπέθανε.

(2) Hermocrates, who lay by the New Wall, was seized with fever. He began to feel pain in the head and loins; tension of the hypochondrium with concavity; tongue at the beginning parched; deafness at once; no sleep; no great thirst; passed thick, red urine, which formed no sediment on standing; bowels passed burned stools.

Fifth day. Passed thin urine with particles floating in it, no deposit; at night delirium.

Sixth day. Jaundice; general exacerbation; did not regain his reason.

Seventh day. Discomfort; urines thin and as before.

The following days similar.

About the eleventh day he seemed to experience general relief; coma began; he passed thicker, reddish urine with a thin deposit that gradually settled; he recovered his reason.

Fourteenth day. No fever; no sweat; slept; reason completely recovered; urines the same.

About the seventeenth day he had a relapse, and became heated.

On the following days there was acute fever; urines thin; delirium.

Twentieth day. A fresh crisis; no fever; no sweat. Lack of appetite the whole time; recovered his reason, but could not talk; tongue dry; no thirst; slept a little; somnolent.

About the twenty-first day he became heated; bowels loose with copious, thin discharges—also on the following days—acute fever; tongue parched.

Twenty-seventh day. He died.

---

[1] Del. Jouanna

Τούτῳ κώφωσις διὰ τέλεος παρέμενεν, οὖρα παχέα
καὶ ἐρυθρά, οὐ καθιστάμενα, ἢ λεπτὰ[2] καὶ ἄχροα[3]
ἐναιώρημα ἔχοντα· γεύεσθαι οὐκ ἠδύνατο.

γ΄. Ὁ κατακείμενος ἐν τῷ Δεάλκεος[4] κήπῳ κεφαλῆς
βάρος καὶ κρόταφον δεξιὸν ἐπωδύνως[5] εἶχε χρόνον
40 πολύν. μετὰ | δὲ προφάσιος πῦρ ἔλαβε, κατεκλίθη.

Δευτέρη ἐξ ἀριστεροῦ ὀλίγον ἄκρητον ἐρρύη· ἀπὸ
δὲ κοιλίης κόπρανα καλῶς διῆλθεν· οὖρα λεπτὰ ποικίλα, ἐναιώρημα ἔχοντα κατὰ σμικρὰ οἷον κρίμνα,
γονοειδέα.

Τρίτη πυρετὸς ὀξύς, διαχωρήματα μέλανα, λεπτά,
ἔπαφρα· ὑπόστασις πελιδνὴ διαχωρήμασιν· ὑπεκα-
ροῦτο, ἐδυσφόρει περὶ τὰς ἀναστάσιας· οὔροις ὑπόστασις πελιδνή, ὑπόγλισχρος.

Τετάρτη ἤμεσε χολώδεα ξανθὰ ὀλίγα, διαλιπὼν
ὀλίγον ἰώδεα· ἐξ ἀριστεροῦ ὀλίγον ἄκρητον ἐρρύη·
διαχωρήματα ὅμοια, οὖρα ὅμοια, ἐφίδρου περὶ κεφαλὴν καὶ κληῖδας· σπλὴν ἐπήρθη· μηροῦ ὀδύνη κατ᾽
ἴξιν, ὑποχονδρίου δεξιοῦ ξύντασις ὑπολάπαρος· νυκτὸς οὐκ ἐκοιμήθη, παρέκρουσε σμικρά.

Πέμπτη διαχωρήματα πλείω, μέλανα, ἔπαφρα,
ὑπόστασις μέλαινα διαχωρήμασι· νύκτα οὐχ ὕπνωσε,
παρέκρουσεν.

Ἕκτη διαχωρήματα μέλανα, λιπαρά, γλίσχρα, δυ-
σώδεα· ὕπνωσε, κατενόει μᾶλλον.

Ἑβδόμη γλῶσσα ἐπίξηρος, διψώδης, οὐκ ἐκοιμήθη, παρέκρουσεν· οὖρα λεπτά, οὐκ εὔχροα.

In this patient deafness persisted throughout; urines thick and red without deposition, or thin and colorless, with material floating in them. The patient was incapable of taking food.

(3) The man lying in the garden of Dealces had for a long time heaviness in the head, and pain in the right temple. From some exciting cause he was seized with fever, and took to his bed.

Second day. Slight flow of unmixed blood from the left (sc. nostril). The bowels functioned well; urines thin and varied, with small particles like barley-meal or semen floating in them.

Third day. Acute fever; stools dark, thin, frothy, with a livid deposit on them; slight stupor; getting up caused distress; in the urine a livid, rather viscous sediment.

Fourth day. Vomited scanty, bilious, yellow material, which after a short interval was like verdigris; slight flow of unmixed blood from the left (sc. nostril); stools and urines remained the same; sweat about the head and collarbones; spleen raised; pain in the thigh on the same side; tension of the right hypochondrium, somewhat concave; at night no sleep; slight delirium.

Fifth day. Stools more copious, dark, frothy; a dark deposit in them; at night no sleep; delirium.

Sixth day. Stools dark, oily, viscid, foul-smelling; slept; regained more of his reason.

Seventh day. Tongue dry; thirsty; no sleep; delirium; urine thin, not of a good color.

---

2 λεπτὰ V: λευκὰ I   3 καὶ add. I
4 Δεάλκεος Littré: Δεάλδεος VI
5 ἐπωδύνως V: ἐπώδυνον I

Ὀγδόη διαχωρήματα μέλανα ὀλίγα, συνεστηκότα· ὕπνωσε, κατενόει, διψώδης οὐ λίην.

Ἐνάτη ἐπερρίγωσε, πυρετὸς ὀξύς, ἵδρωσε, ψύξις, παρέκρουσε, δεξιῷ ἵλλαινε, γλῶσσα ἐπίξηρος, διψώδης, ἄγρυπνος.

42 Δεκάτη περὶ | τὰ αὐτά.

Ἑνδεκάτη κατενόει πάντα, ἄπυρος, ὕπνωσεν, οὖρα λεπτὰ περὶ κρίσιν.

Δωδεκάτη διέλιπεν, ἄπυρος.

Ὑπέστρεψε[6] τεσσαρεσκαιδεκάτη αὐτίκα·[7] νύκτα οὐκ ἐκοιμήθη, πάντα παρέκρουσεν.

Πεντεκαιδεκάτη οὖρον θολερὸν οἷον ἐκ τῶν καθεστηκότων γίνεται, ὅταν ἀναταραχθῇ, πυρετὸς ὀξύς· πάντα παρέκρουσεν, οὐκ ἐκοιμήθη, γούνατα καὶ κνήμας ἐπώδυνα εἶχεν· ἀπὸ δὲ κοιλίης βάλανον προσθεμένῳ μέλανα κόπρανα διῆλθεν.

Ἑξκαιδεκάτη οὖρα λεπτά, εἶχεν ἐναιώρημα ἐπινέφελον· παρέκρουσεν.

Ἑπτακαιδεκάτη πρωὶ ἄκρεα ψυχρά, περιεστέλλετο· πυρετὸς ὀξύς, ἵδρωσε δι᾿ ὅλου, ἐκουφίσθη, κατενόει μᾶλλον, οὐκ ἄπυρος, διψώδης· ἤμεσε χολώδεα, ξανθά, ὀλίγα, ἀπὸ δὲ κοιλίης κόπρανα διῆλθε, μετ᾿ ὀλίγον δὲ μέλανα, ὀλίγα, λεπτά· οὖρα λεπτά, οὐκ εὔχροα.

Ὀκτωκαιδεκάτη οὐ κατενόει, κωματώδης.

Ἐννεακαιδεκάτη διὰ τῶν αὐτῶν.

Εἰκοστῇ ὕπνωσε, κατενόει πάντα, ἵδρωσεν, ἄπυρος, οὐκ ἐδίψη, οὖρα δὲ λεπτά.

44 Εἰκοστῇ πρώτῃ σμικρὰ παρέκρουσεν, | ὑπεδίψα,

Eighth day. Stools dark, scanty, compact; sleep; reason regained; not very thirsty.

Ninth day. Rigor, acute fever; sweat; chill; delirium; squinting of the right eye; tongue dry; thirsty; sleepless.

Tenth day. About the same.

Eleventh day. Total recovery of reason; no fever; slept, urines thin around the crisis.

Twelfth day. Intermission: afebrile.

Relapsed immediately *on the fourteenth day*; at night no sleep; completely delirious.

Fifteenth day. Urine turbid, like urine with a sediment that has been shaken; acute fever; completely delirious; no sleep; pain in knees and lower legs. On the application of a suppository, dark stools were passed.

Sixteenth day. Urines thin, with a cloudy substance floating in them; delirium.

Seventeenth day. Early in the morning extremities cold; he would wrap himself up; acute fever. Sweated over the whole body; was relieved; regained more of his reason; some fever; thirst; vomited a little bilious, yellow material; motions from the bowels; after a while they became dark, scanty and thin; urines thin, and not of a good color.

Eighteenth day. Lost his reason; somnolent.

Nineteenth day. The same.

Twentieth day. Slept; recovered his reason completely; sweated; no fever; no thirst; urines thin.

Twenty-first day. Slightly delirious; a little thirst; pain

---

6 ὑπέστρεψε GalL: διέστρεψεν VI
7 GalL: add. δὲ VI

ὑποχονδρίου πόνος καὶ περὶ ὀμφαλὸν παλμὸς διὰ τέλεος.

Εἰκοστῇ τετάρτῃ οὔροισιν ὑπόστασις· κατενόει πάντα.

Εἰκοστῇ ἑβδόμῃ ἰσχίου δεξιοῦ ὀδύνη, τὰ δ᾽ ἄλλα ἔσχεν ἐπιεικέστατα, οὔροισιν ὑπόστασις.

Περὶ δὲ εἰκοστὴν ἐνάτην ὀφθαλμοῦ δεξιοῦ ὀδύνη, οὖρα λεπτά.

Τεσσαρακοστῇ διεχώρησε φλεγματώδεα, λευκά, ὑπόσυχνα· ἵδρωσε πολλῷ δι᾽ ὅλου, τελέως ἐκρίθη.

δ΄. Ἐν Θάσῳ Φιλίστις κεφαλὴν ἐπόνεε χρόνον πολὺν καί ποτε καὶ ὑποκαρωθεῖσα κατεκλίθη· ἐκ δὲ πότων[8] ξυνεχέων γενομένων ὁ πόνος παρωξύνθη. νυκτὸς ἐπεθερμάνθη | τὸ πρῶτον.

Τῇ πρώτῃ ἤμεσε χολώδεα, ὀλίγα, ξανθὰ τὸ πρῶτον, μετὰ δὲ ταῦτα ἰώδεα πλείω, ἀπὸ δὲ κοιλίης κόπρανα διῆλθε· νύκτα δυσφόρως.

Δευτέρη κώφωσις, πυρετὸς ὀξύς, ὑποχόνδριον δεξιὸν συνετάθη, ἔρρεπεν ἐς τὰ ἔσω· οὖρα λεπτά, διαφανέα, εἶχεν ἐναιώρημα γονοειδές, σμικρόν· ἐξεμάνη περὶ μέσον ἡμέρης.

Τρίτῃ δυσφόρως.

Τετάρτῃ σπασμοί, παρωξύνθη.

Πέμπτῃ πρωῒ ἀπέθανεν.

ε΄. Χαιρίωνα, ὃς κατέκειτο παρὰ Δηλίην,[9] ἐκ πότου πῦρ | ἔλαβεν. αὐτίκα δὲ κεφαλῆς βάρος ἐπώδυνον, οὐκ ἐκοιμᾶτο, κοιλίη ταραχώδης λεπτοῖσίν, ὑποχολώδεσι.

in the hypochondrium and throbbing about the navel right through to the end.

Twenty-fourth day. Sediment in the urines; completely rational.

Twenty-seventh day. Pain in the right hip; otherwise condition quite favorable; sediment in the urines.

About the twenty-ninth day pain in the right eye; urines thin.

Fortieth day. Passed phlegmy, white and rather frequent stools; copious sweat over the whole body; definitive crisis.

(4) Philistis in Thasos had for a long time pain in the head, and at last fell into a state of stupor and took to her bed. With continued drinking the pain grew worse; at night she became heated for the first time.

First day. She vomited bilious material, scanty, at first yellow, afterward increasingly like verdigris; motions from the bowels; an uncomfortable night.

Second day. Deafness; acute fever; the right hypochondrium became tense; it curved inward. Urines thin, transparent, with a small quantity of material like semen floating in it. About midday began raving.

Third day. Uncomfortable.

Fourth day. Convulsions; exacerbation.

Fifth day. Early in the morning she died.

(5) Chaerion, who lay in the house of Delia, was seized with fever after drinking. At once there was painful heaviness of the head; no sleep; bowels disturbed with thin, somewhat bilious stools.

---

8 πότων Jouanna: τούτων V: πότων πυρετῶν I

9 Δηλίην Jouanna: Δηλίαν VI

Τρίτῃ πυρετὸς ὀξύς, κεφαλῆς τρόμος, μάλιστα δὲ χείλεος τοῦ κάτω· μετ᾽ ὀλίγον δὲ ῥῖγος, σπασμοί, πάντα παρέκρουσε, νύκτα δυσφόρως.

Τετάρτῃ δι᾽ ἡσυχίης, σμικρὰ ἐκοιμήθη, παρέλεγε.

Πέμπτῃ ἐπιπόνως, πάντα παρωξύνθη, λῆρος, νύκτα δυσφόρως, οὐκ ἐκοιμήθη.

Ἕκτῃ διὰ τῶν αὐτῶν.

Ἑβδόμῃ ἐπερρίγωσε, πυρετὸς ὀξύς, ἵδρωσε δι᾽ ὅλου, ἐκρίθη.

Τούτῳ διὰ τέλεος ἀπὸ κοιλίης διαχωρήματα χολώδεα, ὀλίγα, ἄκρητα· οὖρα λεπτά, εὔχροα, ἐναιώρημα ἐπινέφελον ἔχοντα.

Περὶ ὀγδόην οὔρησεν εὐχροώτερα, ἔχοντα ὑπόστασιν λευκὴν ὀλίγην· κατενόει, ἀπύρετος· διέλιπεν.

Ἐνάτῃ ὑπέστρεψε.

Περὶ δὲ τεσσαρεσκαιδεκάτην πυρετὸς ὀξύς.

Ἑκκαιδεκάτῃ ἤμεσε χολώδεα, ξανθά, ὑπόσυχνα.

Ἑπτακαιδεκάτῃ ἐπερρίγωσε, πυρετὸς ὀξύς, ἵδρωσεν, ἄπυρος ἐκρίθη. οὖρα μετὰ ὑποστροφὴν καὶ κρίσιν εὔχροα, ὑπόστασιν ἔχοντα· οὐδὲ παρέκρουσεν ἐν τῇ ὑποστροφῇ.

Ὀκτωκαιδεκάτῃ ἐθερμαίνετο σμικρά, ὑπεδίψα,[10] οὖρα λεπτά, ἐναιώρημα ἐπινέφελον· σμικρὰ παρέκρουσεν.

Περὶ δὲ[11] ἐννεακαιδεκάτην ἄπυρος, τράχηλον ἐπωδύνως εἶχεν, οὔροισιν ὑπόστασις.

Τελέως ἐκρίθη εἰκοστῇ. |

Third day. Acute fever, trembling in the head, particularly of the lower lip; a little later rigor, convulsions, complete delirium; an uncomfortable night.

Fourth day. Quiet; a little sleep; delirium.

Fifth day. Troublesome; general exacerbation; irrational talk; uncomfortable night; no sleep.

Sixth day. The same.

Seventh day. Rigor; acute fever; sweating over the whole body; crisis occurred.

This patient's stools were throughout bilious, scanty and uncompounded. Urines thin, of a good color, with cloudy material floating in them.

About the eighth day he passed urine of an even better color, with a little white sediment; regained his reason; no fever; an intermission.

Ninth day. He relapsed.

About the fourteenth day acute fever.

Sixteenth day. Vomited bilious, yellow matters rather frequently.

Seventeenth day. Rigor; acute fever; sweating; crisis ended the fever.

Urines after the relapse and the crisis were of a good color, with a sediment; no delirium during the relapse.

Eighteenth day. Slight heat; a little thirst; urines thin, with cloudy material floating in them; slight delirium.

About the nineteenth day. Afebrile; pain in the neck; sediment in the urines.

Twentieth day. Definitive crisis.

---

10 ὑπεδίψα V: ἐπεδίψα I
11 Περὶ δὲ ἐννεακαιδεκάτην I: Ἐννεακαιδεκάτη V

50 ϛʹ. Τὴν Εὐρυάνακτος θυγατέρα, παρθένον, πῦρ ἔλαβεν. ἦν δὲ ἄδιψος διὰ τέλεος· γεύματα οὐ προσεδέ-
χετο. ἀπὸ δὲ κοιλίης σμικρὰ διῄει· οὖρα λεπτά, ὀλίγα,
οὐκ εὔχροα. ἀρχομένου δὲ τοῦ πυρετοῦ περὶ ἕδρην
ἐπόνεε.

Ἑκταίη δὲ ἐοῦσα ἄπυρος οὐχ ἵδρωσεν· ἐκρίθη. τὸ
δὲ περὶ τὴν ἕδρην σμικρὰ ἐξεπύησεν, ἐρράγη ἅμα
κρίσει.

Μετὰ δὲ κρίσιν ἑβδομαίη ἐοῦσα ἐρρίγωσε, σμικρὰ
ἐπεθερμάνθη, ἵδρωσεν. ὕστερον δὲ ἄκρεα ψυχρὰ αἰεί.

Περὶ δὲ δεκάτην μετὰ τὸν ἱδρῶτα τὸν γενόμενον
παρέκρουσε καὶ πάλιν ταχὺ κατενόει· ἔλεγον δὲ γευ-
σαμένην βότρυος.

Διαλιποῦσα δὲ δωδεκάτῃ πάλιν πολλὰ παρελήρει,
52 κοιλίη | ἐπεταράχθη χολώδεσιν, ὀλίγοισιν, ἀκρή-
τοισι,[12] λεπτοῖσι, δακνώδεσι, πυκνὰ ἀνίστατο.

Ἀφ᾽ ἧς δὲ παρέκρουσε τὸ ὕστερον, ἀπέθανεν
ἑβδόμῃ.

Αὕτη ἀρχομένου τοῦ νοσήματος ἤλγεε φάρυγγα,
καὶ διὰ τέλεος· ἔρευθος εἶχε, γαργαρεὼν ἀνεσπασμέ-
νος. ῥεύματα πολλά, σμικρά, δριμέα. ἔβησσε πέπονα,
οὐδὲν ἀνῆγεν· ἀπόσιτος πάντων παρὰ πάντα τὸν χρόνον οὐδ᾽ ἐπεθύμησεν οὐδενός. ἄδιψος, οὐδ᾽ ἔπινεν οὐδὲν ἄξιον λόγου. σιγῶσα, οὐδὲν διελέγετο. δυσθυμίη,
ἀνελπίστως ἑωυτῆς[13] εἶχεν. ἦν δέ τι καὶ ξυγγενικὸν
φθινῶδες.

[12] ὀλίγοισι καὶ ἀκρήτοισι I: ἀκρίτοισιν ὀλίγοις V

(6) The unmarried daughter of Euryanax was seized with fever. Throughout her illness she suffered no thirst and had no appetite for food. Slight alvine discharges; urines thin, scanty, and not of a good color. At the beginning of the fever she suffered pain in her seat.

On the sixth day she was afebrile, did not sweat, and had a crisis. The sore in her seat suppurated slightly, and broke open at the crisis.

After the crisis, *on the seventh day*, she had a rigor; she became slightly heated; she sweated. Afterward the extremities were always cold.

About the tenth day, after sweating that occurred, she grew delirious, but then soon regained her reason. They said she had eaten grapes.

After an intermission, *on the twelfth day* she was again very delirious; her bowels were disturbed, with bilious, scanty, uncompounded, thin, irritating stools; she got up often.

She died *the seventh day from the second attack of delirium.*

At the beginning of the illness and through to its end this patient had pain in her throat, with redness; her uvula was drawn back. There were many fluxes, scanty and acrid. She had a cough, but failed to bring up any concocted sputum. She had no appetite for any food the whole time, nor did she desire anything. Also no thirst, and she drank nothing worth mentioning. She was silent, and did not converse at all. Depression, the patient despairing of herself. There was also some inherited tendency to consumption.

ζ'. Ἡ κυναγχικὴ ἡ παρὰ Ἀριστίωνος, ᾗ πρῶτον
54 ἀπὸ | γλώσσης[14] ἤρξατο· ἀσαφὴς φωνή· γλῶσσα ἐρυ-
θρή, ἐπεξηράνθη.

Τῇ πρώτῃ φρικώδης, ἐπεθερμάνθη.

Τρίτῃ ῥῖγος, πυρετὸς ὀξύς· οἴδημα ὑπέρυθρον,
σκληρὸν τραχήλου καὶ ἐπὶ στῆθος ἐξ ἀμφοτέρων·
ἄκρεα ψυχρά, πελιδνά, πνεῦμα μετέωρον· ποτὸν διὰ
ῥινῶν, καταπίνειν οὐκ ἠδύνατο· διαχωρήματα καὶ
οὖρα ἐπέστη.

Τετάρτῃ πάντα παρωξύνθη.

Πέμπτῃ ἀπέθανε. |

56 η'. Τὸ μειράκιον, ὃ κατέκειτο ἐπὶ ψευδέων ἀγορῇ,
πῦρ ἔλαβεν ἐκ κόπων καὶ πόνων καὶ δρόμων παρὰ τὸ
ἔθος.

Τῇ πρώτῃ κοιλίη ταραχώδης χολώδεσι, λεπτοῖσι,
πολλοῖσιν, οὖρα λεπτά, ὑπομέλανα· οὐχ ὕπνωσε, διψώδης.

Δευτέρῃ πάντα παρωξύνθη, διαχωρήματα πλείω,
ἀκαιρότερα· οὐχ ὕπνωσε, τὰ τῆς γνώμης ταραχώδεα·
σμικρὰ ὑφίδρωσε.

Τρίτῃ δυσφόρως, διψώδης, ἀσώδης, πολὺς βληστρισμός, ἀπορίη, παρέκρουσεν· ἄκρεα πελιδνὰ καὶ
ψυχρά, ὑποχονδρίου ἔντασις ὑπολάπαρος ἐξ ἀμφοτέρων.

Τετάρτῃ οὐχ ὕπνωσεν· ἐπὶ τὸ χεῖρον.

Ἑβδόμῃ ἀπέθανεν.

Ἡλικίην περὶ ἔτεα εἴκοσιν. |

58 θ'. Ἡ παρὰ Τισαμένου γυνὴ κατέκειτο, ᾗ τὰ εἰλεώ-

(7) The woman suffering from angina who lay in the house of Aristion began her complaint from the tongue with indistinctness of speech; tongue red, and becoming parched.

First day. Shivered, and grew heated.

Third day. Rigor; acute fever; a reddish, hard swelling in the neck, extending to the chest on either side; extremities cold and livid, breathing elevated; drink (sc. returned) through the nostrils—she could not swallow—passage of stools and urine stopped.

Fourth day. General exacerbation.

Fifth day. She died.

(8) The adolescent who lay by the Liars' Market was seized with fever after fatigue, toil and running, to which he was unaccustomed.

First day. Bowels disturbed with bilious, thin, copious stools; urines thin and darkish; no sleep; thirst.

Second day. General exacerbation; stools more copious and more unfavorable. No sleep; mind disturbed; slight sweating.

Third day. Uncomfortable; thirst; nausea; much tossing; distress; delirium; extremities livid and cold; tension, with inward curvature, of the hypochondrium on both sides.

Fourth day. No sleep; grew worse.

Seventh day. He died.

About twenty years of age.

(9) The woman who lodged with Tisamenus took to

---

14 ἀπὸ γλώσσης I: om. V

δεα δυσφόρως ὥρμησεν. ἔμετοι πολλοί, ποτὸν κατέχειν οὐκ ἠδύνατο. πόνοι περὶ ὑποχόνδρια· καὶ ἐν τοῖσι κάτω κατὰ κοιλίην[15] πόνοι. στρόφοι ξυνεχέες. οὐ διψώδης. ἐπεθερμαίνετο, ἄκρεα ψυχρὰ διὰ τέλεος. ἀσώδης, ἄγρυπνος. οὖρα ὀλίγα, λεπτά. διαχωρήματα ὠμά, λεπτὰ, ὀλίγα. ὠφελέειν οὐκέτι ἠδύνατο, ἀπέθανεν. |

60   ι'. Γυναῖκα ἐξ ἀποφθορῆς νηπίου τῶν περὶ Παντιμίδην τῇ πρώτῃ πῦρ ἔλαβε. γλῶσσα ἐπίξηρος, διψώδης, ἀσώδης, ἄγρυπνος. κοιλίη ταραχώδης λεπτοῖσι, πολλοῖσιν, ὠμοῖσι.

Δευτέρῃ ἐπερρίγωσε, πυρετὸς ὀξύς, ἀπὸ κοιλίης πολλά, οὐχ ὕπνωσε.

Τρίτῃ μείζους οἱ πόνοι.

Τετάρτῃ παρέκρουσεν.

Ἑβδόμῃ ἀπέθανε.

Κοιλίη διὰ παντὸς ὑγρὴ διαχωρήμασι πολλοῖσι, λεπτοῖσιν, ὠμοῖσιν· οὖρα ὀλίγα λεπτά.

ια'. Ἑτέρην ἐξ ἀποφθορῆς περὶ πεντάμηνον, Ἰκέτεω
62 γυναῖκα, | πῦρ ἔλαβεν. ἀρχομένη κωματώδης ἦν, καὶ πάλιν ἄγρυπνος· ὀσφύος ὀδύνη, κεφαλῆς βάρος.

Δευτέρῃ κοιλίη ἐπεταράχθη ὀλίγοισι, λεπτοῖσιν, ἀκρήτοισι τὸ πρῶτον.

Τρίτῃ πλείω, χείρω· νυκτὸς οὐδὲν ἐκοιμήθη.

Τετάρτῃ παρέκρουσε, φόβοι, δυσθυμίαι· δεξιῷ ἴλλαινεν· ἵδρω τὰ[16] περὶ κεφαλὴν ὀλίγῳ ψυχρῷ, ἄκρεα ψυχρά.

bed with a troublesome attack of ileus. Much vomiting; she could not retain what she drank. Pains in the region of the hypochondria; also lower down in the bowel; constant colic. No thirst. She became heated, although her extremities were cold through to the end. Nausea; sleeplessness. Urines scanty and thin. Excreta crude, thin and scanty. Nothing could help her, and she died.

(10) A woman who miscarried of her infant—she belonged to the household of Pantimides—was seized with fever *on the first day*. Tongue dry; thirst; nausea; sleeplessness. Bowels disordered with thin, copious and crude stools.

Second day. Rigor; acute fever; much material from the bowels; no sleep.

Third day. The pains increased.

Fourth day. Delirium.

Seventh day. She died.

The bowels were throughout loose, with copious, thin, crude stools. Urines scanty and thin.

(11) Another woman, the wife of Hicetas, after a miscarriage about the fifth month, was seized with fever. At the beginning she had alternations of somnolence and sleeplessness; pain in the loins; heaviness in the head.

Second day. Bowels disordered with scanty, thin stools, which at first were uncompounded.

Third day. Stools more copious and worse; no sleep at night.

Fourth day. Delirium; fears; depression. Squinting of the right eye; slight cold sweat about the head; extremities cold.

---

15 Add. οἱ V   16 ἴδρω τὰ Langholf (p. 110, 2): ἰδρῶτα VI

Πέμπτῃ πάντα παρωξύνθη, πολλὰ παρέλεγε καὶ πάλιν ταχὺ κατενόει· ἄδιψος, ἄγρυπνος, κοιλίη πολλοῖσιν ἀκαίροισι διὰ τέλεος· οὖρα ὀλίγα, λεπτά, ὑπομέλανα· ἄκρεα ψυχρά, ὑποπέλιδνα.

Ἕκτῃ διὰ τῶν αὐτῶν.

Ἑβδόμῃ ἀπέθανε.

ιβ'. Γυναῖκα, ἣ[17] κατέκειτο ἐπὶ ψευδέων ἀγορῆς, τεκοῦσαν | τότε πρῶτον ἐπιπόνως ἄρσεν πῦρ ἔλαβεν. αὐτίκα ἀρχομένη διψώδης, ἀσώδης, καρδίην ὑπήλγει· γλῶσσα ἐπίξηρος, κοιλίη ἐπεταράχθη λεπτοῖσιν ὀλίγοισιν· οὐχ ὕπνωσε.

Δευτέρῃ σμικρὰ ἐπερρίγωσε, πυρετὸς ὀξύς, σμικρὰ περὶ κεφαλὴν ἵδρωσε ψυχρῷ.

Τρίτῃ ἐπιπόνως· ἀπὸ κοιλίης ὠμά, λεπτὰ πολλὰ διήει.

Τετάρτῃ ἐπερρίγωσε, πάντα παρωξύνθη· ἄγρυπνος.

Πέμπτῃ ἐπιπόνως.

Ἕκτῃ διὰ τῶν αὐτῶν· ἀπὸ κοιλίης ἦλθε πολλά.

Ἑβδόμῃ ἐπερρίγωσε, πυρετὸς ὀξύς, δίψα, πολὺς βληστρισμός· περὶ δείλην ἵδρωσε δι᾽ ὅλου ψυχρῷ, ψύξις, ἄκρεα ψυχρά, οὐκέτι ἀνεθερμαίνετο· καὶ πάλιν ἐς νύκτα ἐπερρίγωσεν, ἄκρεα οὐκ ἀνεθερμαίνετο· οὐχ ὕπνωσε, σμικρὰ παρέκρουσε, καὶ πάλιν ταχὺ κατενόει.

Ὀγδόῃ περὶ μέσον ἡμέρης ἀνεθερμάνθη, διψώδης, κωματώδης, ἀσώδης, ἤμεσε χολώδεα σμικρὰ ὑπό-

Fifth day. General exacerbation; much wandering in her talk, followed by a rapid recovery of her reason; no thirst; no sleep; stools copious and unfavorable through to the end; urines scanty, thin and darkish; extremities cold and somewhat livid.

Sixth day. Same.

Seventh day. She died.

(12) A woman who lay by the Liars' Market, when she had for the first time and with difficulty given birth to a male child, was seized with fever. From the very first she had thirst, nausea, slight pain in the cardia, and a dry tongue; bowels disordered with thin and scanty discharges; no sleep.

Second day. Slight rigor; acute fever; a little cold sweating around the head.

Third day. Discomfort; crude, thin, copious discharges from the bowels.

Fourth day. Rigor; general exacerbation; sleepless.

Fifth day. Discomfort.

Sixth day. The same. Copious discharges from the bowels.

Seventh day. Rigor; acute fever; thirst; much tossing; toward evening cold sweat over the whole body; chill; extremities cold and could not be warmed back up. Again at night she had a rigor; the extremities could not be warmed; no sleep; slight delirium, but then she quickly recovered her reason.

Eighth day. About midday she recovered her heat; thirst; somnolence; nausea; vomited bilious, scanty, yel-

---

[17] ἢ Π₅ V: ἥτις I

ξανθα. νύκτα δυσφόρως, οὐκ ἐκοιμήθη, οὔρησε πολὺ ἀθρόον· οὐκ εἶδον.

66   Ἐνάτῃ | ξυνέδωκε πάντα, κωματώδης· πρὸς δείλην σμικρὰ ἐπερρίγωσεν, ἤμεσε μικρὰ χολώδεα.

Δεκάτῃ ῥῖγος, πυρετὸς παρωξύνθη, οὐχ ὕπνωσεν οὐδέν· πρωὶ οὔρησε πολὺ ὑπόστασιν ἔχον· ἄκρεα ἀνεθερμάνθη.

Ἑνδεκάτῃ ἤμεσε χολώδεα, ἰώδεα. ἐπερρίγωσεν οὐ μετὰ πολύ, καὶ πάλιν ἄκρεα ψυχρά· ἐς δείλην ἱδρώς, ῥῖγος, ἤμεσε πολλά, νύκτα ἐπιπόνως.

Δωδεκάτῃ ἤμεσε πολλὰ μέλανα δυσώδεα· λυγμὸς πολύς, δίψος ἐπιπόνως.

Τρισκαιδεκάτῃ μέλανα, δυσώδεα πολλὰ ἤμεσε· ῥῖγος· περὶ δὲ μέσον ἡμέρης ἄφωνος.

Τεσσαρεσκαιδεκάτῃ αἷμα διὰ ῥινῶν· ἀπέθανε.

Ταύτῃ διὰ τέλεος κοιλίη ὑγρή· φρικώδης· ἡλικίη περὶ ἔτεα ἑπτακαίδεκα.

<Κατάστασις>[18]

2. Ἔτος νότιον ἔπομβρον· ἄπνοια διὰ τέλεος· αὐχμῶν
68   δὲ | γενομένων τοὺς ὑπόπροσθεν χρόνους ἐπ᾽ ἐνιαυτόν·[19] ἐν νοτίοισι περὶ ἀρκτοῦρον ὕδατα πολλά. φθινόπωρον σκιῶδες, ἐπινέφελον· ὑδάτων πλήθεα. χειμὼν νότιος, ὑγρός, μαλθακός· μετὰ δὲ ἡλίου τροπὰς ὕστερον πολλῷ, πλησίον ἰσημερίης, ὀπισθοχειμῶνες.

---

[18] Κατάστασις add. GalL
[19] ἐπ᾽ ἐνιαυτόν VI²: ἐνιαυτός I

lowish material. An uncomfortable night; no sleep; she had a copious discharge of urine: I did not see it.

Ninth day. General abatement of the symptoms; somnolence. Toward evening slight rigor; vomited scanty, bilious material.

Tenth day. Rigor; exacerbation of the fever; no sleep at all. In the early morning she had a copious discharge of urine with sediment; extremities warmed up.

Eleventh day. Vomited verdigris-colored, bilious material. A rigor shortly afterward, and the extremities became cold again; in the evening sweat, rigor and copious vomiting; a painful night.

Twelfth day. Vomited copious, dark, foul-smelling material; much hiccupping; troublesome thirst.

Thirteenth day. Vomited copious dark, foul-smelling material; rigor. About midday she lost her speech.

Fourteenth day. Epistaxis; she died.

In this patient the bowels were throughout loose, and she had shivering fits. Age about seventeen years.

Catastasis

2. The year was southerly and rainy, with no winds at any time. While during the immediately preceding period droughts had prevailed for a year, about (sc. the rising of) Arcturus there were heavy rains with southerly winds. Autumn dark and cloudy, with an abundance of rain. The winter southerly, humid, mild; much later, long after the solstice, in fact near the equinox, there was a bout of late-winter weather, and at the actual equinoctial period there

καὶ ἤδη περὶ ἰσημερίην, βόρεια, χιονώδεα, οὐ πολὺν
χρόνον. ἦρ πάλιν νότιον, ἄπνοον· ὕδατα πολλὰ διὰ
τέλεος μέχρι κυνός. θέρος αἴθριον, θερμόν· πνίγεα με-
γάλα· ἐτησίαι σμικρά, διεσπασμένως ἔπνευσαν. πάλιν δὲ περὶ ἀρκτοῦρον ἐν βορείοισιν ὕδατα πολλά.

70   Γενομένου δὲ τοῦ ἔτεος νοτίου καὶ ὑγροῦ καὶ | μαλ-
θακοῦ, κατὰ μὲν χειμῶνα διῆγον ὑγιηρῶς, πλὴν τῶν
φθινωδέων, περὶ ὧν γεγράψεται.

3. Πρωὶ δὲ τοῦ ἦρος ἅμα τοῖσι γενομένοισι ψύχε-
σιν, ἐρυσίπελας πολλοῖσι μὲν μετὰ προφάσιος, τοῖσι
δ᾽ οὔ· κακοήθεα, πολλοὺς ἔκτεινε. πολλοὶ φάρυγγας
ἐπόνησαν· φωναὶ κακούμεναι, καῦσοι, φρενιτικοί,
στόματα ἀφθώδεα· αἰδοίοισι φύματα, ὀφθαλμίαι, ἄνθρακες. κοιλίαι ταραχώδεες, ἀπόσιτοι, διψώδεες οἱ
μέν, οἱ δ᾽ οὔ· οὖρα ταραχώδεα πολλὰ κακά. κωματώδεες ἐπὶ πολλοῖσι καὶ πάλιν ἄγρυπνοι· ἀκρισίαι[20]
πολλαί, δύσκριτα, ὕδρωπες, φθινώδεες πολλοί. τὰ μὲν
ἐπιδημήσαντα νοσήματα ταῦτα. ἑκάστου δὲ τῶν ὑπογεγραμμένων εἰδέων ἦσαν οἱ κάμνοντες καὶ ἔθνησκον
πολλοί. συνέπιπτε δ᾽ ἐφ᾽ ἑκάστοισι τούτων ὧδε.

4. Πολλοῖσι μὲν τὸ ἐρυσίπελας μετὰ προφάσιος ἐπὶ
72 τοῖσι τυχοῦσι | καὶ πάνυ ἐπὶ σμικροῖσι τρωματίοισιν
ἐφ᾽ ὅλῳ τῷ σώματι, μάλιστα δὲ τοῖσι περὶ ἑξήκοντα
ἔτεα περὶ κεφαλήν, εἰ καὶ σμικρὸν ἀμεληθείη. πολλοῖσι δὲ καὶ ἐν θεραπείῃ ἐοῦσι. μεγάλαι φλεγμοναὶ
ἐγίνοντο, καὶ τὸ ἐρυσίπελας πολὺ ταχὺ πάντοθεν

[20] ἀκρισίαι GalL: ἀκρασίαι VI

were northerly winds with snow—but not for long. The spring southerly again, with no winds; strong rains through until (the heliacal rising of) Sirius. The summer was clear and warm, with waves of stifling heat. The Etesian winds blew faint and intermittent. Once more, near Arcturus, there were northerly winds with heavy rains.

The year having proved southerly, wet and mild, in the winter the general health was good except for the consumptives, who will be described in due course.

3. Early in the spring, at the time the cold spells occurred, many cases of erysipelas broke out, in many from an identifiable cause, but in others without one. The disease was malignant and killed many of its sufferers. Many were affected in their throat: voices impaired; ardent fevers; phrenitis; aphthae in the mouth; growths on the genitalia; inflammations of the eyes; carbuncles. Disordered bowels; loss of appetite; thirst in some cases, though not in all; urines disordered, copious, bad. Somnolence in many, and again sleeplessness. No crisis appeared in the majority; obscure crises; dropsies; many consumptives. Such were the diseases prevalent through the populace: there were examples of each of the kinds described, and fatal cases were many. The symptoms in each of them were as follow.

4. Erysipelas from a cause occurred in many who had had some chance, and especially very minor, wounds anywhere in the whole body; especially when the patients were about sixty years old and the wound was in their head—however little their negligence may have been, and even in many who were in treatment. Severe inflammations arose, and the erysipelas would very quickly erode

ἐπενέμετο. τοῖσι μὲν οὖν πλείστοισιν αὐτῶν ἀποστάσιες ἐς ἐμπυήματα συνέπιπτον· σαρκῶν καὶ νεύρων
καὶ ὀστέων ἐκπτώσιες μεγάλαι. ἦν δὲ καὶ τὸ ῥεῦμα τὸ
74 ξυνιστάμενον οὐ | πύῳ ἴκελον, ἀλλὰ σηπεδών τις
ἄλλη, καὶ ῥεῦμα πολὺ καὶ ποικίλον. οἷσι μὲν οὖν περὶ
κεφαλὴν τούτων τι ξυμπίπτοι γίνεσθαι, μάθησίς τε
ὅλης τῆς κεφαλῆς ἐγίνετο καὶ τοῦ γενείου, καὶ ὀστέων
ψιλώματα καὶ ἐκπτώσιες, καὶ πολλὰ ῥεύματα. ἐν πυ-
ρετοῖσί τε ταῦτα καὶ ἄνευ πυρετῶν. ἦν δὲ ταῦτα
φοβερώτερα ἢ κακίω. οἷσι γὰρ ἐς ἐμπύημα ἤ τινα
τοιοῦτον ἀφίκοιτο πεπασμόν, οἱ πλεῖστοι τούτων ἐσώ-
ζοντο. οἷσι δὲ ἡ μὲν φλεγμονὴ καὶ τὸ ἐρυσίπελας
ἀπέλθοι, τοιαύτην δὲ ἀπόστασιν μηδεμίαν ποιήσαιτο,
τούτων ἀπώλλυντο πολλοί. ὁμοίως δὲ καὶ εἴ πῃ ἄλλῃ
τοῦ σώματος πλανηθείη, ξυνέπιπτε ταῦτά.²¹ πολλοῖσι
μὲν γὰρ βραχίων καὶ πῆχυς ὅλος περιερρύη. οἷσι δ᾽
ἐπὶ τὰ πλευρά, ταῦτα ἐκακοῦτο ἢ τῶν ἔμπροσθέν τι ἢ
τῶν ὄπισθεν. οἷσι δ᾽ ὅλος ὁ μηρὸς ἢ τὰ περὶ κνήμην
ἀπεψιλοῦτο· καὶ ποὺς ὅλος. ἦν δὲ πάντων χαλεπώτατα
τῶν τοιούτων, ὅτε περὶ ἥβην καὶ αἰδοῖα γενοίατο.

Τὰ μὲν περὶ ἕλκεα καὶ μετὰ προφάσιος τοιαῦτα·
πολλοῖσι δὲ ἐν πυρετοῖσι καὶ πρὸ πυρετοῦ καὶ ἐπὶ
πυρετοῖσι ξυνέπιπτεν. ἦν δὲ καὶ τούτων, ὅσα μὲν ἀπόστασιν ποιήσαιτο διὰ τοῦ ἐκπυῆσαι ἢ κατὰ κοιλίην
ταραχή τις ἐπίκαιρος ἢ χρηστῶν οὔρων διάδοσις γέ-
76 νοιτο, | διὰ τούτων λελύσθαι· οἷσι δὲ μηδὲν τούτων

---

²¹ ταὐτά Potter: ταῦτα VI

its way forward in all directions. Now most of the patients had apostases ending in empyemas: flesh, sinews and bones were greatly eroded. The sanies which formed was not like pus, but was a different sort of putrefaction, with a copious and varied product. Now in any patient in whom the lesion occurred in the head, there was loss of hair from all the head and from the chin; the bones became exposed and broke away, and there were copious fluxes. Sometimes there was fever, and other times none. These symptoms were terrifying rather than dangerous, for whenever they resulted in an empyema or some similar form of concoction, the patients usually recovered; but wherever the inflammation and the erysipelas disappeared without producing any such apostasis, many patients died. Similarly, if the disease invaded some random part of the body, the symptoms were the same; for in many the upper arm or the whole forearm withered away; in cases where it settled in the sides, these decayed toward the front or the back. In cases where the entire thigh or the area around the lower leg was involved, they were eroded; also an entire foot. But the most severe of all such cases were those arising in the region of the pubes and the genitalia.

Regarding erysipelas occurring around lesions and with a clear cause, they were like this: in many cases, however, it occurred in the accompaniment of fevers, or before a fever, or subsequent to fevers. Among these too there were cases that produced an apostasis by forming an empyema, or by an auspicious disturbance through the bowels, or a favorable passage of urines, and were resolved in this way. But in cases where none of these things hap-

ξυμπίπτοι, ἀσήμως δὲ ἀφανιζομένων, θανατώδεα γίνεσθαι. πολὺ μὲν οὖν πλείστοισι ξυνέπιπτε τὰ περὶ
τὸ ἐρυσίπελας τοῦ ἦρος. παρείπετο δὲ καὶ διὰ τοῦ
θέρεος καὶ ὑπὸ φθινόπωρον.

5. Πολλὴ δὲ ταραχή τισι καὶ τὰ περὶ φάρυγγα
φύματα, καὶ φλεγμοναὶ γλώσσης, καὶ τὰ παρ᾽ ὀδόν-
τας ἀποστήματα. φωναί τε πολλοῖσιν ἐπεσήμαινον
80 κακούμεναι καὶ †κατίλλουσαι,† | πρῶτον μὲν τοῖσι
φθινώδεσιν ἀρχομένοισιν, ἀτὰρ καὶ τοῖσι καυσώδεσι
καὶ τοῖσι φρενιτικοῖσιν.

6. Ἤρξαντο μὲν οὖν οἱ καῦσοι καὶ τὰ φρενιτικὰ
πρωῒ τοῦ ἦρος μετὰ τὰ γενόμενα ψύχεα, καὶ πλεῖστοι
τηνικαῦτα διενόσησαν· ὀξέα δὲ τούτοισι καὶ θανατώδεα ξυνέπιπτεν. ἦν δὲ ἡ κατάστασις τῶν γενομένων
καύσων ὧδε· ἀρχόμενοι κωματώδεες, ἀσώδεες, φρικώ-
82 δεες· πυρετὸς ὀξύς, οὐ διψώδεες λίην, οὐ | παράληροι·
ἀπὸ ῥινῶν ἔσταξε σμικρόν. οἱ παροξυσμοὶ τοῖσι πλείστοισιν ἐν ἀρτίῃσι· περὶ δὲ τοὺς παροξυσμοὺς λήθη
καὶ ἄφεσις καὶ ἀφωνίη. ἄκρεά τε τούτοισιν αἰεὶ μὲν
ψυχρότερα ποδῶν καὶ χειρῶν, πολὺ δὲ περὶ τοὺς
παροξυσμοὺς μάλιστα· πάλιν τε βραδέως καὶ οὐ
καλῶς ἀνεθερμαίνοντο καὶ πάλιν κατενόουν καὶ διελέ-
γοντο. κατεῖχε δὲ ἢ τὸ κῶμα ξυνεχές, οὐχ ὑπνῶδες, ἢ
μετὰ πόνων ἄγρυπνοι. κοιλίαι ταραχώδεες τοῖσι πλείστοισι τούτων, διαχωρήμασιν ὠμοῖσι, λεπτοῖσι, πολ-
λοῖσιν· οὐρά[22] πολλὰ λεπτὰ κρίσιμον οὐδὲ χρηστὸν
οὐδὲν ἔχοντα· οὐδὲ ἄλλο κρίσιμον οὐδὲν τοῖσιν οὕτως
ἔχουσιν ἐφαίνετο· οὔτε γὰρ ἡμορράγει καλῶς οὔτε τις

pened, and the symptoms disappeared without a sign, death resulted. Now it was in the spring that by far the greatest number of cases of erysipelas occurred, but they did continue through the summer and in the autumn.

5. Much trouble was caused in some patients by growths in the throat, inflammations of the tongue, and abscesses next the teeth. In many of these the patients' voices gave an indication, being impaired and †muffled,† first in consumptives at the beginning (sc. of the disease), but also in cases of ardent fevers and phrenitis.

6. Now the ardent fevers and phrenitis began early in the spring after the cold spells that occurred, and very many fell sick at that time, suffering acute and fatal symptoms. The form of the ardent fevers that occurred was as follows: at the beginning somnolence, nausea, shivering, acute fever, not much thirst, no delirium, slight epistaxis. The exacerbations were in most cases on even days, and about the time of the exacerbations there was loss of memory with prostration and speechlessness. The extremities of these patients were always too cold, the feet and the hands, most especially about the times of exacerbation; slowly and imperfectly they recovered their heat, becoming rational again and conversing. Either a coma held them permanently, without actual sleep, or they were awake and in pain. Bowels disordered in the majority of these cases, with crude, thin, copious stools; urines copious, thin, with no critical or favorable sign. Nor did any other critical sign appear in these patients: for there occurred neither a favorable hemorrhage nor any other of

---

²² Add. τὰ V

ἄλλη τῶν εἰθισμένων ἀπόστασις ἐγένετο κρίσιμος.
ἔθνησκόν τε ἕκαστος ὡς τύχοι, πεπλανημένως τὰ
πολλά, περὶ τὰς κρίσιας, ἐκ πολλοῦ δέ τινες ἄφωνοι·
ἱδρῶντες πολλοί. τοῖσι μὲν ὀλεθρίως ἔχουσι ξυνέπιπτε ταῦτα· παραπλήσια δὲ καὶ τοῖσι φρενιτικοῖσιν.
ἄδιψοι δὲ πάνυ οὗτοι ἦσαν· οὐδ᾽ ἐξεμάνη τῶν φρενι-
τικῶν οὐδείς, ὥσπερ ἐπ᾽ ἄλλοισιν, ἀλλ᾽ ἄλλη τινὶ
καταφορῇ νωθρῇ βαρέως ἀπώλλυντο. |

84   7. Ἦσαν δὲ καὶ ἄλλοι πυρετοί, περὶ ὧν γεγράψεται.
στόματα πολλοῖσιν ἀφθώδεα, ἑλκώδεα. ῥεύματα περὶ
αἰδοῖα πολλά, ἑλκώματα, φύματα ἔξωθεν, ἔσωθεν· τὰ
περὶ βουβῶνας. ὀφθαλμίαι ὑγραί, μακραί, χρόνιαι·
μετὰ πόνων, ἐπιφύσιες βλεφάρων ἔξωθεν, ἔσωθεν,
πολλῶν φθείροντα τὰς ὄψιας, ἃ σῦκα ἐπονομάζουσιν.
ἐφύετο δὲ καὶ ἐπὶ τῶν ἄλλων ἑλκέων πολλὰ καὶ ἐν
αἰδοίοισιν. ἄνθρακες πολλοὶ κατὰ θέρος καὶ ἄλλα, ἃ
σὴψ καλέεται. ἐκθύματα μεγάλα. ἕρπητες πολλοῖσι
μεγάλοι.

8. Τὰ δὲ κατὰ κοιλίην πολλοῖσι πολλὰ καὶ βλα-
βερὰ συνέβαινε. πρῶτον μὲν τεινεσμοὶ πολλοῖσιν
ἐπιπόνως, πλείστοισι δὲ παιδίοισι, καὶ πᾶσιν ὅσα
πρὸ ἥβης, καὶ ἀπώλλυντο τὰ πλεῖστα τούτων. λειεν-
86   τερικοὶ πολλοί. δυσεντεριώδεες, οὐδ᾽ οὗτοι λίην | ἐπι-
πόνως. τὰ δὲ χολώδεα καὶ λιπαρὰ καὶ λεπτά· καὶ
ὑδατώδεα· πολλοῖσι μὲν αὐτὸ τὸ νόσημα ἐς τοῦτο
κατέσκηψεν ἄνευ²³ πυρετῶν καὶ ἐν πυρετοῖσι. μετὰ

²³ Add. τε I

the usual critical apostases. They died, each one as it chanced, irregularly, usually at the crises, some long after the loss of speech, and many with sweating. These were the symptoms attending the fatal cases (sc. of ardent fever), and in fact the cases of phrenitis were similar, although such patients suffered from no thirst at all, and no one with phrenitis displayed the mad delirium that attacked in ardent fever, but rather they passed away overpowered by a dull oppression of stupor.

7. There were other fevers as well, which I shall describe later. Many patients had aphthae and sores in the mouth; fluxes about the genitalia were frequent; sores and growths on the outside and the inside in the region of the groin. Inflammations of the eyes, watery, long, and chronic, with pains. Growths accompanied by pains developed on the eyelids, on both the outside and the inside, in many cases impairing vision: they call these "figs." There were also often growths on top of other sores, particularly in the genitalia. Many anthraxes appeared in the summer, and other serious eruptions which are called a "pustule," being large excrescences. Many had large shingles.

8. The bowel troubles in many cases were frequent and harmful. First the tenesmus experienced by many was troublesome—mostly in babies and children before puberty—and most of these died. Many lienteries; and cases of dysentery, but these were not very troublesome: stools bilious, greasy, and thin; also dropsies. In many patients their previous disease changed into this (sc. disease of the bowels), sometimes without fever and sometimes with fevers. Painful colics and tormina (sc. of the bowels),

88 πόνων | στρόφοι καὶ ἀνειλήσιες κακοήθεες. διέξοδοί
τε τῶν πολλῶν ἐνόντων καὶ ἐπισχόντων. τά τε διεξιόντα πόνους οὐ λύοντα τοῖσί τε προσφερομένοισι
δυσκόλως ὑπακούοντα· καὶ γὰρ αἱ καθάρσιες τοὺς
πλείστους προσέβλαπτον. τῶν δὲ οὕτως ἐχόντων πολλοὶ μὲν ὀξέως ἀπώλλυντο, ἔστι δ᾽ οἷσι καὶ μακρότερα
διῆγεν. ὡς δ᾽ ἐν κεφαλαίῳ εἰρῆσθαι, πάντες, καὶ οἱ τὰ
μακρὰ νοσέοντες καὶ οἱ τὰ ὀξέα, ἐκ τῶν κατὰ κοιλίην
ἀπέθνησκον μάλιστα. πάντας γὰρ κοιλίη συναπή-
νεγκεν. |

90   9. Ἀπόσιτοι δ᾽ ἐγένοντο πάντες μὲν[24] καὶ ἐπὶ πᾶσι
τοῖσι προγεγραμμένοισιν, ὡς ἐγὼ οὐδὲ πώποτε ἐνέτυ-
χον, πολὺ δὲ μάλιστα οὗτοι· καὶ οἱ ἐκ τούτων καὶ ἐκ
τῶν ἄλλων δὲ οἳ καὶ ὀλεθρίως ἔχοιεν. διψώδεες οἱ μέν,
οἱ δὲ οὔ, τῶν ἐν πυρετοῖσι· καὶ τοῖσιν ἄλλοισιν οὐδεὶς
ἀκαίρως, ἀλλ᾽ ἦν κατὰ ποτὸν διαιτᾶν ὡς ἤθελες.

10. Οὖρα δὲ πολλὰ μὲν τὰ διεξιόντα ἦν, οὐκ ἐκ τῶν
προσφερομένων ποτῶν, ἀλλὰ πολλὸν ὑπερβάλλοντα.
πολλὴ δέ τις καὶ τῶν οὔρων κακότης ἦν τῶν ἀπιόν-
των. οὔτε γὰρ πάχος οὔτε πεπασμός, οὔτε κάθαρσις
χρηστή [εἶχεν]·[25] ἐπὶ πολλοῖσι γὰρ αἱ κατὰ κύστιν
καθάρσιες χρησταὶ γινόμεναι ἀγαθόν, ἐσήμαινον δὲ
τοῖσι πλείστοισι ξύντηξιν καὶ ταραχὴν καὶ πόνους
καὶ χρόνους καὶ ἀκρισίας.

11. Κωματώδεες δὲ μάλιστα οἱ φρενιτικοὶ καὶ οἱ
92 καυσώδεες | ἦσαν· ἀτὰρ καὶ ἐπὶ τοῖσιν ἄλλοισι νο-
σήμασι πᾶσι τοῖσι μεγίστοισιν, ὅ τι μετὰ πυρετοῦ

which were malignant. Evacuations occurred, but much
of what was inside remained there, so that the evacuations
failed to relieve the pains, since the bowels yielded only
with difficulty to the purgatives administered: cleanings,
in fact, did additional harm in most cases. Of patients in
this condition many died rapidly, though a few held out
longer. To sum up, all patients—both those who had been
ill for longer, and the acute cases—died chiefly from bowel
complaints: it was the bowels that were responsible for
carrying everyone off.

9. Loss of appetite, to a degree that I had never met
with before, attended all the diseases described above, but
most especially the last ones, both patients of this kind (sc.
with bowel conditions) and of other kinds, who were fa-
tally stricken. Thirst afflicted some of the fever patients,
but not others; of the other cases, none was affected excessively (sc. by thirst), but as far as drink was concerned you
could diet them as you pleased.

10. The urines passed were copious, not in proportion
to, but far exceeding, the drink administered; great also
was the malignancy of the urines passed. For it had neither
the proper consistency, nor concoction, nor a beneficial
cleaning power. Although in many cases cleanings through
the bladder, if they were favorable, were a good sign,
in most patients they portended wasting, trouble, pains,
chronicity, and that no crisis would supervene.

11. Somnolence attended in particular phrenitis and
the ardent fevers, without, however, sparing all the other
diseases of the most severe febrile sort. In general most

---

24 ἐγένοντο . . . μὲν V: πάντες μὲν ἐγένοντο I
25 VI: del. Diller, *Gnomon* 13 (1927), p. 275: om. GalL

γίνοιτο. διὰ παντὸς δὲ τοῖσι πλείστοισιν ἢ βαρὺ
κῶμα παρείπετο ἢ μικροὺς καὶ λεπτοὺς ὕπνους κοιμᾶ-
σθαι.

12. Πολλὰ δὲ καὶ ἄλλα πυρετῶν ἐπεδήμησεν εἴδεα,
τριταίων, τεταρταίων, νυκτερινῶν, ξυνεχέων, μακρῶν,
πεπλανημένων, ἀσωδέων, ἀκαταστάτων. ἅπαντες δὲ
οὗτοι μετὰ πολλῆς ἐγίνοντο ταραχῆς· κοιλίαι τε γὰρ
τοῖσι πλείστοισι ταραχώδεες, φρικώδεες· ἱδρῶτες οὐ
κρίσιμοι· καὶ τὰ τῶν οὔρων ὡς ὑπογέγραπται. μακρὰ
δὲ τοῖσι πλείστοισι τούτων· οὐδὲ γὰρ αἱ γινόμεναι
τούτοισιν ἀποστάσιες ἔκρινον ὥσπερ ἐπὶ τοῖσιν ἄλλοισι. δύσκριτα μὲν οὖν²⁶ πᾶσι πάντα ἐγίνετο καὶ
ἀκρισίαι καὶ χρόνια, πολὺ δὲ μάλιστα τούτοισιν.
ἔκρινε δὲ τούτων ὀλίγοισι περὶ ὀγδοηκοστήν. τοῖσι δὲ
πλείστοισιν ἐξέλιπεν ὡς ἔτυχεν.²⁷ ἔθνησκον δὲ τούτων
ὀλίγοι ὑπὸ ὕδρωπος ὀρθοστάδην. πολλοῖσι δὲ καὶ ἐπὶ
τοῖσιν ἄλλοισι νοσήμασιν οἰδήματα παρώχλει, πολὺ
δὲ μάλιστα τοῖσι φθινώδεσι.

13. Μέγιστον δὲ καὶ χαλεπώτατον καὶ πλείστους
ἔκτεινε τὸ φθινῶδες. πολλοὶ γάρ τινες ἀρξάμενοι
94 κατὰ χειμῶνα πολλοὶ μὲν | κατεκλίθησαν, οἱ δὲ αὐτῶν
ὀρθοστάδην ὑπεφέροντο· πρωὶ δὲ τοῦ ἦρος ἔθνησκον
οἱ πλεῖστοι τῶν κατακλιθέντων· τῶν δὲ ἄλλων ἐξέλιπον μὲν αἱ βῆχες οὐδενί, ὑφίεσαν δὲ κατὰ θέρος. ὑπὸ
δὲ τὸ φθινόπωρον κατεκλίθησαν πάντες καὶ πάλιν²⁸
ἔθνησκον. μακρὰ δὲ τούτων οἱ πλεῖστοι διενόσεον.
ἤρξατο μὲν οὖν τοῖσι πλείστοισι τούτων ἐξαίφνης ἐκ
τούτων κακοῦσθαι· φρικώδεες πυκνά. πολλάκις πυρε-

patients were either sunk in a heavy coma, or experienced fitful and shallow sleep.

12. Many other forms of fever were also present in the populace—tertians, quartans, night fevers, fevers continuous, protracted, erratic, nauseous, irregular. All these cases were accompanied by great suffering: for in most cases the bowels were disturbed, shivering fits supervened, the sweats portended no crisis, and the character of the urine was as I have described. Most of these cases were protracted, for the apostases that took place did not prove critical as in other cases: all the crises in these conditions were difficult, or absent, or protracted, but most especially in the patients last described. A few of these did have a crisis about the eightieth day, but in most recovery occurred at random. A few of them died of dropsy, while still up; many sufferers from the other diseases too were troubled in addition with swellings, most particularly the consumptives.

13. The severest and most troublesome disease, which killed the greatest numbers, was consumption. Many cases began in the winter, and several of these patients took to their bed, although some stayed up through the disease. Early in the spring most of those who had gone to bed died, while none of the others lost their cough, although it did become lighter in the summer. During autumn all took to bed, and again many died. Most of these were ill for a long time. Now the condition of most of these began suddenly to grow worse, showing the following symptoms—frequent shivering; often continuous, acute fevers;

---

²⁶ οὖν V: om. I   ²⁷ ἐξέλιπεν ὡς ἔτυχεν V: ὡς ἔτυχεν ἐξέλιπεν I   ²⁸ πάλιν V: πολλοὶ I

τοὶ ξυνεχέες, ὀξέες· ἱδρῶτές τε ἄκαιροι πολλοί, ψυχροὶ διὰ τέλεος· πολλὴ ψύξις, καὶ μόγις πάλιν
ἀναθερμαινόμενοι· κοιλίαι ποικίλως ἐφιστάμεναι καὶ
πάλιν ταχὺ καθυγραινόμεναι,[29] καὶ τῶν περὶ πνεύμονα
πάντων διάδοσις κάτω· πλῆθος οὔρων οὐ χρηστῶν·
ξυντήξιες κακαί. αἱ δὲ βῆχες ἐνῆσαν μὲν διὰ τέλεος
πολλαὶ καὶ πολλὰ ἀνάγουσαι πέπονα καὶ ὑγρά, μετὰ
πόνων δὲ οὐ λίην· ἀλλ᾽ εἰ καὶ ἐπόνεον,[30] πάνυ πρηέως
πᾶσιν ἡ κάθαρσις τῶν ἀπὸ πνεύμονος ἐγίνετο. φάρυγγες οὐ λίην δακνώδεες, οὐδὲ ἁλμυρίδες οὐδὲν ἠνώ-
χλεον· τὰ μέντοι γλίσχρα καὶ λευκὰ καὶ ὑγρὰ καὶ
ἀφρώδεα πολλὰ ἀπὸ κεφαλῆς ᾖει. πολὺ δὲ μέγιστον
κακὸν παρείπετο καὶ τούτοισι καὶ τοῖσιν ἄλλοισι τὰ
περὶ τὴν ἀποσιτίην, καθάπερ ὑπογέγραπται· οὐδὲ
γὰρ πότοισι[31] μετὰ τροφῆς ἡδέως εἶχον, ἀλλὰ πάνυ
96 διῆγον | ἀδίψως· βάρος σώματος· κωματώδεες· τοῖσι
πλείστοισιν αὐτῶν οἴδημα, καὶ ἐς ὕδρωπα περι-
ίσταντο· φρικώδεες, παράληροι περὶ θάνατον.

14. Εἶδος δὲ τῶν φθινωδέων ἦν τὸ λεῖον, τὸ ὑπόλευ-
κον, τὸ φακῶδες, τὸ ὑπέρυθρον, τὸ χαροπόν· λευ-
98 κοφλεγματίαι, | πτερυγώδεες· καὶ γυναῖκες οὕτω. τὸ
μελαγχολικὸν καὶ ὕφαιμον· οἱ καῦσοι καὶ τὰ φρενιτικά, καὶ τὰ δυσεντεριώδεα τούτων ἥπτετο. τεινεσμοὶ
νέοισι φλεγματώδεσιν· αἱ μακραὶ διάρροιαι καὶ τὰ
δριμέα διαχωρήματα καὶ λιπαρὰ πικροχόλοισιν.

[29] Add. περὶ δὲ τὴν τελευτὴν πᾶσι βιαίως καθυγραινόμεναι GalL Hb marg.

unseasonable sweats, numerous, and cold throughout; a severe chill which was difficult to warm back up; bowels variously blocked, and then quickly becoming fluent again; a downward spread of the humors around the lung; copious unfavorable urines; malignant wasting. The coughs throughout were frequent, bringing up copious, concocted and liquid sputa, but without excessive effort; but even if there was a slight strain, in all cases the cleaning from the lungs took place very gently. The throat was not greatly irritated, nor did salty humors cause any distress at all. Viscid, white, moist, frothy, material descended in great quantity from the head. But by far the worst symptom that attended both these cases and the others was the distaste for food, as has been mentioned. Patients also had no pleasure in taking drinks with their food, but they remained entirely without thirst. Heaviness in the body; somnolence. In most of them there were swellings, which developed into dropsy. Shivering fits and delirium as they approached death.

14. The physical characteristics of the consumptives were: skin smooth, whitish, lentil-colored, reddish; bright eyes; a leucophlegmatic condition; winged scapulae. The same in women. As to those with a melancholic or a somewhat sanguine complexion, they were attacked by ardent fevers, phrenitis and dysenteric troubles. Tenesmus affected young, phlegmatic people; chronic diarrhea and acrid, greasy stools afflicted persons with bitter bile.

---

30 ἐπόνεον V: ὑπεπόνεον I

31 πότοισι V: πότων I

15. (1) Ἦν δὲ πᾶσι τοῖσιν ὑπογεγραμμένοισι χαλεπώτατον μὲν τὸ ἔαρ καὶ πλείστους ἀπέκτεινε, τὸ δὲ θέρος ῥήϊστον, καὶ ἐλάχιστοι ἀπώλλυντο. τοῦ δὲ φθινοπώρου καὶ ὑπὸ πληϊάδα πάλιν ἔθνησκον, οἱ πολλοὶ τεταρταῖοι.[32]

100   (2) Δοκεῖ δέ μοι | προσωφελῆσαι κατὰ λόγον τὸ γενόμενον θέρος· τὰς γὰρ θερινὰς νούσους χειμὼν ἐπιγενόμενος λύει, καὶ τὰς χειμερινὰς θέρος ἐπιγενόμενον μεθίστησι. καίτοι αὐτό γε ἐπὶ ἑωυτοῦ τὸ γενόμενον θέρος οὐκ εὐσταθὲς ἐγένετο· καὶ γὰρ ἐξαίφνης θερμὸν καὶ νότιον καὶ ἄπνοον· ἀλλ' ὅμως πρὸς τὴν ἄλλην κατάστασιν μεταλλάξαν ὠφέλησε.

16. Μέγα δὲ μέρος ἡγεῦμαι τῆς τέχνης εἶναι τὸ δύνασθαι σκοπεῖν καὶ περὶ τῶν γεγραμμένων ὀρθῶς·
102 ὁ γὰρ γνοὺς καὶ | χρεόμενος τούτοις οὐκ ἄν μοι δοκέει μέγα σφάλλεσθαι ἐν τῇ τέχνῃ. δεῖ δὲ καταμανθάνειν ἀκριβῶς[33] τὴν κατάστασιν τῶν ὡρέων ἑκάστην καὶ τὸ νόσημα· ἀγαθὸν ὅ τι κοινὸν ἐν τῇ καταστάσει ἢ ἐν τῇ νούσῳ, κακὸν ὅ τι καίριον[34] ἐν τῇ καταστάσει ἢ ἐν τῇ νούσῳ· μακρὸν ὅ τι νόσημα καὶ θανάσιμον, μακρὸν ὅ τι καὶ περιεστικόν· ὀξὺ ὅ τι θανάσιμον, ὀξὺ ὅ τι καὶ περιεστικόν. τάξιν τῶν κρισίμων ἐκ τούτων σκοπεῖσθαι καὶ προλέγειν ἐκ τούτων εὐπορεῖται. εἰδότι περὶ τούτων ἔστιν εἰδέναι οὓς καὶ ὅτε καὶ ὡς δεῖ διαιτᾶν.

---

[32] τεταρταῖοι V: ἐς τεταρταῖον I
[33] ἀκριβῶς trans. after ὡρέων V

15. (1) In all the cases described, spring was the most difficult season, and carried off the greatest numbers; summer was the easiest season, in which fewest died. In autumn and during the Pleiades, on the other hand, there were again deaths, many on the fourth day.

(2) It seems to me that on principle the arrival of summer should have been helpful, for the arrival of winter resolves the diseases of summer, and the arrival of summer dissipates those of winter. Yet the summer itself that arrived was not stable: in fact it suddenly became hot, southerly, and windless. But nevertheless the change with regard to the rest of the catastasis did prove beneficial.

16. A large part, I believe, of the medical art consists also in being able to examine correctly its writings; for a person who knows and makes use of these would not, in my opinion, be likely to go astray in the art. You must understand precisely the constitution of each of the seasons and the disease: what is a common good in the constitution or in the disease, and what is the critical evil in the constitution or in the disease. Which disease is long and deadly, which long and tending to recovery, which is acute and deadly, and which is acute and tending to recovery. The order of the critical days is to be examined starting from these principles, and medical prediction is facilitated by them. For a person who is knowledgeable about these things, it is possible to know which patients, when, and how you must treat by regimen.

---

34 κακὸν ὅ τι καίριον V: καὶ ὅ τι κοινὸν I

17. αʹ. Ἐν Θάσῳ τὸν Πάριον, ὃς κατέκειτο ὑπὲρ Ἀρτεμισίου, πυρετὸς ἔλαβεν ὀξύς, κατ᾽ ἀρχὰς συνεχής, 104 καυσώδης· δίψος· | ἀρχόμενος κωματώδης καὶ πάλιν ἄγρυπνος· κοιλίη ταραχώδης ἐν ἀρχῇσιν, οὖρα λεπτά.

Ἕκτῃ οὔρησεν ἐλαιῶδες, παρέκρουσεν.

Ἑβδόμῃ παρωξύνθη πάντα, οὐδὲν ἐκοιμήθη, ἀλλὰ οὐρά τε ὅμοια καὶ τὰ τῆς γνώμης ταραχώδεα· ἀπὸ δὲ κοιλίης χολώδεα, λιπαρὰ διῆλθεν.

Ὀγδόῃ σμικρὸν ἀπὸ ῥινῶν ἔσταξεν, ἤμεσεν ἰώδεα ὀλίγα· σμικρὰ ἐκοιμήθη.

Ἐνάτῃ διὰ τῶν αὐτῶν.

Δεκάτῃ πάντα ξυνέδωκεν.

Ἑνδεκάτῃ ἵδρωσε οὐ[35] δι᾽ ὅλου· περιέψυξε, ταχὺ δὲ πάλιν ἀνεθερμάνθη.

Τεσσαρεσκαιδεκάτῃ πυρετὸς ὀξύς, διαχωρήματα χολώδεα, λεπτά, πολλά· οὔροισιν ἐναιώρημα· παρέκρουσεν.

Ἑπτακαιδεκάτῃ ἐπιπόνως· οὔτε γὰρ ὕπνοι, ὅ τε πυρετὸς ἐπέτεινεν.

106 Εἰκοστῇ ἵδρωσε δι᾽ ὅλου· ἄπυρος, | διαχωρήματα χολώδεα, ἀπόσιτος, κωματώδης.

Εἰκοστῇ τετάρτῃ ὑπέστρεψε.

Τριακοστῇ τετάρτῃ ἄπυρος, κοιλίη οὐ συνίστατο, καὶ πάλιν ἀνεθερμάνθη.

Τεσσαρακοστῇ ἄπυρος· κοιλίη ξυνέστη χρόνον οὐ

Cases c1–c16

17. (1) In Thasos the Parian who lay sick above the temple of Artemis was seized with a fever that was acute at the beginning, continuous and ardent. Thirst. At the beginning somnolence, and then again sleeplessness. Bowels disordered at the beginning; urines thin.

Sixth day. He passed oily urine; delirium.

Seventh day. General exacerbation; no sleep; urines similar, and disorders of the mind; stools bilious and fatty.

Eighth day. Slight epistaxis; vomited a little material like verdigris; a little sleep.

Ninth day. Same.

Tenth day. General improvement.

Eleventh day. Sweated, but not over the whole body; grew chilly, but was quickly warmed up again.

Fourteenth day. Acute fever; stools bilious, thin, copious; material floating in the urines; delirium.

Seventeenth day. Troublesome, since he did not sleep, and his fever grew more intense.

Twentieth day. Sweated over the whole body; no fever; stools bilious; aversion to food; somnolent.

Twenty-fourth day. He relapsed.

Thirty-fourth day. No fever; bowels not blocked, he warmed up again.

Fortieth day. No fever; bowels blocked, but not very

πολύν·[36] ἀπόσιτος, σμικρὰ πάλιν ἐπύρεξε, καὶ διὰ
παντὸς πεπλανημένως· ἄπυρος τὰ μέν, τὰ δ᾽ οὔ· εἰ
γάρ τι διαλίποι καὶ διακουφίσαι, ταχὺ[37] πάλιν ὑπ-
έστρεφε. σιταρίοισί τε πολλοῖσι καὶ φαύλοισι προσ-
εχρῆτο· ὕπνοι κακοί. περὶ τὰς ὑποστροφὰς παρέκρουσεν, οὖρα πάχος μὲν ἔχοντα οὔρει τηνικαῦτα,
ταραχώδεα δὲ καὶ πονηρά. καὶ τὰ κατὰ κοιλίην συν-
ιστάμενα καὶ πάλιν διαλυόμενα. πυρέτια ξυνεχέα.
διαχωρήματα πολλά, λεπτά.

Ἐν εἴκοσι καὶ ἑκατὸν ἀπέθανε.

108   Τούτῳ κοιλίη συνεχέως | ἀπὸ τῆς πρώτης ὑγρὴ
χολώδεσιν, ὑγροῖσι πολλοῖσιν ἦν, ἢ ξυνισταμένη
ζέουσι καὶ ἀπέπτοισιν· οὖρα διὰ τέλεος κακά· κωματώδης τὰ πλεῖστα, μετὰ πόνων ἄγρυπνος, ἀπόσιτος
ξυνεχέως.

β΄. Ἐν Θάσῳ τὴν κατακειμένην παρὰ τὸ ψυχρὸν
110 ὕδωρ, ἐκ | τόκου θυγατέρα τεκοῦσαν καθάρσιος οὐ
γενομένης, πυρετὸς ὀξὺς φρικώδης τριταίην ἔλαβεν.
ἐκ χρόνου δὲ πολλοῦ πρὸ τοῦ τόκου πυρετώδης ἦν,
κατακλινής, ἀπόσιτος. μετὰ δὲ τὸ γενόμενον ῥῖγος
ξυνεχέες, ὀξέες, φρικώδεες οἱ πυρετοί.

Ὀγδόη πολλὰ παρέκρουσε, καὶ τὰς ἐχομένας, καὶ
ταχὺ πάλιν[37] κατενόει· κοιλίη ταραχώδης πολλοῖσι
λεπτοῖσιν, ὑδατοχόλοις· ἄδιψος.

Ἑνδεκάτη κατενόει· κωματώδης δ᾽ ἦν· οὖρα πολλὰ
λεπτὰ καὶ μέλανα, ἄγρυπνος.

Εἰκοστῇ σμικρὰ περιέψυξε καὶ ταχὺ πάλιν ἀνεθερ-

long; aversion to food; became slightly feverish again, always irregularly, the fever being sometimes absent, sometimes present: for if the fever intermitted and was relieved, a relapse followed soon afterward. He used many little bits of food, and that of a very bad kind; sleep poor. Around the relapses delirium. At that time he passed urines that were thick, disturbed, and troublesome. The bowels were sometimes constricted, but then again relaxed. Continuous slight fevers. Stools thin and copious.

Hundred and twentieth day. He died.

In this case the bowels from the first day on were either continuously moist with bilious, loose, copious stools, or contracted with fermenting, undigested stools. Urines throughout bad; mostly somnolent; sleeplessness with pains; continued aversion to food.

(2) In Thasos the woman who lay by the Cold Water was seized, on the third day after giving birth to a daughter but having no lochial discharge, by acute fever and shivering. For a long time before the birth she had suffered from fever, being confined to bed, and been averse to food. After the appearance of rigor, the fevers were continuous, acute, and attended with shivering.

Eighth day. Much delirium—also the following days—which was quickly followed by recovery of her reason; bowels disturbed with copious, thin, watery and bilious stools; no thirst.

Eleventh day. She recovered her reason, but was somnolent. Urines copious, thin and dark; no sleep.

Twentieth day. Slight chills, but she quickly warmed up

---

36 πολύν V: συχνόν I   37 ταχὺ V: om. I

259

μάνθη· σμικρὰ παρέλεγεν, ἄγρυπνος· τὰ κατὰ κοιλίην ἐπὶ τῶν αὐτῶν· οὖρα ὑδατώδεα πολλά.

Εἰκοστῇ ἑβδόμῃ ἄπυρος, κοιλίη ξυνέστη, οὐ πολλῷ δὲ χρόνῳ ὕστερον ἰσχίου δεξιοῦ ὀδύνη ἰσχυρὴ χρόνον πολύν· πυρετοὶ πάλιν παρείποντο· οὖρα ὑδατώδεα.

Τεσσαρακοστῇ τὰ μὲν περὶ τὸ ἰσχίον ἐπεκούφισε, βῆχες δὲ ξυνεχέες ὑγραὶ πολλαί· κοιλίη ξυνέστη, ἀπόσιτος· οὖρα ἐπὶ τῶν αὐτῶν. οἱ δὲ πυρετοὶ | τὸ μὲν ὅλον οὐκ ἐκλείποντες, πεπλανημένως δὲ παροξυνόμενοι, τὰ μέν, τὰ δ' οὔ.

Ἑξηκοστῇ αἱ μὲν βῆχες ἀσήμως ἐξέλιπον· οὔτε γάρ τις πτυάλων πεπασμὸς ἐγένετο οὔτε ἄλλη τῶν εἰθισμένων ἀπόστασις· σιηγὼν δὲ ἡ ἐκ τῶν ἐπὶ δεξιὰ κατεσπάσθη· κωματώδης· παρέλεγε πάλιν καὶ ταχὺ κατενόει· πρὸς δὲ τὰ γεύματα ἀπονενοημένως εἶχεν· ἡ σιηγὼν μὲν ἐπανῆκε, ἡ κοιλίη δὲ χολώδεα σμικρὰ διέδωκεν· ἐπύρεξεν ὀξυτέρως, φρικώδης—καὶ τὰς ἐχομένας—ἄφωνος καὶ πάλιν διελέγετο.

Ὀγδοηκοστῇ ἀπέθανε.

Ταύτῃ τὰ τῶν οὔρων διὰ τέλεος ἦν μέλανα καὶ λεπτὰ καὶ ὑδατώδεα. κῶμα παρείπετο, ἀπόσιτος, ἄθυμος, ἄγρυπνος, ὀργαί, δυσφορίαι· τὰ περὶ τὴν γνώμην μελαγχολικά.

γ'. Ἐν Θάσῳ Πυθίωνα, ὃς κατέκειτο ὑπεράνω τοῦ Ἡρακλείου, ἐκ πόνων καὶ κόπων καὶ διαίτης γενομένης ἀμελέος, ῥῖγος μέγα καὶ πυρετὸς ὀξὺς ἔλαβε.

again; slight wandering of her talk; no sleep; bowels the same; urines watery and copious.

Twenty-seventh day. No fever; bowels blocked; not long afterward severe pain in the right hip for a long time. Fevers again present; urines watery.

Fortieth day. Condition in the hip relieved; continuous coughing, with watery, copious sputa; bowels stopped; aversion to food; urines as before. The fevers, without entirely intermitting, were exacerbated irregularly, sometimes increasing and sometimes not doing so.

Sixtieth day. The coughing ceased without any (sc. critical) sign—there were neither any concocted sputa, nor any other of the usual apostases. Her jaw on the right side convulsed; somnolent; wandering in her talk, but reason quickly recovered; desperately averse to food; her jaw relaxed; passed small, bilious stools; fever grew more acute, with shivering—also the following days—she lost her voice, but then again she conversed.

Eightieth day. She died.

In this patient the urines right to the end were dark, thin and watery. Coma was present; aversion to food, despondency, sleeplessness, irritability, and restlessness; her turn of mind was melancholic.

(3) In Thasos Pythion, who lay above the temple of Heracles, after labor, fatigue and careless living, was seized by a violent rigor and acute fever. Tongue dry; thirst; bilious;

114 γλῶσσα ἐπίξηρος· διψώδης, χολώδης, | οὐχ ὕπνωσεν· οὖρα ὑπομέλανα, ἐναιώρημα μετέωρον, οὐχ ἵδρυτο.

Δευτέρῃ περὶ μέσον ἡμέρης ψύξις ἀκρέων, τὰ περὶ χεῖρας καὶ κεφαλὴν μᾶλλον· ἄναυδος, ἄφωνος, βραχύπνοος ἐπὶ πολὺν χρόνον· ἀνεθερμάνθη, δίψα· νύκτα <δι’>³⁸ ἡσυχίης, ἵδρωσε περὶ κεφαλὴν σμικρά.

Τρίτῃ ἡμέρην δι’ ἡσυχίης, ὀψὲ δὲ περὶ ἡλίου δυσμὰς ὑπεψύχθη σμικρά, ἄση, ταραχή· νυκτὸς ἐπιπόνως, οὐδὲν ὕπνωσεν, ἀπὸ δὲ κοιλίης σμικρὰ ξυνεστηκότα κόπρανα διῆλθε.

Τετάρτῃ πρωῒ δι’ ἡσυχίης, περὶ δὲ μέσον ἡμέρης πάντα παρωξύνθη· ψύξις, ἄναυδος, ἄφωνος, ἐπὶ τὸ χεῖρον· ἀνεθερμάνθη μετὰ χρόνον, οὔρησε μέλανα ἐναιώρημα ἔχοντα· νύκτα δι’ ἡσυχίης, ἐκοιμήθη.

116 Πέμπτῃ ἔδοξε | κουφίσαι· κατὰ δὲ κοιλίην βάρος μετὰ πόνου, διψώδης· νύκτα ἐπιπόνως.

Ἕκτῃ πρωῒ μὲν δι’ ἡσυχίης· δείλης δὲ οἱ πόνοι μέζους, παρωξύνθη· ἀπὸ δὲ κοιλίης ὀψὲ κλυσματίῳ καλῶς διῆλθε· νυκτὸς ἐκοιμήθη.

Ἑβδόμῃ ἡμέρη<ν>³⁹ ἀσώδης, ὑπεδυσφόρει, οὔρησεν ἐλαιῶδες· νυκτὸς ταραχὴ πολλή, παρέλεγεν, οὐδὲν ἐκοιμήθη.

Ὀγδόῃ πρωῒ μὲν ἐκοιμήθη σμικρά, ταχὺ δὲ ψύξις, ἀφωνίη, λεπτὸν πνεῦμα καὶ μινυθῶδες· ὀψὲ δὲ πάλιν ἀνεθερμάνθη, παρέκρουσεν· ἤδη δὲ πρὸς ἡμέρην σμικρὰ ἐκουφίσθη, διαχωρήματα ἄκρητα σμικρὰ χολώδεα.

Ἐνάτῃ κωματώδης, ἀσώδης, ὅτε διεγείροιτο· οὐ

no sleep; urines rather dark, with material suspended in them, which did not settle.

Second day. About midday coldness of the extremities, especially in the hands and head; lost his speech and his voice; short of breath for a long time; warmed up; thirst; a quiet night; slight sweats about the head.

Third day. A quiet day, but later, about sunset, he had a slight chill; nausea; distress; troubled night without sleep; small, compact stools passed.

Fourth day. Early morning peaceful, but about midday general exacerbation; chill; speechless and voiceless; deterioration; he warmed up after a time; passed dark urine with material floating in it; night peaceful; he slept.

Fifth day. Seemed to be relieved, but there was heaviness in the bowels with pain; thirst; troubled night.

Sixth day. Early morning peaceful; toward evening the pains increased; exacerbation; but later a small clyster caused a good movement of the bowels. At night he slept.

Seventh day. Nausea during the day; patient rather uneasy; he passed oily urine; at night much distress; wandering in his talk; no sleep at all.

Eighth day. Early in the morning a little sleep; but quickly there was a chill; loss of voice; respiration thin and weak; in the evening he warmed up again; was delirious; toward morning slightly better; stools uncompounded, small in amount, bilious.

Ninth day. Somnolent; nausea whenever he woke up.

---

38 Add. Aldina: abs. from VI
39 ἡμέρην Jouanna: ἡμέρῃ VI

λίην διψώδης· περὶ δὲ ἡλίου δυσμὰς ἐδυσφόρει, παρέλεγε, νύκτα κακήν.

Δεκάτῃ πρωῒ ἄφωνος, πολλὴ ψύξις, πυρετὸς ὀξύς, πολὺς ἱδρώς· ἔθανεν.

Ἐν ἀρτίῃσιν οἱ πόνοι τούτῳ.

δʹ. Ὁ φρενιτικὸς τῇ πρώτῃ κατακλιθεὶς ἤμεσεν ἰώδεα πολλά, λεπτά· πυρετὸς φρικώδης, πολὺς ἱδρὼς ξυνεχὴς δι᾽ ὅλου· κεφαλῆς καὶ τραχήλου βάρος μετ᾽ ὀδύνης· οὖρα λεπτά, | ἐναιωρήματα σμικρά, διεσπασμένα, οὐχ ἵδρυτο. ἀπὸ δὲ κοιλίης ἐξεκόπρισεν ἀθρόα· πολλὰ παρέκρουσεν· οὐδὲν ὕπνωσε.

Δευτέρῃ πρωῒ ἄφωνος, πυρετὸς ὀξύς, ἵδρωσεν· οὐ διέλιπε· παλμοὶ δι᾽ ὅλου τοῦ σώματος· νυκτὸς σπασμοί.

Τρίτῃ πάντα παρωξύνθη.

Τετάρτῃ ἀπέθανεν.

εʹ. Ἐν Λαρίσσῃ φαλακρὸς μηρὸν δεξιὸν ἐπόνησεν ἐξαίφνης· τῶν δὲ προσφερομένων οὐδὲν ὠφέλει.

Τῇ πρώτῃ πυρετὸς ὀξύς, καυσώδης, ἀτρεμέως εἶχεν, οἱ δὲ πόνοι παρείποντο.

Δευτέρῃ τοῦ μηροῦ μὲν ὑφίεσαν οἱ πόνοι, ὁ δὲ πυρετὸς ἐπέτεινεν, ὑπεδυσφόρει, οὐκ ἐκοιμᾶτο, ἄκρεα ψυχρά, οὔρων πλῆθος διῄει οὐ χρηστῶν.

Τρίτῃ τοῦ μηροῦ μὲν ὁ πόνος ἐπαύσατο, παρακοπὴ δὲ τῆς γνώμης καὶ ταραχὴ καὶ πολὺς βληστρισμός.

Τετάρτῃ περὶ μέσον ἡμέρης ἔθανεν. |

ϛʹ. Ἐν Ἀβδήροισι Περικλέα πυρετὸς ἔλαβεν ὀξύς, ξυνεχὴς μετὰ πονου· πολλὴ δίψα, ἄση, ποτὸν κατέ-

Not too thirsty. About sunset he was uncomfortable; wandered in his talk; a bad night.

Tenth day. In the early morning he lost his voice; severe chill; acute fever; abundant sweat; he died.

In this case the pains were on even days.

(4) A patient with phrenitis vomited much thin material resembling verdigris on the first day he took to his bed; fever with shivering; abundant continuous sweating over his whole body; heaviness of the head and neck, with pain; urines thin, with a little scattered material floating in them, which did not settle. He discharged material from the bowels in a mass; much delirium; no sleep at all.

Second day. In the early morning voiceless; acute fever; sweating; no intermission; throbbing all over the body; at night convulsions.

Third day. General exacerbation.

Fourth day. He died.

(5) In Larissa a bald man suddenly experienced pain in his right thigh. None of the remedies applied did any good.

First day. Acute fever of the ardent type; the patient was quiet, but the pains persisted.

Second day. The pains in the thigh subsided, but the fever heightened; the patient was quite restless and did not sleep; extremities cold; a good amount of urine passed but to no benefit.

Third day. The pain in the thigh ceased, but there was derangement of the intellect, with distress and much tossing.

Fourth day. He died about midday.

(6) In Abdera Pericles was seized with acute fever, continuous and painful; much thirst; nausea; he could not

χειν οὐκ ἠδύνατο· ἦν δὲ ὑπόσπληνός τε καὶ καρηβα-
ρικός.

Τῇ πρώτῃ ἡμορράγησεν ἐξ ἀριστεροῦ· πολὺς μέντοι ὁ πυρετὸς ἐπέτεινεν· οὔρησε πολύ, θολερόν, λευ-
κόν· κείμενον οὐ καθίστατο.

Δευτέρη πάντα παρωξύνθη· τὰ μέντοι οὖρα παχέα
μὲν ἦν, ἱδρυμένα δὲ μᾶλλον· καὶ τὰ περὶ τὴν ἄσην
ἐκούφισεν, ἐκοιμήθη.

Τρίτῃ πυρετὸς ἐμαλάχθη, οὔρων πλῆθος, πέπονα,
πολλὴν ὑπόστασιν ἔχοντα, νύκτα δι' ἡσυχίης.

Τετάρτῃ περὶ μέσον ἡμέρης ἵδρωσε πολλῷ θερμῷ
δι' ὅλου· ἄπυρος, ἐκρίθη· οὔθ' ὑπέστρεψεν. |

122   ζ'. Ἐν Ἀβδήροισι τὴν παρθένον, ἣ κατέκειτο ἐπὶ
τῆς ἱρῆς ὁδοῦ, πυρετὸς καυσώδης ἔλαβεν· ἦν δὲ διψώδης καὶ ἄγρυπνος. κατέβη δὲ τὰ γυναικεῖα πρῶτον
αὐτῇ.

Ἕκτῃ ἄση πολλή, ἔρευθος, φρικώδης, ἀλύουσα.

Ἑβδόμῃ διὰ τῶν αὐτῶν· οὖρα λεπτὰ μέν, εὔχροα
δέ· τὰ περὶ τὴν κοιλίην οὐκ ἠνώχλει.

Ὀγδόῃ κώφωσις, πυρετὸς ὀξύς, ἄγρυπνος, ἀσώδης, φρικώδης· κατενόει, οὖρα ὅμοια.

Ἐνάτῃ διὰ τῶν αὐτῶν, καὶ τὰς ἐπομένας οὕτως· ἡ
κώφωσις παρέμεινε.

Τεσσαρεσκαιδεκάτῃ τὰ τῆς γνώμης ταραχώδεα· ὁ
πυρετὸς συνέδωκεν.

Ἑπτακαιδεκάτῃ διὰ ρινῶν ἐρρύη πολύ, ἡ κώφωσις
σμικρὰ συνέδωκε.

keep down what he drank. There was slight enlargement of the spleen and heaviness in the head.

First day. Epistaxis from the left (sc. nostril); the fever, however, increased greatly. He passed urine that was copious, turbid and white, but on standing did not produce a sediment.

Second day. General exacerbation; the urines, however, were thick and deposited more; the nausea was relieved; the patient slept.

Third day. The fever slackened; abundance of urine, with plentiful, concocted sediment; night quiet.

Fourth day. About midday an abundant, hot, sweating over the whole body; afebrile; crisis occurred; no relapse.

(7) In Abdera the girl who lay near the Sacred Way was seized with a fever of the ardent type. She was thirsty and sleepless. Menstruation occurred for the first time.

Sixth day. Much nausea; ruddiness; shivering; restlessness.

Seventh day. Same. Urines thin but of a good color; no trouble with her bowels.

Eighth day. Deafness; acute fever; sleeplessness; nausea; shivering; she retained her reason; urines similar.

Ninth day. Same, and also on the following days: the deafness persisted.

Fourteenth day. Disturbance of the mind; the fever subsided.

Seventeenth day. Copious epistaxis; the deafness improved a little.

Καὶ τὰς ἑπομένας ἄση· κωφότης ἐνῆν καὶ παράληρος.

Εἰκοστῇ ποδῶν ὀδύνη· κωφότης, παράληρος ἀπέλιπεν· ἡμορράγησε σμικρὰ διὰ ῥινῶν, ἵδρωσεν, ἄπυρος.

Εἰκοστῇ τετάρτῃ ὁ πυρετὸς ὑπέστρεψε, κώφωσις πάλιν, ποδῶν ὀδύνη παρέμεινε, παρακοπή.

Εἰκοστῇ ἑβδόμῃ ἵδρωσε πολλῷ, ἄπυρος, ἡ κώφωσις ἐξέλιπεν, ἡ τῶν ποδῶν ὑπέμενεν ὀδύνη, τὰ δ' ἄλλα τελέως ἐκρίθη. |

124   η΄. Ἐν Ἀβδήροισιν Ἀναξίωνα, ὃς κατέκειτο παρὰ τὰς Θρηϊκίας πύλας, πυρετὸς ὀξὺς ἔλαβε· πλευροῦ δεξιοῦ ὀδύνη ξυνεχής, ἔβησσε ξηρά, οὐδ' ἔπτυε τὰς πρώτας· διψώδης, ἄγρυπνος, οὖρα δὲ εὔχρω πολλὰ λεπτά.

Ἕκτῃ παράληρος· πρὸς δὲ τὰ θερμάσματα οὐδὲν ἐνεδίδου.

Ἑβδόμῃ ἐπιπόνως· ὁ γὰρ πυρετὸς ἐπέτεινεν, οἵ τε πόνοι οὐ ξυνεδίδοσαν, αἵ τε βῆχες ἠνώχλεον, δύσπνοός τε ἦν.

Ὀγδόῃ ἀγκῶνα ἔταμον·[40] ἐρρύη πολλὸν οἷον δεῖ· ξυνέδωκαν μὲν οἱ πόνοι, αἱ μέντοι βῆχες ξηραὶ παρείποντο.

Ἑνδεκάτῃ ξυνέδωκαν οἱ πυρετοί, σμικρὰ περὶ κεφαλὴν ἵδρωσεν, αἵ τε βῆχες καὶ τὰ ἀπὸ πνεύμονος ὑγρότερα.

Ἑπτακαιδεκάτῃ ἤρξατο σμικρὰ πέπονα πτύειν· ἐκουφίσθη. |

On the following days nausea; deafness was still present, and also delirium.

Twentieth day. Pain in the feet; deafness; the delirium ceased; slight epistaxis; she sweated; no fever.

Twenty-fourth day. The fever returned; deafness again; pain in the feet persisted; delirium.

Twenty-seventh day. She passed abundant sweat; no fever; the deafness remitted; a little of the pain in the feet remained, but in other respects there was a definitive crisis.

(8) In Abdera Anaxion, who lay by the Thracian Gate, was seized with acute fever. Continuous pain in the right side; a dry cough, with no sputum on the first days. Thirst; sleeplessness; urines of a good color, copious and thin.

Sixth day. Delirium; warm applications brought no relief.

Seventh day. Troublesome, for the fever grew stronger and the pains were not relieved, while the coughing caused distress and the patient had difficulty breathing.

Eighth day. I bled him at the elbow. There was an abundant, proper flow of blood: the pains were relieved, although the dry coughing persisted.

Eleventh day. The fever abated; slight sweating about the head; the cough and the sputa became looser.

Seventeenth day. He began to expectorate a little, concocted sputa; was relieved.

---

126   Εἰκοστῇ ἵδρωσεν, ἄπυρος, μετὰ δὲ κρίσιν ἐκου-
φίσθη· διψώδης δὲ ἦν καὶ τῶν ἀπὸ πνεύμονος οὐ χρησταὶ αἱ καθάρσιες.

Ἑβδόμη καὶ εἰκοστῇ ὁ πυρετὸς ὑπέστρεψεν, ἔβησ-
σεν, ἀνῆγε πέπονα πολλά· οὔροισιν ὑπόστασις πολλὴ
λευκή, ἄδιψος ἐγένετο, ὕπνοι.

Τριακοστῇ τετάρτῃ ἵδρωσε δι᾽ ὅλου, ἄπυρος,
ἐκρίθη πάντα. |

128   θ΄. Ἐν Ἀβδήροισιν Ἡρόπυθος κεφαλὴν ὀρθοστάδην ἐπιπόνως εἶχεν· οὐ πολλῷ δὲ χρόνῳ ὕστερον κατεκλίθη. ᾤκει πλησίον τῆς ἄνω ἀγωγῆς. πυρετὸς ἔλαβε
καυσώδης, ὀξύς.

Ἔμετοι τὸ κατ᾽ ἀρχὰς πολλῶν χολωδέων, διψώδης,
πολλὴ δυσφορίη· οὖρα λεπτὰ μέλανα, ἐναιώρημα
μετέωρον ὁτὲ μέν, ὁτὲ δ᾽ οὔ· νύκτα ἐπιπόνως, πυρετὸς
ἄλλοτε ἀλλοίως παροξυνόμενος, τὰ πλεῖστα ἀτάκτως.

Περὶ δὲ τεσσαρεσκαιδεκάτην κώφωσις, οἱ πυρετοὶ
ἐπέτεινον, οὖρα διὰ τῶν αὐτῶν.

Εἰκοστῇ πολλὰ παρέκρουσε καὶ τὰς ἑπομένας.

Τεσσαρακοστῇ διὰ ῥινῶν ἠμορράγησε πολύ, καὶ
κατενόει μᾶλλον· ἡ κώφωσις ἐνῆν μέν, ἧσσον δέ· οἱ
πυρετοὶ ξυνέδωκαν· ἠμορράγει τὰς ἑπομένας πυκνὰ
κατ᾽ ὀλίγον.

Περὶ δὲ ἑξηκοστὴν αἱ μὲν αἱμορραγίαι ἀπεπαύ-
σαντο, ἰσχίου δὲ δεξιοῦ ὀδύνη ἰσχυρὴ καὶ οἱ πυρετοὶ
ἐπέτεινον.

Οὐ πολλῷ δὲ χρόνῳ ὕστερον πόνοι τῶν κάτω πάν-
των· ξυνέπιπτε δὲ ἢ τοὺς πυρετοὺς εἶναι μέζους καὶ

Twentieth day. He sweated and became afebrile; after the crisis he was relieved; but he was thirsty and the cleanings from his lungs were not favorable.

Twenty-seventh day. The fever returned; he coughed bringing up concocted sputa in a good amount; in the urines there was copious, white sediment; thirst disappeared; he slept.

Thirty-fourth day. Sweated over the whole body; no fever; complete crisis.

(9) In Abdera Heropythus had pain in the head while he was still up, but not much later he took to his bed. He lived close to the Upper Road. An ardent fever seized him which was acute.

He vomited at the beginning copious, bilious material; thirst; great discomfort; urines thin and dark, sometimes containing material suspended high up in it, but not at other times. Troublesome night: the fever rose now in one way, now in another, but for the most part irregularly.

About the fourteenth day. Deafness; the fever grew worse; urines the same.

Twentieth day. Much delirium, also on the following days.

Fortieth day. Copious epistaxis; regained more of his reason; some deafness, but less than before; the fever was relieved. Epistaxis on the following days, frequent but only a little at a time.

About the sixtieth day the bleedings (sc. from the nose) ceased, but there was violent pain in his right hip, and the fever increased.

Not long afterward, pains arose in all the lower parts; it happened that either the fever was then higher and the

271

τὴν κώφωσιν πολλήν, ἢ ταῦτα μὲν ὑφιέναι καὶ κου-
φίζειν, τῶν δὲ κάτω περὶ ἰσχία μέζους εἶναι τοὺς πόνους.

Ἤδη δὲ περὶ ὀγδοηκοστὴν ξυνέδωκε μὲν πάντα,
ἐξέλιπε δὲ οὐδέν· οὐρά τε γὰρ εὔχρω καὶ πλείους ὑποστάσιας ἔχοντα κατέβαινεν, οἱ παράληροί τε μείους
ἦσαν.

Περὶ δὲ ἑκατοστὴν κοιλίη πολλοῖσι χολώδεσιν ἐπεταράχθη, καὶ ᾔει χρόνον οὐκ ὀλίγον πολλὰ τοιαῦτα,
καὶ πάλιν δυσεντεριώδεα μετὰ πόνου· τῶν δὲ ἄλλων
ῥαστώνη. τὸ δὲ σύνολον οἵ τε πυρετοὶ ἐξέλιπον καὶ ἡ
130   κώφωσις | ἐπαύσατο.

Ἐν ἑκατοστῇ εἰκοστῇ τελέως ἐκρίθη.

ί. Ἐν Ἀβδήροισι Νικόδημον ἐξ ἀφροδισίων καὶ
πότων πῦρ ἔλαβεν. ἀρχόμενος δὲ ἦν ἀσώδης καὶ καρ-
διαλγικός, διψώδης· γλῶσσα ἐπεκαύθη· οὖρα λεπτὰ
μέλανα.

Δευτέρη ὁ πυρετὸς παρωξύνθη· φρικώδης, ἀσώδης,
οὐδὲν ἐκοιμήθη· ἤμεσε χολώδεα ξανθά· οὖρα ὅμοια·
νύκτα δι' ἡσυχίης, ὕπνωσε.

Τρίτη ὑφῆκε πάντα, ῥαστώνη· περὶ δὲ ἡλίου δυσμὰς πάλιν ὑπεδυσφόρει· νύκτα ἐπιπόνως.

Τετάρτη ῥῖγος, πυρετὸς πολύς, πόνοι πάντων· οὖρα
λεπτά, ἐναιώρημα· νύκτα πάλιν δι' ἡσυχίης. |

132   Πέμπτη ἐνῆν μὲν πάντα, ῥαστώνη δὲ ἦν.

Ἕκτη τῶν αὐτῶν πόνοι πάντων· οὔρησεν, ἐναι-
ώρημα· παρέκρουσε πολλά.

Ἑβδόμη ῥαστώνη.

deafness great, or else these symptoms became lighter and less severe, but the pains in the lower parts about the hips increased.

But from *about the eightieth day* on all the symptoms were relieved, although none disappeared. The urines passed were of a good color and had more deposits, while the wanderings in his talk were less frequent.

About the hundredth day the bowels were disturbed with copious, bilious stools, and such continued for no little time; then followed the symptoms of dysentery accompanied by pain, while the other symptoms abated. In summary, the fevers remitted and the deafness disappeared.

Hundred and twentieth day. Definitive crisis.

(10) In Abdera Nicodemus after sexual pleasures and drunkenness was seized with fever. At the beginning he had nausea and pain in the cardia; thirst; tongue parched; urines thin and dark.

Second day. The fever peaked; shivering; nausea; no sleep; vomitus bilious and yellow; urine the same; a quiet night; he slept.

Third day. All symptoms less severe; relief. But about sunset he was again somewhat uncomfortable; night troublesome.

Fourth day. Rigor; much fever; pains throughout the body; urines thin, with material floating in them; the night, on the other hand, was quiet.

Fifth day. All symptoms present, although mitigated.

Sixth day. Same general pains; he urine with material floating in it; much delirium.

Seventh day. Relief.

Ὀγδόῃ τὰ ἄλλα ξυνέδωκε πάντα.

Δεκάτῃ καὶ τὰς ἑπομένας ἐνῆσαν μὲν οἱ πόνοι, ἦσσον δὲ πάντες· οἱ δὲ παροξυσμοὶ καὶ οἱ πόνοι τούτῳ διὰ τέλεος ἐν ἀρτίῃσιν ἦσαν μᾶλλον.

Εἰκοστῇ οὔρησε λευκόν, πάχος εἶχε, κείμενον οὐ καθίστατο· ἵδρωσε πολλῷ, ἔδοξεν ἄπυρος γενέσθαι· δείλης δὲ πάλιν ἐθερμάνθη, καὶ τῶν αὐτῶν πόνοι· φρίκη, δίψα, σμικρὰ παρέκρουσεν.

Εἰκοστῇ τετάρτῃ οὔρησε πολὺ λευκόν, πολλὴν ὑπόστασιν ἔχον. ἵδρωσε πολλῷ θερμῷ δι᾽ ὅλου, ἄπυρος ἐκρίθη. |

134 ια'. Ἐν Θάσῳ γυνὴ δυσήνιος ἐκ λύπης μετὰ προφάσιος ὀρθοστάδην ἐγένετο ἄγρυπνός τε καὶ ἀπόσιτος· καὶ διψώδης ἦν καὶ ἀσώδης. ᾤκει δὲ πλησίον τῶν Πυλάδου ἐπὶ τοῦ λείου.

Τῇ πρώτῃ ἀρχομένης νυκτὸς φόβοι, λόγοι πολλοί, δυσθυμίη, πυρέτιον λεπτόν. πρωῒ σπασμοὶ πολλοί· ὅτε δὲ διαλίποιεν οἱ σπασμοὶ οἱ πολλοί, παρέλεγεν, ᾐσχρομύθει· πολλοὶ πόνοι, μεγάλοι, ξυνεχέες.

Δευτέρῃ διὰ τῶν αὐτῶν· οὐδὲν ἐκοιμᾶτο, πυρετὸς ὀξύτερος.

Τρίτῃ οἱ μὲν σπασμοὶ ἀπέλιπον· κῶμα δὲ καὶ καταφορή, καὶ πάλιν ἔγερσις· ἀνήϊσσε, κατέχειν οὐκ ἠδύνατο, παρέλεγε πολλά· πυρετὸς ὀξύς· ἐς νύκτα δὲ ταύτην ἵδρωσε πολλῷ θερμῷ δι᾽ ὅλου· ἄπυρος, ὕπνωσε, πάντα κατενόει, ἐκρίθη.

Περὶ δὲ τρίτην ἡμέρην οὖρα μέλανα λεπτά, ἐναι-

Eighth day. All the other symptoms were less severe.

Tenth day and following days. The pains were present, but all were less severe. The exacerbations and the pains in the case of this patient always tended to occur on the even days.

Twentieth day. Urines white, having a thick consistency; on standing they deposited no sediment. Copious sweating; he seemed to lose his fever, but toward evening he grew warm again, with pains in the same parts; shivering; thirst; slight delirium.

Twenty-fourth day. Passed much white urine with abundant sediment. Hot sweating over the whole body; the fever passed away in a crisis.

(11) In Thasos a woman disturbed after suffering grief for a good reason, without taking to bed became sleepless and adverse to food; she suffered thirst and nausea. Her habitation was near the place of Pylades on the plain.

First day. As night began she experienced fears, much rambling, depression and slight feverishness. Early in the morning frequent convulsions; when these had for the most part ceased, her talk wandered and she uttered obscenities; many pains, severe and continuous.

Second day. Same; no sleep; fever more acute.

Third day. The convulsions ceased, but were succeeded by coma and oppression, and then again by wakefulness. She would jump up, and could not keep herself still; she wandered a great deal in her talk; fever acute; on this night a copious, hot sweating over her whole body; no fever; she slept; she regained her reason perfectly, and had a crisis.

About the third day urines dark and thin, with mostly

275

ὥρημα δὲ ἐπὶ πολὺ στρογγύλον, οὐχ ἵδρυτο, περὶ δὲ κρίσιν γυναικεῖα πολλὰ κατέβη. |

136 ιβ'. Ἐν Λαρίσσῃ παρθένον πυρετὸς ἔλαβε καυσώδης, ὀξύς· ἄγρυπνος, διψώδης, γλῶσσα λιγνυώδης, ξηρή· οὖρα εὔχρω μέν, λεπτὰ δέ.

Δευτέρῃ ἐπιπόνως, οὐχ ὕπνωσε.

Τρίτῃ πολλὰ διῆλθεν ἀπὸ κοιλίης ὑδατόχλοα,[41] καὶ τὰς ἑπομένας· ᾔει τοιαῦτα εὐφόρως.

Τετάρτῃ οὔρησε λεπτὸν ὀλίγον· εἶχεν ἐναιώρημα μετέωρον, οὐχ ἵδρυτο· παρέκρουσεν ἐς νύκτα.

Ἕκτῃ διὰ ῥινῶν λάβρον ἐρρύη πολύ· φρίξασα ἵδρωσε πολλῷ θερμῷ δι' ὅλου· ἄπυρος, ἐκρίθη. ἐν δὲ τοῖσι πυρετοῖσι καὶ ἤδη κεκριμένων γυναικεῖα κατέβη πρῶτον τότε· παρθένος γὰρ ἦν.

Ἦν δὲ διὰ παντὸς ἀσώδης, φρικώδης, ἔρευθος προσώπου, ὀμμάτων ὀδύνη· καρηβαρική. ταύτῃ οὐχ ὑπέστρεψεν, ἀλλ' ἐκρίθη. οἱ πόνοι ἐν ἀρτίῃσιν.

ιγ'. Ἀπολλώνιος ἐν Ἀβδήροισιν ὀρθοστάδην ὑπε-
138 φέρετο χρόνον | πολύν. ἦν δὲ μεγαλόσπλαγχνος, καὶ περὶ ἧπαρ ξυνήθης ὀδύνη χρόνον πολὺν παρείπετο· καὶ δὴ τότε καὶ ἰκτερώδης ἐγένετο· φυσώδης, χροιῆς τῆς ὑπολεύκου. φαγὼν δὲ καὶ πιὼν ἀκαιρότερον βόειον ἐθερμάνθη σμικρά· τὸ πρῶτον κατεκλίθη. γάλαξι δὲ χρησάμενος ἐφθοῖσι καὶ ὠμοῖσι πολλοῖσιν, αἰγείοισι καὶ μηλείοισι, καὶ διαίτῃ κακῇ, πάντων βλά-

---

[41] ὑδατόχλοα V: ὑδατόχροα I

round particles floating in them, which did not settle. Near the crisis copious menstruation came down.

(12) In Larissa a girl was seized with a fever of the ardent type in an acute form. Sleeplessness; thirst; tongue sooty and parched; urines of good color, but thin.

Second day. Troublesome; no sleep.

Third day. She passed copious stools, watery and greenish, also the following days; this was how the stools passed, and the patient bore it well.

Fourth day. She passed a little thin urine with material suspended in its upper part, that did not settle; delirium at night.

Sixth day. Violent and abundant epistaxis; after a shivering fit, an abundant hot sweating over her whole body followed; no fever; this was a crisis. During the fevers and after the crisis menstruation came down for the first time, for she was just a girl.

Throughout she suffered nausea and shivering; redness of her face; pain in the eyes; heaviness of the head. In this patient there was no relapse, but a true crisis. The pains were on even days.

(13) Apollonius in Abdera bore his ailment for a long time while still up. He had enlarged viscera, and continuous pain in the region of the liver was present for a long time. Moreover, at this time he became jaundiced; flatulent; complexion whitish. After dining unseasonably on beef and drinking cow's milk, he warmed up a little; he took to his bed for the first time. Then, after employing milk both boiled and raw in great amounts—of both goats and ewes—and adopting a thoroughly bad regimen, he

βαι αἱ μεγάλαι·[42] οἵ τε γὰρ πυρετοὶ παρωξύνθησαν,
κοιλίη τε τῶν προσενεχθέντων οὐδὲν διέδωκεν ἄξιον
λόγου, οὐρά τε λεπτὰ καὶ ὀλίγα διῄει· ὕπνοι οὐκ ἐνῆ-
σαν· ἐμφύσημα κακόν, πολὺ δίψος, κωματώδης, ὑποχονδρίου δεξιοῦ ἔπαρμα ξὺν ὀδύνῃ, ἄκρεα πάντοθεν
ὑπόψυχρα· σμικρὰ παρέλεγε, λήθη πάντων ὅ τι λέγοι·
παρεφέρετο.

Περὶ δὲ τεσσαρεσκαιδεκάτην, ἀφ' ἧς ῥιγώσας ἐπεθερμάνθη[43] καὶ κατεκλίθη ἐξεμάνη· βοή, ταραχή, λόγοι πολλοί, καὶ πάλιν ἵδρυσις, καὶ τὸ κῶμα τηνι-
καῦτα προσῆλθε. μετὰ δὲ ταῦτα κοιλίη ταραχώδης |
140   πολλοῖσι χολώδεσιν, ἀκρήτοισιν, ὠμοῖσιν· οὐρα μέλανα, σμικρά, λεπτά· πολλὴ δυσφορίη· τὰ τῶν δια-
χωρημάτων ποικίλως· ἢ γὰρ μέλανα καὶ σμικρὰ καὶ
ἰώδεα ἢ λιπαρὰ καὶ ὠμὰ καὶ δακνώδεα· κατὰ δὲ χρόνους ἐδόκεε καὶ γαλακτώδεα διδόναι.

Περὶ δὲ τετάρτην καὶ εἰκοστὴν διὰ παρηγορίης· τὰ
μὲν ἄλλα ἐπὶ τῶν αὐτῶν, σμικρὰ δὲ κατενόησεν· ἐξ
οὗ δὲ κατεκλίθη, οὐδενὸς ἐμνήσθη· πάλιν δὲ ταχὺ
παρενόει, ὥρμητο πάντα ἐπὶ τὸ χεῖρον.

Περὶ δὲ τριηκοστὴν πυρετὸς ὀξύς· διαχωρήματα
πολλὰ λεπτά, παράληρος, ἄκρεα ψυχρά, ἄφωνος.

Τριηκοστῇ τετάρτῃ ἀπέθανε.

Τούτῳ διὰ τέλεος, ἐξ οὗ καὶ ἐγὼ οἶδα, κοιλίη ταραχώδης, οὐρα λεπτὰ μέλανα, κωματώδης, ἄγρυπνος,
ἄκρεα ψυχρά, παράληρος διὰ τέλεος.

suffered many injuries of all kinds: the fevers were exacerbated, the bowels passed practically nothing of the food ingested, and the urines were thin and scanty; no sleep; grievous distension with wind; much thirst; somnolence; swelling of the right hypochondrium with pain; extremities in all parts rather cold; a little delirious talk; forgetfulness of everything he talked about; disorientation.

About the fourteenth day, after he had had a rigor, he became warm and took to his bed; he raged madly; shouting, distress, much rambling, followed by calm again; coma came on then. Afterward the bowels were disordered with copious, bilious, uncompounded and crude stools; urines dark, scanty and thin. Great discomfort. The evacuations varied in their appearance: they were either dark, scanty and like verdigris, or greasy, crude and irritating; at times he even seemed to pass what looked like milk.

About the twenty-fourth day the patient became more comfortable; the other symptoms were as before but he did recover a little of his reason; from the time since he had taken to bed he could remember nothing; he quickly became delirious again. Everything took a turn for the worse.

About the thirtieth day acute fever; copious, thin stools; delirium; cold extremities; voiceless.

Thirty-fourth day. He died.

This patient throughout, as long as I had knowledge of the case, suffered from disordered bowels; urines thin and dark; somnolence; sleeplessness; extremities cold; delirious throughout.

---

42 αἱ βλάβαι μεγάλαι I: βλάβη μεγάλη V
43 ἐπεθερμάνθη V: ἀπεθερμάνθη I

ιδ΄. Ἐν Κυζίκῳ γυναικὶ[44] θυγατέρας τεκούσῃ διδύμους[45] καὶ δυστοκησάσῃ καὶ οὐ πάνυ καθαρθείσῃ.

Τῇ πρώτῃ πυρετὸς φρικώδης ὀξύς· κεφαλῆς καὶ
τραχήλου βάρος μετ᾽ ὀδύνης· ἄγρυπνος ἐξ ἀρχῆς,
σιγῶσα δὲ καὶ σκυθρωπὴ καὶ οὐ πειθομένη· οὖρα
λεπτὰ καὶ ἄχρω· διψώδης, ἀσώδης τὸ πολύ· κοιλίη
πεπλανημένως ταραχώδης καὶ πάλιν ξυνισταμένη.

Ἕκτῃ ἐς νύκτα πολλὰ παρέλεγεν, οὐδὲν ἐκοιμήθη.

Περὶ δὲ ἑνδεκάτην ἐοῦσα ἐξεμάνη καὶ πάλιν κατ-
ενόει· οὖρα μέλανα, | λεπτά, καὶ πάλιν διαλείποντα
ἐλαιώδεα· καὶ κοιλίη πολλοῖσι, λεπτοῖσι, ταραχώδεσι.

Τεσσαρεσκαιδεκάτῃ σπασμοὶ πολλοί, ἄκρεα ψυχρά, οὐδὲν ἔτι κατενόει, οὖρα ἐπέστη.

Ἑξκαιδεκάτῃ ἄφωνος.

Ἑπτακαιδεκάτῃ ἀπέθανεν.

ιε΄. Ἐν Θάσῳ Δεάλκεος γυναῖκα, ἣ κατέκειτο ἐπὶ
τοῦ Ἡλείου,[46] πυρετὸς φρικώδης ὀξὺς ἐκ λύπης ἔλαβεν. ἐξ ἀρχῆς δὲ περιεστέλλετο καὶ διὰ τέλεος· αἰεὶ
σιγῶσα ἐψηλάφα, ἔτιλλεν, ἔγλυφεν, ἐτριχολόγει· δάκρυα καὶ πάλιν γέλως, οὐκ ἐκοιμᾶτο· ἀπὸ κοιλίης ἐρεθισμῷ[47] οὐδὲν διῄει· σμικρὰ ὑπομιμνησκόντων ἔπινεν·
οὖρα λεπτὰ σμικρά, οὐκ ἔχοντα ὑπόστασιν·[48] πυρετοὶ
πρὸς χεῖρα λεπτοί· ἀκρέων ψύξις.

44 γυναικὶ I: om. V
45 διδύμους I: διδύμας V
46 Ἡλείου Wenkebach (II) in *app. crit.* GalL (Ar.) "temple
of the Sun": λείου VI

(14) In Cyzicus a woman gave birth with a difficult labor to twin daughters, and her lochial cleaning was far from complete.

First day. Acute fever with shivering; heaviness of the head and neck, with pain. Sleepless from the first; silent, sulky and disobedient. Urines thin and colorless; thirsty; nausea much of the time; bowels irregularly disturbed and then blocked again.

Sixth day. Much wandering in her talk at night; no sleep.

About the eleventh day she lost her mind, but then became rational again; urines dark, thin, and then intermittently oily; copious, thin, disordered stools.

Fourteenth day. Many convulsions; extremities cold; no further recovery of reason; urines suppressed.

Sixteenth day. Voiceless.

Seventeenth day. She died.

(15) In Thasos the wife of Dealces, who lay near the temple of the Sun, was seized after a grief with an acute fever and shivering. From the beginning she would wrap herself up, and so throughout to the end; she was always silent; she would fumble about, pluck, scratch, and pick hairs; she would weep tears, and then laugh; she did not sleep. Though stimulated, the bowels passed nothing; she drank a little when she was reminded to. Urines thin and scanty with no deposit; fever slight to the touch; coldness of the extremities.

---

47 ἐρεθισμῷ Ermerins: ἐρεθισμός VI
48 οὐκ ἔχοντα ὑπόστασιν GalT: οὐχ ἑκοῦσα V: om. I

144   Ἐνάτῃ πολλὰ παρέλεγε καὶ πάλιν | ἱδρύνθη· σιγῶσα.

Τεσσαρεσκαιδεκάτῃ πνεῦμα ἀραιόν, μέγα, διὰ χρόνου καὶ πάλιν βραχύπνοος.

Ἑπτακαιδεκάτῃ ἀπὸ κοιλίης ἐρεθισμῷ ταραχώδεα, 146 ἔπειτα δὲ αὐτὰ τὰ ποτὰ διῄει, | οὐδὲν συνίστατο· ἀναισθήτως εἶχε πάντων· δέρματος περίτασις, καρφαλέον.

Εἰκοστῇ λόγοι πολλοὶ καὶ πάλιν ἱδρύνθη· ἄφωνος, βραχύπνοος.

Εἰκοστῇ πρώτῃ ἀπέθανε.

Ταύτῃ διὰ τέλεος πνεῦμα ἀραιόν, μέγα· ἀναισθήτως πάντων εἶχεν· αἰεὶ περιεστέλλετο· ἢ λόγοι πολλοὶ ἢ σιγῶσα διὰ τέλεος.

ις'. Ἐν Μελιβοίῃ νεηνίσκος ἐκ πότων καὶ ἀφροδισίων πολλῶν πολὺν χρόνον θερμανθεὶς κατεκλίθη· φρικώδης δὲ καὶ ἀσώδης ἦν, καὶ ἄγρυπνος καὶ ἄδιψος.

Ἀπὸ δὲ κοιλίης τῇ πρώτῃ πολλὰ κόπρανα διῆλθε σὺν περιρρόῳ πολλῷ, καὶ τὰς ἐπομένας, ὑδατόχλοα πολλὰ διῄει· οὖρα λεπτὰ ὀλίγα ἄχρω· πνεῦμα ἀραιόν, μέγα διὰ χρόνου· ὑποχονδρίου ἔντασις ὑπολάπαρος, παραμήκης ἐξ ἀμφοτέρων· καρδίης παλμὸς διὰ τέλεος ξυνεχής· οὔρησεν ἐλαιῶδες.

148   Δεκάτῃ παρέκρουσεν ἀτρεμέως, ἦν δὲ κόσμιός | τε καὶ σιγῶν· δέρμα καρφαλέον καὶ περιτεταμένον· διαχωρήματα ἢ πολλὰ καὶ λεπτά, ἢ χολώδεα λιπαρά.

Ninth day. Much wandering in her talk, followed by a return of reason; silent.

Fourteenth day. Respirations rare and large, with long intervals, and then again shortness of breath.

Seventeenth day. Bowels under a stimulus passed disordered material, then her drinks passed through just as they were, without any coagulation. The patient was insensible to everything; her skin was taut all around, and dry.

Twentieth day. Much rambling followed by the recovery of reason; voiceless; shortness of breath.

Twenty-first day. She died.

The breathing of this patient throughout was rare and large; she was insensible to everything; she constantly wrapped herself up; either much rambling or silence right through to the end.

(16) In Meliboea a youth took to his bed after being for a long time heated by excessive drunkenness and sexual pleasures. He had shivering fits, nausea, and sleeplessness, but no thirst.

First day. Copious stools, passed with an abundant coating around them—also the following days—watery and of a greenish color. Urines thin, scanty and colorless; respirations rare and large, with long intervals; tightness of the hypochondrium on both sides lengthwise, slightly concave; continuous throbbing in the cardia through to the end; passage of oily urine.

Tenth day. Delirious but quiet: he was orderly and silent; skin dry and taut all around; stools either copious and thin, or bilious and greasy.

Τεσσαρεσκαιδεκάτῃ πάντα[49] παρωξύνθη, πολλὰ παρέλεγεν.

Εἰκοστῇ ἐξεμάνη, πολὺς βλήστρισμός· οὐδὲν οὔρει· σμικρὰ ποτὰ κατείχετο.

Εἰκοστῇ τετάρτῃ ἀπέθανε.

[49] Add. παρεκρούσθη I

Fourteenth day. General exacerbation; much wandering talk.

Twentieth day. Wildly out of his mind; much tossing; urine suppressed; slight quantities of drink were kept down.

Twenty-fourth day. He died.

OATH

INTRODUCTION

The earliest reference to the *Oath* is found in a dedicatory letter to Gaius Julius Callistus prefaced to Scribonius Largus' pharmalogical compendium *Compositiones* (AD 47/8):

(4) Anyone who is rightly bound to medicine by oath will not give even enemies an evil drug—although as a soldier and a good citizen he will pursue them with every means when the situation calls for it—for medicine does not judge people by their wealth or standing, but rather promises to extend its help equally to all who need assistance, and to harm no one.

(5) Hippocrates, the founder of our profession, passed down the principles of our discipline in an oath which forbids a physician to give or reveal to a pregnant woman any drug which causes abortion of the fetus; in this way Hippocrates sought forcefully to cultivate humanity in the souls (*animos*) of his students; and how much more evil will those who hold it wrong to destroy a human being that is but an uncertain hope think it to harm one fully formed? He holds it of great importance to preserve the name and reputation of medicine through a respectful and reverent soul, and to practice accord-

ing to his intention: for medicine is the science of healing, not of harming.[1]

This document confirms that by the middle of the first century the *Oath*'s attribution to Hippocrates and its role in contemporary medical deontology were firmly established.

Erotian includes the *Oath* in his census of the collection among "works pertaining to a discussion of medicine," and one of his glosses may refer to it.[2] Soranus of Ephesus refers to the oath in his *Gynecology* and quotes a variant line from it: "I will not give an abortive to anyone" (οὐ δώσω δὲ οὐδενὶ φθόριον).[3] Galen, however, has left no reference to the *Oath* among his Greek writings, although Arabic writers report that Galen wrote a commentary to it and quote from that text.[4]

Three papyri from Oxyrhynchus[5]—74 4970 (II c. AD); 3 437 (II/III c.); 31 2547 (III c.) = Π₉—testify to the *Oath*'s circulation in Egypt in the second and third centuries, and a wide range of writers through the medieval period make reference to the document.[6] Since the Renaissance, the *Oath* has attracted an ever-increasing attention,[7] to the point that today it is the most widely known Hippocratic text.

---

[1] *Compositiones*, Epistula dedicatoria, §§4–5.

[2] Cf. Nachmanson, p. 441.

[3] I, 60 (Ilberg, p. 45; Temkin, p. 63.)

[4] Cf. Ullmann, pp. 32f. and 62; Sezgin, pp. 28 and 69.

[5] See Marganne, p. 251; Marganne/Mertens, p. 17; Jouanna *Serm.*, pp. xv–xvi and lxxxviii–xciv.

[6] *Testimonien* vol. III, pp. 267–71.

[7] Cf., e.g., Th. Rütten, "François Tissard and His 1508 Edition of the Hippocratic *Oath*," in Eijk, pp. 465–91.

Within the formal structure of an oath, beginning with the invocation of divine witnesses, and ending with the imprecation of opposite fates upon the oath-taker according to his fidelity, two sections lay out details of:

1. a contractual relation being entered between a learner and his teacher;
2. the ethical framework within which the learner pledges to practice medicine.

Although the *Oath* is for the most part clear in its literal meaning, there are passages whose import seems at variance with medical practice as it is presented in other Hippocratic texts (e.g., the prohibition of surgery). Its historical context and practical significance, however, are difficult to know. As W. H. S. Jones remarks in the first (1923) edition of the Loeb *Hippocrates* (vol. 1, p. 291):

> What is the date of the *Oath*? Is it mutilated or interpolated? Who took the oath, all practitioners or only those belonging to a guild? What binding force had it beyond its moral sanction? Above all, was it ever a reality or merely a "counsel of perfection"? To all these questions the honest inquirer can only say that for certain he knows nothing.

The hundred years that have passed since Jones' statement have brought valuable scholarship clarifying many points concerning the *Oath*, but most of the questions he raises remain without definitive answers.

The text of the Oath is transmitted in the two primary sources M and V, and the two further Greek manuscripts

Amb[a] and Vind, while the papyrus Π_9 contains a fragmentary fifteen-line passage of the text. An Arabic version of the *Oath* quoted by Uṣaibiʿa in his *History of Medicine*[8] is also useful in assessing the original Greek text.

Besides finding a place in all the collected editions and translations of the Hippocratic Collection, including Zwinger, Mack, Adams, Daremberg, Petrequin, Heiberg, Chadwick, and Diller *Schr.*, many special studies and editions of the work have appeared in print.[9] Particularly valuable in preparing the present edition have been:

Jones, W. H. S. *The Doctor's Oath*. Cambridge, 1924.
Edelstein, L. *The Hippocratic Oath. Text, Translation and Interpretation*. Baltimore, 1943.
Jouanna, J. *Hippocrate. Le Serment . . . La Loi*. Budé I (2). Paris, 2018.

[8] Cf. Jouanna *Serm.*, pp. cxli–cliv.
[9] Cf. Littré, vol. 4, pp. 626–27; Jouanna *Serm.*, pp. 47–49.

ΟΡΚΟΣ

IV 628
Littré Ὀμνύω Ἀπόλλωνα ἰητρὸν καὶ Ἀσκληπιὸν καὶ Ὑγείαν
καὶ Πανάκειαν καὶ θεοὺς πάντας τε καὶ πάσας, ἵστο-
ρας ποιεύμενος, ἐπιτελέα ποιήσειν κατὰ δύναμιν καὶ
κρίσιν ἐμὴν ὅρκον τόνδε καὶ ξυγγραφὴν τήνδε.

  Ἡγήσασθαι δὲ[1] τὸν διδάξαντά με τὴν τέχνην
ταύτην ἴσα γενέτῃσιν ἐμοῖσι, καὶ βίου κοινώσασθαι,
καὶ χρεῶν χρῇζοντι μετάδοσιν ποιήσασθαι· καὶ γέ-
630 νος τὸ ἐξ αὐτοῦ ἀδελφεοῖς | ἴσον ἐπικρινέειν ἄρρεσι·
καὶ διδάξειν τὴν τέχνην ταύτην, ἢν χρῇζωσι μανθά-
νειν, ἄνευ μισθοῦ καὶ ξυγγραφῆς, παραγγελίης τε
καὶ ἀκροήσιος καὶ τῆς λοίπης ἁπάσης μαθήσιος
μετάδοσιν ποιήσασθαι[2] υἱοῖσί τε ἐμοῖσι καὶ τοῖσι τοῦ
ἐμὲ διδάξαντος, καὶ μαθητῇσι ξυγγεγραμμένοισί τε
καὶ ὡρκισμένοις νόμῳ ἰητρικῷ, ἄλλῳ δὲ οὐδενί.

  Διαιτήμασί τε[3] χρήσομαι ἐπ' ὠφελείῃ καμνόντων
κατὰ δύναμιν καὶ κρίσιν ἐμήν· ἐπὶ δηλήσει δὲ καὶ
ἀδικίῃ εἴρξειν.[4] οὐ δώσω δὲ οὐδὲ φάρμακον οὐδενὶ
αἰτηθεὶς θανάσιμον, οὐδὲ ὑφηγήσομαι ξυμβουλίην

---

[1] δὲ M: τε V   [2] καὶ γένος (8)—ποιήσασθαι M: om. V
[3] Add. πᾶσι Π₉ Ambᵃ Ar.
[4] Add. κατὰ γνώμην ἐμήν Π₉ Ambᵃ Vind Ar.

OATH

I swear by Apollo the healer, by Asclepius, by Health, by Panacea, and by all the gods and goddesses, making them my witnesses, to carry out, according to my ability and judgment, this oath and this contract.

To hold the one teaching me this art equal to my parents; to share my livelihood with him, and should he be in want of money to give him a share of my own; to consider his offspring equal to my brothers, and to teach them this art, if they wish to learn it, without payment or contract; to share freely written precepts, oral instructions, and all other learning with my sons, with sons of my teacher, and with students bound by contract and oath according to the medical law, but with no one else.

I will employ[1] dietetic means for the benefit of patients, according to my ability and judgment, and protect them against harm and injustice.[2] I will not give a deadly drug to anyone if asked, nor will I suggest such a course;

[1] Some manuscripts add "all" here.
[2] Daremberg (p. 8) considered the second part of this sentence "very embarrassing." See also Edelstein, pp. 20–24.

τοιήνδε· ὁμοίως δὲ οὐδὲ γυναικὶ πεσσὸν φθόριον
δώσω. ἁγνῶς δὲ καὶ ὁσίως διατηρήσω βίον ἐμὸν καὶ
τέχνην ἐμήν. οὐ τεμέω δὲ οὐδὲ μὴν λιθιῶντας, ἐκχω-
ρήσω δὲ ἐργάτῃσιν ἀνδράσι πρήξιος τῆσδε. ἐς οἰκίας
δὲ ὁκόσας ἂν ἐσίω, ἐσελεύσομαι ἐπ' ὠφελείῃ καμνόντων, ἐκτὸς ἐὼν πάσης ἀδικίης ἑκουσίης καὶ φθορίης,
τῆς τε ἄλλης καὶ ἀφροδισίων ἔργων ἐπί τε γυναικείων
σωμάτων καὶ ἀνδρείων, ἐλευθέρων τε καὶ δούλων. ἃ
δ' ἂν ἐν θεραπείῃ ἢ ἴδω ἢ ἀκούσω, ἢ καὶ ἄνευ θεραπείης κατὰ βίον ἀνθρώπων, ἃ μὴ χρή ποτε ἐκλαλέ-
632 εσθαι ἔξω, σιγήσομαι, ἄρρητα ἡγεύμενος | εἶναι τὰ
τοιαῦτα.

Ὅρκον μὲν οὖν μοι τόνδε ἐπιτελέα ποιέοντι, καὶ μὴ
ξυγχέοντι, εἴη ἐπαύρασθαι καὶ βίου καὶ τέχνης δοξα-
ζομένῳ παρὰ πᾶσιν ἀνθρώποις ἐς τὸν ἀεὶ χρόνον·
παραβαίνοντι δὲ καὶ ἐπιορκοῦντι, τἀναντία τούτων.

similarly, I will not give an abortive pessary to a woman. With integrity and reverence will I conduct my life and my art. I will not make an incision, even on sufferers from the stone, but will defer to those who practice this craft. Into whichever houses I enter, I will go for the benefit of patients, keeping myself free of any intentional injustice or corruption, particularly in sexual matters, involving both female and male bodies, both of the free and of slaves. Whatever I see or hear in the course of my practice, or also apart from it in my life among people, which should never be talked about openly, I will keep secret, holding such things to be unspeakable.

Now if I carry out this oath and do not break it, may I win honor for my life and my art among all people for all time; but if I break it and perjure myself, just the opposite.

PRECEPTS

INTRODUCTION

Precepts has left no trace of its existence in ancient literature, and few testimonies from the middle ages, the most
important of which is a scholion to the first chapter of the
work added apparently some time in the first half of the
fifteenth century in the margin of the fourteenth-century
Hippocratic manuscript Vaticanus Urbinas Graecus 68
(= U).[1] The Greek text of *Precepts* is transmitted directly
in the single independent manuscript M and its roughly
twenty descendants.[2]

The contents of the work might be summarized as follows:

1–2   The role of theory and experience in medicine.
3–4   Fees should be allowed neither to trouble the pa-
       tient nor to influence the practitioner's willing-
       ness to help.
5–6   False physicians practice badly and then reject
       assistance to the detriment of their patients:
       good physicians consult colleagues when they
       become aware of their limitations.
  7   Physicians should give their patients confidence.
       Health is a natural state that can be achieved
       by means of a proper regimen.

[1] See Ecca, pp. 66–71, 316–70, esp. 319f.
[2] See Ecca, pp. 33–65, 72 (*stemma codicum*).

8–9   Physicians should not attempt to impress their patients by adopting exotic accouterments or giving them theoretical discourses.

10   Random practical and theoretical points relating to health and disease.

The text of *Precepts*, which Littré considered the most difficult of the whole collection to understand,[3] is difficult for a variety of different reasons:

Vocabulary: many individual words in the treatise are otherwise unknown in extant Greek literature, that is, they are *hapax legomena* occurring uniquely in *Precepts*. Other words have a wide range of possible meanings, making them understandable in any particular passage only from their context: in *Precepts*, with its aphoristic style, such a context is often very limited or even absent. Still other words appear to be technical terms invented to express special concepts employed in particular philosophical and/or medical schools (e.g., Epicurean, Stoic, and Empiricist), which may have influenced the author of *Precepts*.[4]

Syntax: sentences or their parts often contain grammatical constructions difficult or impossible to construe, on occasion due to incompleteness or an excessive compression of the text.

Logic: in the continuous text, the author's train of thought is frequently difficult or impossible to follow, since no relationship between successive statements seems apparent.

[3] Vol. 9, p. 247.
[4] Cf. Ecca, pp. 9–22.

As an explanation for this unclarity, scholars have suggested the state of Greek at the author's relatively late date (perhaps second century AD), his unfamiliarity in general with the Greek language—Jones identifies evidence of Latin grammatical constructions in his prose[5]—or his level of educational and/or linguistic competence. No amount of physical damage of the text over the course of its transmission can alone account for its present form.

Since the Renaissance, editors and translators of collected Hippocratic works, including Zwinger and Heiberg, have made attempts to turn the often nebulous text of *Precepts* into a coherent narrative, by emending the Greek and/or guessing at its meaning—with various degrees of success.

Three more recent dissertations devoted in part or wholly to *Precepts* are Fleischer, Moisan, and Ecca. These have all made significant improvements to both the text and its understanding, although many serious problems and obscurities still remain. This treatise must be approached with particular care, and for more serious work, Ecca's very informed and thorough commentary should be consulted.

The present edition and translation depend strongly on the scholarly work of my predecessors, including W. H. S. Jones' 1923 Loeb *Hippocrates* volume 1. The chapter divisions adopted in the text are those of Ecca, with Littré's noted in parentheses.

[5] Jones *Loeb*, vol. 1, p. 310; Ecca, pp. 22f.

ΠΑΡΑΓΓΕΛΙΑΙ

IX 250
Littré

1. Χρόνος ἐστὶν ἐν ᾧ καιρός, καὶ καιρὸς ἐν ᾧ χρόνος οὐ πολύς· ἄκεσις χρόνῳ, ἔστι δὲ ἡνίκα καὶ καιρῷ. δεῖ γε μὴν ταῦτα εἰδότα μὴ λογισμῷ πρότερον πιθανῷ προσέχοντα ἰητρεύειν, ἀλλὰ τριβῇ μετὰ λόγου. ὁ γὰρ λογισμὸς μνήμη τίς ἐστι ξυνθετικὴ τῶν μετ᾽ αἰσθήσιος ληφθέντων. ἐφαντασιώθη γὰρ ἐναργέως ἥ τε αἴσθησις προπαθὴς καὶ ἀναπομπὸς ἐοῦσα ἐς διά-
252 νοιαν τῶν ὑποκειμένων· ἡ δὲ | παραδεξαμένη πολ-λάκις, οἷς ὅτε ὁκοίως ταῦτα τηρήσασα, καὶ ἐς ἑωυτὴν καταθεμένη ἐμνημόνευσε. ξυγκαταινέω μὲν οὖν καὶ τὸν λογισμόν, ἤνπερ ἐκ περιπτώσιος ποιῆται τὴν ἀρ-χήν καὶ τὴν καταφορὴν ἐκ τῶν φαινομένων μεθοδεύῃ. ἐκ γὰρ τῶν ἐναργέως ἐπιτελεομένων ἢν[1] τὴν ἀρχὴν ποιήσηται ὁ λογισμὸς ἐν διανοίης δυνάμει ὑπάρχων εὑρίσκεται, παραδεχομένης αὐτῆς ἕκαστα παρ᾽ ἄλ-λων. ὑποληπτέον οὖν τὴν φύσιν [τὴν][2] ὑπὸ τῶν πολ-λῶν καὶ παντοίων πρηγμάτων κινηθῆναί τε καὶ δι-δαχθῆναι βίης ὑπεούσης, ἡ δὲ διάνοια παρ᾽ αὐτῆς λαβοῦσα, ὡς προεῖπον, ὕστερον ἐς ἀληθείην ἤγαγεν.

---

[1] ἢν I: om. M   [2] τὴν M: del. M[2]

PRECEPTS

1. Time is that within which the opportune moment exists: the opportune moment is that within which not much time exists. Healing is a matter of time, but it is sometimes also a matter of the opportune moment. Knowing this, one must practice medicine not according to plausible *a priori* theories, but based on experience combined with reason. For a theory is a composite memory of things apprehended by means of sense perception: images appear distinctly to the sense perception, which being the first instance to be affected, then transmits the underlying reality to conscious thought; conscious thought, frequently receiving and observing these things as to what they are, when, and how, stores them up in itself and remembers them. Now I approve of theorizing, too, if it takes its beginning from experience, and makes its deductions in accordance with visible evidence. For if theorizing lays its foundation in clear fact, it evidently belongs to the workings of thought, which itself receives each of its impressions from external sources. So we must conceive of our nature as being moved and instructed, through some underlying force, by a great variety of things; thought, as I have said, takes over the perceptions of the things from nature, and then leads us to truth. But if (sc. theorizing)

εἰ δὲ μὴ ἐξ ἐναργέος ἐφόδου, ἐκ δὲ πιθανῆς ἀναπλά-
σιος λόγου, πολλάκις βαρείην καὶ ἀνιηρὴν ἐπήνεγκε
διάθεσιν. οὗτοι δὲ ἀνοδίην χειρίζουσι. τί γὰρ ἂν ἦν
κακόν, εἰ τὰ ἐπίχειρα[3] ἐκομίζοντο οἱ τὰ τῆς ἰητρικῆς
ἔργα κακῶς δημιουργέοντες; νῦν δὲ τοῖσιν ἀναιτίοισιν
ἐοῦσι τῶν καμνόντων, ὁκόσοισιν οὐχ ἱκανὴ ἐφαίνετο
ἐοῦσα τοῦ νοσέειν βίη, εἰ μὴ ξυνέλθοι τῇ τοῦ ἰητροῦ
ἀπειρίῃ. περὶ τούτων μὲν οὖν ἅλις ἔστω διειλεγμένα.

2. Τῶν δ' ὡς λόγου μούνου ξυμπεραινομένων μὴ
εἴη ἐπαύρασθαι, τῶν δὲ ὡς ἔργου ἐνδείξιος· σφαλερὴ
γὰρ καὶ εὔπταιστος ἡ μετ' ἀδολεσχίης ἰσχύρισις. διὸ
254 καὶ καθόλου δεῖ ἔχεσθαι τῶν γινομένων, καὶ | περὶ
ταῦτα μὴ ἐλαχίστως γίνεσθαι, ἢν μέλλῃ ἕξειν ῥηϊ-
δίην καὶ ἀναμάρτητον ἕξιν ἣν δὴ ἰητρικὴν προσαγο-
ρεύομεν. κάρτα γὰρ μεγάλην ὠφελίην περιποιήσει
τοῖσί γε νοσέουσι καὶ τοῖσι τούτων δημιουργοῖσι. μὴ
ὀκνέειν δὲ καὶ παρὰ ἰδιωτέων ἱστορέειν, ἤν τι δοκοίη
ξυνοίσειν ἐς καιρὸν θεραπείης. οὕτω γὰρ δοκέω τὴν
ξύμπασαν τέχνην ἀναδειχθῆναι διὰ τὸ ἐξ ἑκάστου τὸ
τέλος τηρηθῆναι καὶ ἐς ταὐτὸ ξυναλισθῆναι. προσέχειν οὖν δεῖ περιπτώσει τε τῇ ὡς ἐπὶ τὸ πολύ, καὶ
μετ' ὠφελίης καὶ ἠρεμαιότητος μᾶλλον ἢ ἐπαγγελίης
καὶ ἀπολογίης τῆς μετ' ἀπρηξίης. (3 L.) χρήσιμος δὲ
καὶ ποικιλίη[4] τῶν προσφερομένων τῷ νοσέοντι καὶ οὐ[5]
προορισμός, ὅτι μόνον τι προσενεχθὲν ὠφελήσει· οὐ

---

[3] ἐπίχειρα Paris. Gr. 2145 (XV c.): ἐπιχείρια M
[4] ποικιλίη Ecca: ποικίλος codd.   [5] οὐ Ecca: ὁ codd.

begins not from a clear impression, but from a plausible fiction, it often leads to a difficult and troublesome situation: people who practice in this way are off the track. Now what harm would be done if only the bad practitioners of the art themselves received their due payment? But as it is, innocent patients, too, are affected, for whom the violence of their disorder would not apparently have been too great, had it not been for the additional effect of their physician's inexperience. Now enough has been said on this subject.

2. Conclusions that are based merely on words are not to be employed, but rather those founded on demonstrated fact. For mere affirmation and talk are deceptive and undependable. For this reason you must hold completely to the facts, and not be negligent about them, if you are to acquire that ready and infallible habit which we call "the art of medicine." For this bestows a very great advantage upon both those who are ill and those who are treating them. Also do not hesitate to question laymen, if this seems likely to contribute to finding some correct measure in treatment. For in this way I believe the complete art has been made clear, by observing the final end which is revealed in each particular situation, and then combining all these into a single whole. So you must pay as much attention as possible to what you experience, in an attitude of helpfulness and serenity, rather than dealing in promises and the excuses that accompany ill-success. Also useful is to have a variety of treatments to administer to the patient, rather than using what was determined upon previously, in the belief that only what was given

γὰρ ἰσχυρίσιος δεῖ· πάντα γὰρ τὰ πάθη διὰ πολλὰς
περιστάσιας καὶ μεταβολὰς μονῇ τινι προσκαθίζει.

3. (4 L.) Παραινέσιος δ᾽ ἂν καὶ τοῦτ᾽ ἐπιδεηθείη τῆς
θεωρίης. εἰ γὰρ ἄρξαιο περὶ μισθαρίων (ξυμβάλλει
γάρ τι τῷ ξύμπαντι) τῷ μὲν ἀλγέοντι τοιαύτην δια-
νόησιν ἐμποιήσεις τὴν ὅτι [οὐκ]⁶ ἀπολιπὼν | αὐτὸν
πορεύσῃ—μὴ ξυνθέμενος—καὶ⁷ ὅτι ἀμελήσεις καὶ
οὐχ ὑποθήσῃ τινὰ τῷ παρεόντι. ἐπιμελεῖσθαι οὖν δεῖ
περὶ στάσιος μισθοῦ· ἄχρηστον γὰρ ἡγεύμεθα ἐν-
θύμησιν ὀχλεομένου τὴν τοιαύτην, πολὺ δὲ μᾶλλον ἐν
ὀξεῖ νοσήματι· νούσου γὰρ ταχυτὴς καιρὸν μὴ δι-
δοῦσα ἐς ἀναστροφὴν οὐκ ἐποτρύνει τὸν καλῶς ἰη-
τρεύοντα ζητεῖν τὸ λυσιτελές, ἔχεσθαι δὲ δόξης μᾶλλον. κρέσσον οὖν σῳζομένοισιν ὀνειδίζειν ἢ ὀλεθρίως
ἔχοντας προμύσσειν.

(5 L.) Καίτοι ἔνιοι νοσέοντες [ἀξιοῦσι]⁸ τὸ ξενοπρε-
πὲς ἢ τὸ εὔδηλον προκρίνοντες, ἀξιοῦσι⁹ μὲν ἀμε-
λείης, οὐ μέντοι γε κολάσιος. [προκρίνοντες]¹⁰ διὸ
τοιούτοισιν ἀντιτάξῃ εἰκότως μεταβολῆς ἐπὶ σάλου

---

⁶ οὐκ del. Coray
⁷ καὶ M: δὲ Ecca
⁸ ἀξιοῦσι del. Ermerins
⁹ ἀξιοῦσι Ecca: ἄξιοι codd.
¹⁰ προκρίνονες M: del. M²

before will help again. There is no need for assertions: for it is through many turns and changes that all diseases settle into some sort of consistent form.

3. This piece of advice will also need consideration. For if you begin on the subject of fees (for it contributes something to the whole), you will give the patient the idea that you will go away and leave him—in the case of not coming to an agreement—and that you will neglect him and not prescribe anything for the present situation.[1] So you must be careful about setting a fee, for I consider such a worry in a troubled patient to be harmful, particularly if the disease is acute. Indeed the rapidity of a disease that offers no opportunity for reconsideration compels a good physician not to seek his profit, but rather urges him to secure his good reputation. In essence it is better to reproach patients you have saved (sc. for not paying you), than to fleece those who are in mortal danger.

And further, some patients through prejudice prefer what is out of the way to what is obvious: while such attitudes may deserve to be ignored, they should not be punished. Therefore you may reasonably oppose these patients, driven as they are by incessant ideas of change. For

[1] Ecca (pp. 115 and 132), reading a somewhat different text, translates: "For if you began with a small fee—you would thereby come to an agreement about everything that needed to be settled—you would give the patient the idea that you are progressing (sc. in your therapeutic plan) without leaving him behind; but if you made no agreement, (sc. you would give him the idea) that you are about to neglect him and not prescribe anything at the moment."

258 πονηρευομένοισιν. | τίς γάρ, ὦ πρὸς Διός, ἠδελφισμένως ἰητρεύοι πίστει καὶ[11] ἀτεραμνίη ὥστε ἐν ἀρχῇ ἀνακρίνοντα πᾶν πάθος μὴ οὐχ ὑποθέσθαι τινὰ ξυμφέροντα ἐς θεραπείην, ἀποθεραπεῦσαί τε τὸν νοσέοντα καὶ μὴ παριδεῖν τὴν ἐπικαρπίην, μὴ ἄνευ τῆς ἐπισκευαζούσης ἐς μάθησιν ἐπιθυμίης;

4. (6 L.) Παρακελεύομαι δὲ μὴ λίην ἀπανθρωπίην ἐσάγειν, ἀλλ᾽ ἀποβλέπειν ἔς τε περιουσίην καὶ οὐσίην· ὁτὲ δὲ προῖκα, ἀναφέρων μνήμην προτέρης[12] εὐχαριστίης ἢ παρεοῦσαν εὐδοκίην. ἢν δὲ καιρὸς εἴη χορηγίης ξένῳ τε ἐόντι καὶ ἀπορέοντι, μάλιστα ἐπαρκέειν τοῖσι τοιούτοισιν· ἢν γὰρ παρῇ φιλανθρωπίη, πάρεστι καὶ φιλοτεχνίη. ἔνιοι γὰρ νοσέοντες ᾐσθημένοι τὸ περὶ ἑωυτοὺς πάθος μὴ ἐὸν ἐν ἀσφαλείῃ, καὶ τῇ τοῦ ἰητροῦ ἐπιεικείῃ εὐδοκέουσι, μεταλλάσσοντες ἐς ὑγιείην. εὖ δ᾽ ἔχει νοσεόντων ἐπιστατέειν, ἕνεκεν ὑγιείης, ὑγιαινόντων τε φροντίζειν, ἕνεκεν ἀνοσίης· φροντίζειν καὶ ὑγιαζόντων[13] ἕνεκεν εὐσχημοσύνης.

5. (7 L.) Οἱ μὲν οὖν ἐόντες ἐν βυθῷ ἀτεχνίης τῶν προλελεγμένων οὐκ ἂν αἰσθάνοιντο. καὶ γὰρ οὗτοι ἀνίητροι ἐόντες εὐέλεγκτοι[14] ἐκ ποδὸς ὑψεύμενοι, τύχης γε μὴν δεόμενοι, ὑπό τινων εὐπόρων ἀσθενῶν,[15] ἔνδοσιν ἀναλαμβάνονται,[16] ἑκατέρῃσιν[17] ἐπὶ τύχῃ-

---

[11] καὶ Fleischer (p. 36): ἢ codd.
[12] -τέρης Fleischer (p. 38): -τέρην Μ
[13] ὑγιαζόντων Zwinger in marg.: ὑγιαινόντων Μ
[14] εὐέλεγκτοι Ecca: ἔλεγχοι codd.
[15] ἀσθενῶν Zwinger: καὶ στενῶν codd.

who, by Zeus, would be practicing medicine in a brotherly manner with credit and integrity, unless at the beginning, after making a preliminary examination of every illness, he prescribed the means that would help toward a cure, he healed the patient completely, and he disregarded his reward, and he was not without a desire to prepare (sc. his patient) to learn?

4. I urge you not to be altogether indifferent to your patient's circumstances, but to consider carefully his wealth and property. Sometimes give (sc. your services) without payment, calling to mind a previous benefaction or a present honor. And if there is an opportunity of serving a stranger in financial straits, give full assistance to all such. For where there is love of man, there is also love of the art. For some patients, though they have grasped that their condition is perilous, recover their health simply through appreciating the excellence of their physician. It is right to manage the ill for the sake of their recovery, and to care for the healthy to prevent disease; but also to support healers themselves so that they can maintain their dignity.

5.[2] Now those who are sunk in an abyss of ignorance regarding the art could never appreciate what has been said. For such men are only pseudophysicians, easy to detect, suddenly exalted yet needing good luck. For they take up the management of some wealthy patients, and

[2] Jones *Loeb*, p. 323, n. 2: "The greater part of this chapter is hopeless."

---

16 -λαμβάνονται M[2]: -λαμβάνοντες M
17 ἑκατέρῃσιν Ecca: ἑκάτεροι M

σιν[18]—εὐδοκιμέοντες καὶ διαπιπτόντες ἐπὶ τὸ χεῖρον—
260 | καταχλιδεῦσι καταμεμεληκότες τὰ τῆς τέχνης ἀνυ-
πεύθυνα, ἐφ᾽ οἷς ἂν ἰητρὸς ἀγαθὸς ἀκμάζοι, ὁμότεχνος
καλεόμενος. (ὁ δὲ τὰς ἀκεσίας ἀναμαρτήτους ῥηϊδίως
ἐπιτελέων οὐδὲν ἂν τούτων παραβαίη [ὅ παντὶ][19] σπά-
νει τοῦ δύνασθαι· οὐ γὰρ ἄπιστος ἐστὶν ὡς <οἱ>[20] ἐν
ἀδικίῃ.) πρὸς γὰρ θεραπείην οὐ γίνονται σκοπέοντες
διάθεσιν φθεγγώδεα, φυλασσόμενοι ἑτέρων ἰητρῶν
ἐπεισαγωγήν, ἐνόντες ἐν μισοπονηρίῃ βοηθήσιος. οἵ
τε νοσέοντες ἀνιώμενοι νήχονται ἐπὶ ἑκατέρῃ μοχθη-
ρίῃ μὴ ἐγκεχειρηκότες ἑωυτοὺς ἕως τέλους τῇ ἐν τῇ
τέχνῃ πλείονι θεραπείῃ· ἄνεσις γὰρ νούσου τινὸς
κάμνοντι παρέχει μεγάλην ἀλεωρήν· διὸ δεόμενοι τὴν
ὑγιεινὴν διάθεσιν οὐκ ἐθέλουσι τὴν αὐτὴν χρῆσιν αἰεὶ
262 προσδέχεσθαι, ὁμοιοῦντες ἰητροῦ ποικιλίῃ. | πολυ-
τελείης γὰρ ἀπορέουσιν οἱ νοσέοντες, κακοτροπίῃ
προσκυνεῦντες καὶ ἀχαριστέοντες ξυντυχεῖν. δυνατοὶ
ἐόντες εὐπορέειν, διαντλίζονται περὶ μισθαρίων, ἀτρεκέως ἐθέλοντες ὑγιέες εἶναι εἵνεκεν ἐργασίης τόκων
ἢ γεωργίης, ἀφροντιστέοντες περὶ αὐτῶν λαμβάνειν.
(8 L.) παρασημασίης τοιαύτης ἅλις ἔστω· ἄνεσις γὰρ
καὶ ἐπίτασις νοσέοντος ἐπινέμησιν ἰητρικὴν κέκτην-
ται.

6. Οὐκ ἀσχήμων δέ, οὐδ᾽ ἤν τις ἰητρὸς στενοχω-
ρέων τῷ παρεόντι ἐπί τινι νοσέοντι καὶ ἐπισκοτεόμε-

---

[18] τύχῃσιν Ecca: τεύχεσιν M
[19] ὅ παντὶ M: del. Ermerins   [20] οἱ add. Zwinger

in both turns of fortune—when they win a reputation and when they make mistakes—they make a spectacle of themselves. They neglect the irreproachable methods of the art, in which a good physician excels, thereby earning the title "brother of the art," since by accomplishing his cures easily without making a mistake he transgresses none of these methods through want of ability, being undependable like those who are unjust: for the unjust do not attempt treatment when they see an alarming condition, and avoid calling in other physicians, because they detest help. Then their patients in their distress drift on a sea of twofold wretchedness, for not having entrusted themselves to the end to the fuller treatment that is given by the art: for the remission of a disease affords a sick man much relief. Thus although patients want to recover, they are not willing always to submit consistently to the same treatment, imitating their physician's changes. For indeed patients impoverish themselves by paying high fees, when they are worshiping incompetent practitioners, and are not grateful for what they receive; but even when they are in the position of being well off, they trouble themselves about small fees, and really only wish to be well for the sake of managing their investments or farm incomes, yet without a thought about caring for themselves. Let this suffice concerning my recommendation: remission or aggravation in a patient depends upon the appropriate apportioning of medical assistance.

6. A physician does not violate etiquette even if, being in difficulties on occasion over a patient and in the dark

νος τῇ ἀπειρίῃ κελεύοι καὶ ἑτέρους ἐσάγειν, εἵνεκα
τοῦ ἐκ κοινολογίης ἱστορῆσαι τὰ περὶ τὸν νοσέοντα,
καὶ συνεργοὺς γενέσθαι ἐς εὐπορίην βοηθήσιος. ἐκ[21]
γὰρ κακοπαθείης παρεδρίης ἐπιτείνοντος τοῦ πάθεος,
δι' ἀπορίην τὰ πλεῖστα ἐκκλίνουσι τῷ παρεόντι. |

264   θαρρητέον οὖν ἐν καιρῷ τοιούτῳ· οὐδέποτε γὰρ ἐγὼ
τὸ τοιοῦτο ὁριεῦμαι, ὅτι ἡ τέχνη κέκριται[22] περὶ τούτου. μηδέποτε φιλονικέειν προσκυρέοντας ἑωυτοῖσι
⟨ἢ⟩[23] κατασιλλαίνειν· ὃ γὰρ ἂν μεθ' ὅρκου ἐρέω, οὐδέποτε ἰητροῦ λογισμὸς φθονήσειεν ἂν ἑτέρῳ· ἀκιδ-
νὸς γὰρ ἂν φανείη· ἀλλὰ μᾶλλον οἱ ἀγχιστεύοντες
ἀγοραίης ἐργασίης πρήσσουσι ταῦτα εὐμαρέως. καίτοι γε οὐ ψευδέως κατανενόηται· πάσῃ γὰρ εὐπορίη
ἀπορίη ἔνεστι.

7. (9 L.) Μετὰ τούτων δὲ πάντων μέγα ⟨ξυν⟩τεκ-
μήριον[24] ἂν εἴη [ξὺν][25] τῇ οὐσίῃ τῆς τέχνης, εἴ τις
καλῶς ἰητρεύων προαγορεύσιος τοιαύτης μὴ ἀπο-
σταίη, κελεύων τοῖσι νοσέουσι μηδὲν ὀχλεῖσθαι κατὰ
διάνοιαν ἐν τῷ σπεύδειν ἀφικέσθαι ἐς καιρὸν σωτη-
ρίης· ἡγεύμεθα γὰρ ἀχρηστίην. καὶ προστασσόμενός
γε οὐ διαμαρτήσει· αὐτοὶ μὲν γὰρ οἱ νοσέοντες διὰ
τὴν ἀλγεινὴν διάθεσιν ἀπαυδέοντες ἑωυτούς τε μεταλ-
λάσσουσι τῆς ζωῆς· ὁ δ' ἐγκεχειρισμένος τὸν νοσέοντα, ἐὰν ἀποδείξῃ τὰ τῆς τέχνης ἐξευρήματα, σῴζων

---

[21] ἐκ M²: εἰ M   [22] κέκριται I: κέκρηται M   [23] ἢ add.
Heiberg: καὶ add. Littré   [24] ξυντεκμήριον Fleischer
(pp. 40f.): ἂν τεκμήριον codd.   [25] ξὺν M: del. Ermerins

through inexperience, he suggests calling in others, in order to learn by consultation the patient's situation, and in order that there will be colleagues in finding effective help. For when a disease takes firm hold, as the evil grows most physicians, in their helplessness, turn away from the help available: but on such occasions one must act with decision. For never would I claim that the physician's art deserved to be condemned in such a situation. Physicians who meet in consultation must never quarrel, or mock one another—this I am willing to swear under oath—and never should a physician's reasoning be motivated by envy of another, for that would be a sign of moral weakness: in reality it is those connected with activities in the marketplace that have no scruples about such things. This is not a wrong way of thinking, for within every capability resides some incapability.[3]

7. Together with all these things, a further piece of strong evidence for the existence of his art would be if a physician, while skillfully treating the patient, did not refrain from an exhortation in which he tells patients not to strain their mind in their eagerness to reach the hour of recovery: for we believe that to do so (i.e., to refrain) would have no positive effect, and at least anyone who follows my recommendation will not be going far astray: for the patients themselves, in their miserable condition, give up and depart this life. But whoever has taken the sick man in hand, if he displays the discoveries of the art, by

[3] Ecca (p. 254) understands this sentence as a statement of the inevitability of the individual practitioner's limitations, which makes consultation with colleagues imperative.

266 οὐκ | ἀλλοιῶν²⁶ φύσιν, ἀποίσει τὴν παρεοῦσαν ἐπι-
καρπίην ἢ τὴν παραυτίκα ἀπιστίην. ἡ γὰρ τοῦ ἀνθρώπου εὐεξίη φύσις τίς ἐστι φύσει περιπεποιημένη
κίνησιν οὐκ ἀλλοτρίην, ἀλλὰ λίην τε εὐαρμοστεῦ-
σαν,²⁷ πνεύματί τε καὶ θερμασίη καὶ χυμῶν κατεργα-
σίη, πάντη²⁸ τε καὶ πάσῃ διαίτῃ καὶ τοῖσι ξύμπασι
δεδημιουργημένη, ἢν μή τι ἐκ γενετῆς ἢ ἀπ᾽ ἀρχῆς
ἔλλειμμα ᾖ· εἰ δ᾽ ἂν²⁹ γένηταί τι, ἐξιτήλου ἐόντος,
πειρᾶσθαι ἐξομοιοῦν τῇ ὑποκειμένῃ· παρὰ γὰρ φύσιν
τὸ μινύθημα καὶ διὰ χρόνου.

8. (10 L.) Φευκτέη³⁰ δὲ καὶ θρύψις³¹ ἐπικρατίδων διὰ
προσκύρησιν ἀκέσιος, ὀδμή τε περίεργος· διὰ γὰρ
ἱκανὴν ἀξυνηθείην³² διαβολὴν ἔκτησαι, διὰ δὲ τὴν
ὀλίγην, εὐσχημοσύνην· ἐν γὰρ μέρει πόνος ὀλίγος, ἐν
πᾶσι ἱκανός. εὐχαρίην δὲ οὐ περιαιρέω· ἀξίη γὰρ ἰητρικῆς προστασίης. (11 L.) προσθέσιος δὲ δι᾽ ὀρ-
γάνων καὶ σημαντικῶν ἐπιδείξιος, καὶ τῶν τοιουτο-
τρόπων μνήμην παρεῖναι. (12 L.) ἢν δὲ καὶ εἵνεκεν
268 ὁμίλου θέλῃς ἀκρόασιν ποιήσασθαι, οὐκ | ἀπ᾽ ἀκλε-
οῦς³³ ἐπιθυμίης, μὴ μέντοι γε μετὰ μαρτυρίης ποιητι-
κῆς· ἀδυναμίην γὰρ ἐμφαίνει φιλοπονίης· ἀπαρνέο-
μαι γὰρ εἰς χρῆσιν ἑτέρην φιλοπονίην μετὰ πόνου
ἱστορεομένην, διὸ ἐν μόνῃ ἑωυτῇ αἵρεσιν ἔχουσαν³⁴
χαρίεσσαν· περιποιήσει γὰρ κηφῆνος μετὰ παραπομ-
πῆς ἑτοιμοκοπίην.

²⁶ ἀλλοιῶν Cornarius in marg.: ἀλλοίην codd.
²⁷ -τεῦσαν Paris. Gr. 2145 (XV c.): -τεῦσα M   ²⁸ πάντη I:
πάντι τε παντη M   ²⁹ εἰ δ᾽ ἂν Moisan: ἢ δ᾽ ἂν M

preserving and not altering (sc. the patient's) constitution, will harvest the success that comes with this, or a momentary loss of trust. For the healthy condition of a human being is a constitution that has naturally attained a movement, not alien, but perfectly adapted to the breath, warmth and processing of the humors in every way, and achieved by the whole regimen and by everything combined, unless there be some congenital or early deficiency. But if there is such a deficiency, as it becomes weaker try to assimilate it to the patient's underlying nature: for a defect, even of long standing, is unnatural.

8. You must also avoid adopting luxurious headgear and refined perfume in order to gain a patient. For excess of strangeness wins you ill-repute, while a little is considered in good taste, just as pain in one part is a trifle, while in every part it is serious. Yet I do not forbid your trying to please, for it is not unworthy of a physician's dignity. Pay attention in the employment of instruments and the pointing out of significant symptoms, and so forth. And if for the sake of a crowded audience, but not from inglorious ambition, you wish to hold a lecture, at least it should not be accompanied by citations from the poets, for that would expose the irrelevance of misplaced effort. For I reject in (sc. medical) practice any industry not pertinent to the art, that is laboriously searched for, and ultimately serves an attractive purpose only for itself. For you will have the officiousness of a drone, with the accompaniment of a crowd.

---

30 -τέη I: -ταίη M   31 θρύψις Triller (Littré): τρῦψις codd.
32 ἀξυνηθείην Kühn (Littré): ἀξυνεσίην codd.
33 ἀπ' ἀκλεοῦς Ecca: ἀλαλκέως M
34 ἐχοῦσαν Littré: ἐοῦσαν codd.

9. (13 L.) Εὐκταίη δὲ καὶ διάθεσις ἐκτὸς ἐοῦσα ὀψιμαθίης· παρεόντων μὲν οὐδὲν ἐπιτελέει· ἀπεόντων δὲ μνήμη ἀνέχεται.[35] γίνεται τοίνυν πάμμαχος ἀτυχίη, μετὰ λύμης νεαρῆς, ἀφροντιστεῦσα εὐπρεπῆς, ὁρισμοῖς τε καὶ ἐπαγγελίῃ, ὅρκοις τε παμμεγέθεσι θεῶν εἵνεκεν. ἰητροῦ προστατέοντος νούσου, ἀναγνώσιος ξυνεχείῃ κατηχήσιός τε ἰδιωτέων φιλαλυστέων λόγους ἐκ μεταφορῆς διαζηλευομένων,[36] καὶ πρὶν ἢ νούσῳ καταπορέω ξυνηθροισμένων. τῶν μὲν οὖν τοιούτων ὅπῃ ἂν καὶ ἐπιστατήσαιμι, οὐκ ἂν ἐπὶ θεραπείης ξυλλόγου αἰτήσαιμι δ' ἂν θαρσαλέως βοηθεῖν· ἱστορίης γὰρ εὐσχήμονος σύνεσις ἐν τοιούτοισι διεσπαρμένη. τοιούτων οὖν δι' ἀνάγκην ἀσυνέτων ἐόντων. παρακελεύομαι χρησίμην εἶναι τὴν τρίβην, μεθ' ὑστέρης[ιν][37] | δογμάτων ἱστορίης. τίς γὰρ ἐπιθυμεῖ δογμάτων μὲν πολυσχιδίην [ἀτρεκέως ἐθέλειν][38] ἱστορέειν, μὴ δὲ[39] χειροτριβίης ἀτρεμιότητι; διὸ παραινέω τοιούτοισι λέγουσι μὲν προσέχειν, ποιέουσι δὲ ἐγκόπτειν.

10. (14 L.) Ξυνεσταλμένης[40] διαίτης μὴ μακρὴν ἐγχειρέειν. νοσέοντος χρονίην ἐπιθυμίην ἀνίστησι καὶ ξυγχωρίῃ ἐν χρονίῃ νούσῳ. ἢν τις προσέχῃ τυφλῷ τὸ δέον, ὡς μέγας φόβος φυλακτέος, καὶ χαρὰ δι' ἑνότητα.[41] ἤερος αἰφνιδίη ταραχὴ φυλακτέη. ἀκμὴ ἡλι-

270

35 ἀνέχεται Ecca: ἀνεκτή M
36 -μένων Zwinger: -μενον codd.
37 ὑστέρης Heiberg: ὑστέρῃσιν M

9. Desirable too is a condition free from late learning. For this approach accomplishes nothing in the present, but maintains only a memory of the past. So there follows an incompetence ready for anything mixed with a childish conceit that has no thought for what is seemly, while definitions, declarations, and extravagant oaths in the names of the gods are invoked. When such a physician is in charge of a case, there will be continuous reading and instruction of bewildered laymen, who are amazed by his metaphorical language, and crowd around him before he fails in treating the disease. Even if I were ever in charge of such people, I would never have any confidence in asking for their aid in medical consultation, for in them any grasp of sensible examination is weak, due to their unavoidable lack of intelligence. I urge that experience is useful, when joined with a later consideration of opinions. But who is keen to investigate subtle diversities of opinion, while caring nothing for calm and practiced skill? Wherefore I advise you to listen to these people when they talk, but to oppose them when they act.

10. When regimen has been restricted, you must not extend your management for too long. Indulgence in the course of a chronic disease relieves a patient's persistent desire. If supplying necessary attention to a person who is blind, great fear is to be guarded against, as is also joy, out of empathy. A sudden disturbance of the air is also to be

---

38 ἀτρεκέως ἐθέλειν del. Ecca: ἐθέλειν del. Littré

39 μὴ δὲ Heiberg in *app. crit.*: μήτε codd.

40 -σταλμένης Aldina: -σταμένης M

41 δι᾿ ἑνότητα Ecca: δι᾿ ἧς ἑνότης M

κίης πάντα ἔχει χαρίεντα, ἀπόληξις δὲ τοὐναντίον.
ἀσαφίη δὲ γλώσσης γίνεται ἢ διὰ πάθος, ἢ διὰ τὰ
οὔατα πρίν τε πρότερα ἐξαγγεῖλαι ἕτερα ἐπιβαλεῖν, ἢ
πρὶν τὸ διανενοημένον εἰπεῖν ἕτερα ἐπιδιανοεῖσθαι· τὸ
μὲν ἄνευ πάθους ὁρατοῦ λελεγμένου μάλιστα συμβαίνει φιλοτεχνοῦσιν. ἡλικίη, σμικροῦ ἐόντος τοῦ
ὑποκειμένου, δύναμις ἐνίοτε παμπόλλη. νούσου ἀτα-
ραξίη μῆκος σημαίνει· κρίσις δὲ ἀπόλυσις νούσου.
μικρὴ αἰτίη ἄκεσι λύεται, ἢν μή τι περὶ τόπον καίριον
272 | πάθῃ. διότι ξυμπάθησις ὑπὸ λύπης ἐοῦσα ὀχλέει, ἐξ
ἑτέρου συμπαθείης τινὲς ὀχλεῦνται. καταύδησις λυ-
πεῖ, φιλοπονίης κραταιῆς ὑποπαραίτησις. ἀλυώδης
τρόπος ⟨οὐκ⟩42 ὀνησιφόρος.

42 οὐκ add. Ecca

guarded against. The prime of life contains everything that is lovely, but its decline holds the opposite. Unclear speech comes either from a disease condition, or from the ears, when before the speaker has reported some previous thought, he adds something new, or before he says what he is thinking, he thinks up something new: now this is a thing that happens without any "visible affection" so-called, mostly to those who are immersed in their art. In youth, when some matter is small, a person's strength is sometimes supremely great. Regularity in a disease signifies that it will be a long one, but the appearance of a crisis is the resolution of a disease. A slight cause is relieved by treatment, unless it is something in a vital part that is suffering. Because sympathy with an existing pain causes distress, some people are distressed by sympathy with another person. Loud talking causes pain, being (sc. only) an excuse for exaggerated officiousness. A restless manner is ‹not› beneficial.

NUTRIMENT

INTRODUCTION

Erotian includes the title περὶ τροφῆς in his census of the Hippocratic Collection among the "therapeutic works pertaining to diet" and defines at least one term from the text in his glossary.[1] Galen refers to the work and its title frequently in his writings,[2] and according to his *On My Own Books* wrote a commentary on it in four books.[3] Although Kühn's Galen edition contains a commentary by this name (vol. 15, pp. 224–417), it has been demonstrated to be a Renaissance forgery;[4] of the actual commentary, a short passage survives in Papyrus Flor. 2. 115 (III/V c.).[5] In the Galenic *Explanation of Difficult Words in Hippocrates*, the entry α 44 ἀκροσαπές is taken from *Nutriment* 41.[6] Testimonies to *Nutriment* are also found in several other medical writers, such as Aretaeus of Cappadocia and the Byzantine compilers Oribasius, Alexander of Tralles, and Aëtius of Amida, as well as in the *Attic Nights* of the Latin miscellaneous writer Aulus Gellius.[7]

---

[1] A 148 ἀχώρ ~ *Nutriment* 20; cf. Nachmanson, pp. 439f.

[2] *Testimonien* vol. II,1, pp. 53–56; vol. II,2, pp. 36–48.

[3] IX, 12; see Boudon, p. 161.

[4] See Joly *Aliment*, p. 136, n. 4.

[5] Cf. Marganne, pp. 156–58; *Testimonien* vol. II,1, p. 263.

[6] Perilli, p. 158.

[7] *Testimonien* vol. I, pp. 40–45 (for the Aulus Gellius passages, see pp. 43f.).

The treatise is included by ibn abī Uṣaibiʿa in his canon of twelve Hippocratic texts, and an Arabic translation of the work is edited and translated into English in J. N. Mattock, *Kitāb al-Ġiḏāʾ li Buqrāṭ. Hippocrates . . . on Nutriment* (Cambridge, 1971).[8] Latin translations from the Greek made by Nicolas de Regio in the fourteenth century and Johannes de Conte in the sixteenth century are extant in four manuscripts and one manuscript, respectively.[9]

Nutriment gives the impression of being an almost random collection of material somehow related to its title, presented in a literary form intended to suggest an association with the early Greek philosopher Heraclitus of Ephesus (fl. ca. 500 BC).[10] Estimates of its date of composition range from around 400 BC,[11] to the Hellenistic age,[12] to the first century AD.[13]

Deichgräber in his study mentioned below groups the traditional fifty-five chapters of the work established by Littré into nine sections:

1–10   Nutiment and how it affects the body.
11–20   Actions of the juices on nutriment, and of external and internal applications to the body.

---

[8] For the reception of *Nutriment* in Islamic medicine, see Ullmann, p. 30; Sezgin, pp. 25–27, 41.

[9] See Kibre, pp. 195f.

[10] Text material from *Nutriment* chapters 1, 2, 8, 9, 12, 14, 15, 17, 19, 21, 23, 24, 40, 42, and 45 is printed in DK Herakleitos, C. Imitation, vol. 1, p. 189.

[11] Cf. Jones *Loeb*, vol. 1, p. 339.

[12] Cf. Joly *Aliment*, pp. 131–36.

[13] Cf. H. Diller. "Eine stoisch-pneumatische Schrift im Corpus Hippocraticum," *Archiv für Geschichte der Medizin* 29 (1936): 178–95.

Interspersed among these chapters on nutriment are various more general utterances—whether medical or philosophical: e.g.,

The text of *Nutriment* is probably often intentionally "riddling," as would suit its suggested relationship to Heraclitus, and this may have contributed to the wide variation of the texts transmitted in its two independent manuscript traditions, M and V, and between these and the indirect witnesses. *Nutriment*'s interpretation and translation can be tentative at best.

Besides finding a place in the collected editions and translations, including Zwinger, but not Mack or Grimm,

Nutriment was treated specifically by several Renaissance scholars,[14] and in the twentieth century it is edited by Jones (Loeb), Heiberg (CMG) and Joly (Budé). More recently it has received a detailed examination in:

Deichgräber, K. *Pseudhippokrates Über die Nahrung: Text, Kommentar und Würdigung einer stoisch-heraklitisierenden Schrift aus der Zeit um Christi Geburt*. Wiesbaden, 1973.

[14] See Littré, vol. 9, p. 96f.

ΠΕΡΙ ΤΡΟΦΗΣ

1. Τροφὴ καὶ τροφῆς εἶδος μία καὶ πολλαί· μία μὲν ἢ γένος ἕν, εἶδος δὲ ὑγρότητι καὶ ξηρότητι, καὶ ἐν τούτοις ἰδέαι καὶ ποσόν ἐστιν, καὶ ἔς τινα καὶ ἐς τοσαῦτα.

2. Αὔξει δὲ καὶ ῥώννυσι καὶ σαρκοῖ καὶ ὁμοιοῖ καὶ ἀνομοιοῖ τὰ ἐν ἑκάστοις κατὰ φύσιν τὴν ἑκάστου καὶ τὴν ἐξ ἀρχῆς δύναμιν.

3. Ὁμοιοῖ δὲ ἐς δύναμιν, ὅταν κρατήσῃ ἡ ἐπεισιοῦσα, καὶ ὅταν ἐπικρατέηται[1] ἡ προϋπάρχουσα.

4. Γίνεται δὲ καὶ ἐξίτηλος, ὁτὲ μὲν ἡ προτέρη ἐν
χρόνῳ ἀπολυθεῖσα | ἢ ἐπιπροσθετηθεῖσα, ὁτὲ δὲ ἡ ὑστέρη ἐν χρόνῳ ἀπολυθεῖσα ἢ ἐπιπροσθετηθεῖσα.

5. Ἀμαυροῖ δὲ ἑκατέρας ἐν χρόνῳ καὶ μετὰ χρόνον ἡ ἔξωθεν συνεχὴς ἐπεισκριθεῖσα[2] καὶ ἐπὶ πολλὸν χρόνον στερεμνίως πᾶσι τοῖς μέλεσι διαπλεκεῖσα.

6. Καὶ τὴν μὲν ἰδίην ἰδέην ἐξέβλαστησε· τὴν δὲ προτέρην ἔστιν ὅτε καὶ τὰς προτέρας ἐξημαύρωσεν.

7. Δύναμις δὲ τροφῆς ἀπικνέεται καὶ ἐς ὀστέον καὶ

---

[1] κρατήσῃ . . . ἐπικρατέηται A: κρατέῃ μὲν ἡ ἐπεισιοῦσα, ἐπικρατέῃ δὲ M
[2] ἐπεισκρ. M: ἐπικρ. A

NUTRIMENT

1. Nutriment and the form of nutriment (sc. are) one and many: one, in so far as it is generically one, but (sc. many) in species according to moisture and dryness; and in these there are varieties both in quantity and in the kind and number of effects.

2. It (sc. nutriment) augments, strengthens, clothes with flesh, assimilates, and attenuates what is in each (sc. of the body parts), according to the part's nature and original potency.

3. It assimilates to its own potency, when what enters gains the upper hand and what was there before is overwhelmed.

4. There is also attenuation, sometimes the previous (sc. potency) being effaced or overshadowed with time, sometimes the later one being effaced or overshadowed with time.

5. Nutriment weakens both of these with time, when it has been continually imposed for a time from the outside and become firmly interwoven for a long time with all the limbs.

6. A nutriment also causes its own specific form to sprout forth, obliterating the previous form, and sometimes even the ones from before that.

7. The potency of a nutriment arrives as far as the bone

πάντα τὰ μέρεα αὐτοῦ, καὶ ἐς νεῦρον καὶ ἐς φλέβα
καὶ ἐς ἀρτηρίην καὶ ἐς μῦν καὶ ἐς[3] ὑμένα καὶ σάρκα
καὶ πιμελὴν καὶ αἷμα καὶ φλέγμα καὶ μυελὸν καὶ
ἐγκέφαλον καὶ νωτιαῖον καὶ τὰ ἐντοσθίδια καὶ πάντα
τὰ μέρεα αὐτῶν· καὶ δὴ καὶ ἐς θερμασίην καὶ ὑγρα-
σίην καὶ πνεῦμα.

8. Τροφὴ δὲ τὸ τρέφον, τροφὴ δὲ τὸ οἷον, τροφὴ δὲ
τὸ μέλλον. |

9. Ἀρχὴ δὲ πάντων μία καὶ τελευτὴ πάντων μία,
καὶ ἡ αὐτὴ τελευτὴ καὶ ἀρχή.

10. Καὶ ὅσα κατὰ μέρος ἐν τροφῇ καλῶς καὶ κακῶς
διοικέεται· καλῶς μὲν ὅσα προείρηται, κακῶς δὲ ὅσα
τούτοις τὴν ἐναντίην ἔχει τάξιν.

11. Χυλοὶ ποικίλοι καὶ χρώμασι καὶ δυνάμεσι καὶ
ἐς βλαβὴν καὶ ἐς ὠφελίην, καὶ οὔτε βλάπτειν οὔτε
ὠφελέειν, καὶ πλήθει καὶ ὑπερβολῇ καὶ ἐλλείψει, καὶ
διαπλοκῇ ὧν μέν, ὧν δὲ οὔ.

12. Καὶ πάντων ἐς θερμασίην βλάπτει καὶ ὠφελέει,
ἐς ψύξιν βλάπτει καὶ ὠφελέει, ἐς δύναμιν βλάπτει καὶ
ὠφελέει.

13. Δυνάμιος δὲ ποικίλαι φύσιες.

14. Χυμοὶ φθείροντες καὶ ὅλον καὶ μέρος καὶ ἔξωθεν καὶ ἔνδοθεν, αὐτόματοι καὶ οὐκ αὐτόματοι, ἡμῖν
μὲν αὐτόματοι, αἰτίῃ δὲ οὐκ αὐτόματοι. αἰτίης δὲ τὰ
μὲν δῆλα, τὰ δὲ ἄδηλα, καὶ τὰ μὲν δυνατά, τὰ δὲ
ἀδύνατα.

15. Φύσις ἐξαρκέει πάντα πᾶσι.

and all the bony parts, and to cord, to vein, to artery, to muscle, to membrane, flesh, fat, blood, phlegm, marrow, brain, spine, and the intestines and all the inward parts; naturally it also affects the (sc. body's) heat, moisture, and breath.

8. Nutriment is what nourishes, what can nourish, and what will nourish.

9. The beginning of all things is one, and the end of all things is one, and the same thing is end and beginning.

10. The particular components in nutriment are ordered well or ill; well if as indicated above, ill if they have the opposite arrangement to these.

11. The nutritive juices vary in their colors and potencies, in their tendency to harm or to help, or neither to harm nor to help, in their amount, their excess or defect, and in their combination with some things but not with others.

12. And of all things it (i.e., the nutritive juice) harms and helps in warming, harms and helps in cooling, harms and helps in strengthening.

13. Of potency there are various natures.

14. Humors corrupting both the whole and the part, from without or from within, are both spontaneous and not spontaneous: for us they are spontaneous, but their causation is not spontaneous. The cause is partly clear, partly obscure; partly it is in our power and partly not.

15. Nature is sufficient for all in all.

---

3 ἐς M: om. A

16. Ἐς δὲ ταύτην· ἔξωθεν μὲν κατάπλασμα, κατά-
χρισμα, ἄλειμμα, γυμνότης καὶ ὅλου καὶ μέρεος καὶ
104 σκέπη ὅλου καὶ μέρεος, | θερμασίη καὶ ψύξις κατὰ τὸν
αὐτὸν λόγον· καὶ στύψις καὶ ἕλκυσις⁴ καὶ δηγμὸς καὶ
λίπασμα· ἔνδοθεν δὲ τινά τε τῶν εἰρημένων, καὶ ἐπὶ
τούτοις αἰτίη ἄδηλος καὶ μέρει καὶ ὅλῳ, τινί τε καὶ οὔ
τινι.

17. Ἀποκρίσιες κατὰ φύσιν, κοιλίης, οὔρων, ἱδρῶ-
τος, πτυάλου, μύξης, ὑστέρης, καθ' αἱμορροΐδα, θύ-
μον, λέπρην, φῦμα, καρκίνωμα· ἐκ ῥινῶν, ἐκ πλεύμο-
νος, ἐκ κοιλίης, ἐξ ἕδρης, ἐκ καυλοῦ, κατὰ φύσιν καὶ
παρὰ φύσιν. αἱ διακρίσιες τούτων ἄλλοισι πρὸς ἄλλον λόγον ἄλλοτε καὶ ἀλλοίως. μία φύσις ἐστὶ πάντα
ταῦτα καὶ οὐ μία· πολλαὶ φύσιές εἰσι πάντα ταῦτα
καὶ μία.

18. Φαρμακείη ἄνω καὶ κάτω, οὔτε ἄνω οὔτε κάτω.

19. Ἐν τροφῇ φαρμακείη ἄριστον, ἐν τροφῇ φαρμακείη φλαῦρον· φλαῦρον καὶ ἄριστον πρός τι.

20. Ἕλκος, ἐσχάρη, αἷμα, πύον, ἰχώρ, λέπρη, πίτυ-
ρον, ἄχωρ, λειχήν, ἀλφός, ἔφηλις, ὁτὲ μὲν βλάπτει,
ὁτὲ δὲ ὠφελέει, ὁτὲ δὲ οὔτε βλάπτει οὔτε ὠφελέει.

21. Τροφὴ οὐ τροφή, ἢν μὴ δύνηται· μὴ τροφὴ
106 τροφή, ἢν [μὴ]⁵ οἷόν | τε ᾖ τρέφεσθαι. οὔνομα τροφή,
ἔργον δὲ οὐχί· ἔργον τροφή, οὔνομα δὲ οὐχί.

---

⁴ ἕλκυσις Potter: ἕλκωσις ΜΑ
⁵ δύνηται· μὴ τροφὴ τροφή ἢν [μὴ] Littré: δύνηται τροφὴ
οὐ τροφή ἢν μὴ Α: δύνηται . . . μὴ om. Μ

16. For this.[1] From without: plaster, ointment, salve; exposure of both whole and part, and covering of whole and part; warming and cooling according to the same principle; also astringent, attractive, mordant, and unguent. From within: some of the things named, in which the cause is obscure for both the part and the whole, and in one case but not in another.

17. Secretions in accordance with nature: of the cavity, urines, sweat, sputum, mucus and uterus: through a hemorrhoid, an excrescence, leprosy, a growth, a carcinoma; from the nostrils, the lungs, the cavity, the seat, and the penis—in accordance with nature and contrary to nature. The secretions among these vary according to different principles at different times and in different circumstances. All these things have one nature and not one nature: all these things have many natures and one nature.

18. Purgation upward and downward, neither upward nor downward.

19. In nutriment purgation is excellent, in nutriment purgation is bad; bad or excellent according to circumstances.

20. An ulcer, an eschar, blood, pus, lymph, leprosy, scurf, dandruff, lichen, alphos and freckles sometimes harm and sometimes help, and sometimes they neither harm nor help.

21. Nutriment (sc. is) not nutriment if it has no potency; what is not nutriment (sc. becomes) nutriment if it can nourish. Nutriment in name, but not in deed; nutriment in deed, but not in name.

---

[1] What this ταύτην refers to is unclear. Possibilities are "cause" in ch. 14 or "nature" in ch. 15, but the sense of the whole chapter remains obscure.

22. Ἐς τρίχας τροφὴ⁶ καὶ ἐς ὄνυχας καὶ ἐς τὴν ἐσχάτην ἐπιφανείην ἔνδοθεν ἀφικνέεται· ἔξωθεν τροφὴ ἐκ τῆς ἐσχάτης ἐπιφανείης ἐνδοτάτω ἀπικνέεται.

23. Σύρροια μία, σύμπνοια μία, συμπαθέα πάντα. κατὰ μὲν οὐλομελίην πάντα, κατὰ μέρος δὲ τὰ ἐν ἑκάστῳ μέρει μέρεα πρὸς τὸ ἔργον.

24. Ἀρχὴ μεγάλη ἐς ἔσχατον μέρος ἀπικνέεται· ἐξ ἐσχάτου μέρεος ἐς ἀρχὴν μεγάλην ἀφικνέεται· μία φύσις εἶναι καὶ μὴ εἶναι.

25. Νούσων διαφοραὶ ἐν τροφῇ, ἐν πνεύματι, ἐν θερμασίῃ, ἐν αἵματι, ἐν φλέγματι, ἐν χολῇ, ἐν χυμοῖσιν, ἐν σαρκί, ἐν πιμελῇ, ἐν φλεβί, ἐν ἀρτηρίῃ, ἐν νεύρῳ, μυί, ὑμένι, ὀστέῳ, ἐγκεφάλῳ, νωτιαίῳ μυελῷ, στόματι, γλώσσῃ, στομάχῳ, κοιλίῃ, ἐντέροισι, φρεσί, περιτοναίῳ, ἥπατι, σπληνί, νεφροῖσι, κύστει, μήτρῃ, δέρματι. ταῦτα πάντα καὶ καθ᾽ ἓν καὶ κατὰ μέρος. μέγεθος αὐτῶν μέγα καὶ οὐ μέγα.

26. Τεκμήρια, γαργαλισμός, ὀδύνη, ῥῆξις, γνώμη, ἱδρώς, οὔρων ὑπόστασις, ἡσυχίη, ῥιπτασμός, ὄψιος στάσιες, φαντασίαι, ἴκτερος, λυγμοί, ἐπιληψίη, αἷμα ὁλοσχερές, ὕπνος· καὶ ἐκ τούτων καὶ τῶν ἄλλων τῶν κατὰ φύσιν, καὶ ὅσα ἄλλα τοιουτότροπα ἐς βλάβην καὶ ἐς ὠφελίην ὁρμᾷ. πόνοι ὅλου καὶ μέρεος μεγέθεος σημεῖα, τοῦ | μὲν ἐς τὸ μᾶλλον, τοῦ δὲ ἐς τὸ ἧσσον, καὶ ἀπ᾽ ἀμφοτέρων ἐς τὸ μᾶλλον καὶ ἀπ᾽ ἀμφοτέρων ἐς τὸ ἧσσον.

108

22. Nutriment passes to the hairs, to the finger nails, and to the farthest surface (sc. of the body) from within. From without nutriment passes from the farthest surface to the deepest interior.

23. There is one flowing together, one breathing together: all things are in accord. All things belong to one system, but part by part the parts of each part fulfill their functions.

24. The great beginning extends to the farthest part: from the farthest part there is movement to the great beginning. A single nature, being and not being.

25. Differences of disease: in nutriment, in breath, in heat, in blood, in phlegm, in bile, in humors, in flesh, in fat, in vein, in artery, in cord, muscle, membrane, bone, brain, spinal marrow, the mouth, the tongue, the esophagus, the cavity, the intestines, the diaphragm, the peritoneum, the liver, the spleen, the kidneys, the bladder, the uterus, the skin. All these things both as a unity and part by part: their greatness great and not great.

26. Signs: tickling, pain, tearing, thought, sweating, sediment of the urines, rest, tossing, fixations of the eye, imaginings, jaundice, hiccups, epilepsy, solid blood, sleep: from both these and the other things that are according to nature, and all the others of this kind that tend toward harm or help. Pains of the whole and part are signs of (sc. a disease's) magnitude, of the one, greater severity, of the other less, and from both come signs of greater severity, and from both come signs of less.

---

[6] τροφὴ M: om. A

27. Γλυκύ οὐ γλυκύ, γλυκὺ ἐς δύναμιν οἷον ὕδωρ, γλυκὺ ἐς γεῦσιν οἷον μέλι· σημεῖα ἑκατέρων, ἕλκεα, καὶ ὀφθαλμοὶ καὶ γεύσιες, καὶ ἐν τούτοις τὸ μᾶλλον καὶ ἧσσον· γλυκὺ ἐς τὴν ὄψιν καὶ ἐν χρώμασι καὶ ἐν ἄλλῃσι μίξεσι, γλυκὺ μᾶλλον καὶ ἧσσον.

28. Ἀραιότης σώματος ἐς διαπνοιὴν οἷσι πλέον ἀφαιρέεται ὑγιεινόν· πυκνότης σώματος ἐς διαπνοιὴν οἷσιν ἔλασσον ἀφαιρέεται νοσηλόν· οἱ διαπνεόμενοι καλῶς ἀσθενέστεροι καὶ ὑγιεινότεροι καὶ εὐανάσφαλτοι, οἱ διαπνεόμενοι κακῶς πρὶν ἢ νοσέειν ἰσχυρότεροι, νοσήσαντες δὲ δυσανάσφαλτοι· ταῦτα δὲ καὶ ὅλῳ καὶ μέρει.

29. Πλεύμων ἐναντίην σώματος τροφὴν ἕλκει, τὰ δ' ἄλλα πάντα τὴν αὐτήν.

30. Ἀρχὴ τροφῆς πνεύματος, ῥῖνες, στόμα, βρόγχος, πλεύμων, καὶ ἡ ἄλλη διαπνοή· ἀρχὴ τροφῆς καὶ ὑγρῆς καὶ ξηρῆς, στόμα, στόμαχος, κοιλίη. ἡ δὲ ἀρχαιοτέρη τροφὴ διὰ τοῦ ἐπιγαστρίου, ᾗ ὀμφαλός. |

31. Ῥίζωσις φλεβῶν ἧπαρ, ῥίζωσις ἀρτηριῶν καρδίη· ἐκ τούτων ἀποπλανᾶται ἐς πάντα αἷμα καὶ πνεῦμα, καὶ θερμασίη διὰ τούτων φοιτᾷ.

32. Δύναμις μία καὶ οὐ μία, ᾗ πάντα ταῦτα[7] καὶ τὰ ἑτεροῖα διοικέεται· ἡ μὲν ἐς ζωὴν ὅλου καὶ μέρεος, ἡ δὲ ἐς αἴσθησιν ὅλου καὶ μέρεος.

33. Γάλα τροφή, οἷσι γάλα τροφὴ κατὰ φύσιν, ἄλλοισι δὲ οὐχί. καὶ ἄλλοι[8] δὲ οἷσιν οἶνος τροφή, καὶ

---

7 ταῦτὰ Heidel (Joly, p. 150): ταῦτα ΜΑ

27. Sweet, not sweet; sweet in potency like water, sweet to the taste like honey. Signs of either are sores, eyes and tastes, and in these a greater or less degree. Sweet to the sight, in colors and in other combinations, sweet to a greater or less degree.

28. Porousness of the body for transpiration (sc. is) healthy in those from whom more (sc. breath) is taken; denseness of the body for transpiration (sc. is) unhealthy in those from whom less is taken. Patients who transpire well are weaker, healthier, and recover easily; those who transpire poorly are stronger before they are sick, but on falling ill they have a difficult recovery. This is true for both whole and part.

29. The lungs draw a nutriment which is opposite to that of the body, whereas all the other parts draw the same.

30. Origin of the nutriment breath: the nostrils, mouth, windpipe, lung, and the rest of the respiratory system. Origin of nutriment both wet (i.e., drink) and dry (i.e., food): the mouth, esophagus, cavity. The earlier nutriment (sc. passes) through the epigastrium where the navel is.

31. Root of the veins, the liver; root of the arteries, the heart. Out of these blood and breath are distributed to all the parts, and heat is conducted through them.

32. Potency, single and not single, by which all things, both the same and the different are nourished; one potency for the life of the whole and part, another potency for the sensation of the whole and part.

33. Milk nourishes those for whom it is a natural nutriment, but not others. There are others for whom wine is

---

8 ἄλλοι Heiberg in *app. crit.*: ἄλλοισι MA

ΠΕΡΙ ΤΡΟΦΗΣ

οἷσιν οὐχὶ τροφή· καὶ σάρκες καὶ ἄλλαι ἰδέαι τροφῆς πολλαί, καὶ κατὰ χώρην καὶ κατ' ἐθισμόν.

34. Τρέφεται δὲ τὰ μὲν ἐς αὔξησιν καὶ ἐς τὸ εἶναι, τὰ δὲ ἐς τὸ εἶναι μοῦνον, οἷον γέροντες, τὰ δὲ πρὸ[9] τούτων καὶ ἐς ῥώμην. διάθεσις ἀθλητικὴ οὐ φύσει. ἕξις ὑγιεινὴ κρέσσων ἐν πᾶσι.

35. Μέγα τὸ πόσον εὐστόχως ἐς δύναμιν συναρμοσθέν.

36. Γάλα καὶ αἷμα τροφῆς πλεονασμός.

37. Περίοδοι ἐς πολλὰ σύμφωνοι, ἐς ἔμβρυον ἐς τὴν τούτου τροφήν. αὖτις δ' ἄνω ῥέπει ἐς γάλα καὶ ἐς τροφὴν βρέφεος. |

112   38. Ζωοῦται τὰ μὴ ζῷα, ζωοῦται τὰ ζῷα, ζωοῦται τὰ μέρεα τῶν ζῴων.

39. Φύσιες πάντων ἀδίδακτοι.

40. Αἷμα ἀλλότριον ὠφέλιμον, αἷμα οἰκεῖον βλαπτικόν·[10] αἷμα ἀλλότριον βλαβερόν, αἷμα ἴδιον ξυμφέρον.[11] χυμοὶ ἀλλότριοι ξυμφέροντες, χυμοὶ ἴδιοι βλαβεροί· χυμοὶ ἀλλότριοι βλαβεροί, χυμοὶ ἴδιοι ξυμφέροντες. τὸ σύμφωνον διάφωνον, τὸ διάφωνον σύμφωνον. γάλα ἀλλότριον ἀστεῖον, γάλα ἴδιον φλαῦρον· γάλα ἀλλότριον βλαβερόν, γάλα ἴδιον ὠφέλιμον.

41. Σιτίον νέοις ἀκροσαπές, γέρουσιν ἐς τέλος μεταβεβλημένον, ἀκμάζουσιν ἀμετάβλητον.

[9] πρὸ A: πρὸς M
[10] οἰκεῖον βλαπτικόν A: ἴδιον ὠφέλιμον M

336

a nutriment, and others for whom it is not a nutriment. Also meats and many other kinds of nutriment according to place and habit.

34. Nourishment is sometimes for both growth and existence, at other times for existence alone, as in old men, and at other times for strength rather than for these. An athletic condition is not natural: a healthy state is superior in every way.

35. It is a great thing when quantity is perfectly matched to strength.

36. Milk and blood are a surplus of nutriment.

37. The (sc. blood's) periods are in harmony for many things, for the embryo and for its nutriment. Then next it turns upward, becoming milk for the nourishment of the baby.

38. Life is given to the inanimate, life is given to living creatures, life is given to the parts of living creatures.

39. Natures (sc. are) untaught of all.

40. Foreign blood is beneficial, domestic blood is harmful; foreign blood is harmful, one's own blood is useful. Foreign humors are useful, one's own humors are harmful; foreign humors are harmful, one's own humors are useful. What is in agreement (sc. is) not in agreement, what (sc. is) not in agreement is in agreement. Foreign milk is good, one's own milk is bad; foreign milk is harmful, one's own milk is beneficial.

41. Food for the young (sc. should be) on the point of turning, for the old completely altered, for adults unchanged.

---

11 ξυμφέρον A: βλαβερόν M

42. Ἐς τύπωσιν λε΄ ἤέλιοι, ἐς κίνησιν ο΄, ἐς τε-
114 λειότητα σι΄· | ἄλλοι, ἐς ἰδέην[12] με΄, ἐς κίνησιν ϛ΄, ἐς
ἔξοδον σο΄· ἄλλοι, ν΄ ἐς ἰδέην, ἐς πρῶτον ἅλμα ρ΄, ἐς
τελειότητα τ΄. ἐς διάκρισιν[13] μ΄, ἐς μετάβασιν π΄, ἐς
ἔκπτωσιν σμ΄.

116 Οὐκ ἔστι καὶ ἔστι.[14] γίνεται δὲ ἐν τούτοις καὶ | πλείω
καὶ ἐλάσσω, καὶ καθ᾽ ὅλον καὶ κατὰ μέρος, οὐ πολλὸν
δὲ τὰ <πλείω> πλείω ἢ ἐλάσσω τὰ δὲ ἐλάσσω.[15] τοσαῦτα καὶ ὅσα ἄλλα τούτοισι ὁμοῖα.

43. Ὀστέων τροφὴ ἐκ κατήξιος· ῥινὶ δὶς πέντε,
γνάθῳ καὶ κληῖδι καὶ πλευρῇσι διπλάσιαι, πήχει τρι-
πλάσιαι, κνήμῃ καὶ βραχίονι τετραπλάσιαι, μηρῷ
πενταπλάσιαι. καὶ εἴ τι ἐν τούτοισι δύναται πλέον ἢ
ἔλασσον.

44. Αἷμα ὑγρὸν καὶ αἷμα στερεόν. αἷμα ὑγρὸν
ἀστεῖον, αἷμα ὑγρὸν φλαῦρον· αἷμα στερεὸν ἀστεῖον,
αἷμα στερεὸν φλαυρόν· πρός τι πάντα φλαῦρα καὶ
πάντα ἀστεῖα.

45. Ὁδὸς ἄνω κάτω, μία.[16]

46. Δύναμις τροφῆς κρέσσων ἢ ὄγκος, ὄγκος τροφῆς κρέσσων ἢ δύναμις, καὶ ἐν ὑγροῖσι καὶ ἐν ξη-
ροῖσι.

47. Ἀφαιρέει καὶ προστίθησι τωὐτό· τὸ μὲν ἀφαιρέει, τῷ δὲ προστίθησι τωὐτό.[17]

[12] ἰδέην A: μορφὴν M
[13] διάκρισιν M: ἄκρισιν A
[14] τὰ ὀκτάμηνα add. Aulus Gellius III 16,7

338

42. For formation 35 days; for movement, 70; for perfection, 210. Others (sc. say): for form, 45; for quickening 90; for delivery, 270. Others, 50 for form; for quickening, 100; for perfection, 300. (sc. Others say) for differentiation, 40, for change of position, 80; for expulsion, 240.

(Sc. eight months' children) both are not and are. In these matters there are both more and less—concerning both whole and part—but the more is not much more, and the less not much less. So much of these matters and others similar to them.

43. Nutriment for bones after a break: for the nostril, twice five (sc. days); for the jaw, the collarbone and ribs, twice this; for the forearm, thrice; for the lower leg and upper arm, four times; for the thigh, five times—even if in these there may be a certain amount of more or less.

44. Blood is liquid and blood is solid. Liquid blood is good, liquid blood is bad; solid blood is good, solid blood is bad. It is according to circumstances that all things are bad and all good.

45. The road up and down is one.

46. Potency of nutriment superior to mass; mass of nutriment superior to potency; both in moist things and in dry.

47. It takes away and adds the same thing; first it takes away, but somewhere else it adds the same thing.

---

15 τὰ ⟨πλείω⟩ . . . ἐλάσσω: καὶ πλείω πλείω καὶ ἐλάσσω ἐλάσσω Aulus Gellius III 16,20: πλείω ἢ ἐλάσσω τὰ δὲ ἐλάσσω M: τὰ πλείω A

16 μία M: om. A

17 τὸ μὲν ἀφαιρέει—τωὐτό M: om. A

48. Φλεβῶν διασφύξιες καὶ ἀναπνοὴ πλεύμονος καθ᾽ ἡλικίην, καὶ σύμφωνα καὶ διάφωνα· καὶ νούσου 118 καὶ ὑγιείης σημήϊα, καὶ ὑγιείης | μᾶλλον ἢ νούσου καὶ νούσου μᾶλλον ἢ ὑγιείης·[18] τροφὴ γὰρ καὶ πνεῦμα.

49. Ὑγρὴ τροφὴ εὐμετάβλητος μᾶλλον ἢ ξηρή· ξηρὴ τροφὴ εὐμετάβλητος μᾶλλον ἢ ὑγρή· ἡ δυσαλλοίωτος δυσεξανάλωτος, ἡ εὐπρόσθετος εὐεξανάλωτος.

50. Καὶ ὁκόσοι[19] ταχείης προσθέσιος δέονται, ὑγρὸν ἴημα ἐς ἀνάληψιν δυνάμιος κράτιστον. ὁκόσοι δὲ ἔτι ταχυτέρης, δι᾽ ὀσφρήσιος. ὁκόσοι δὲ βραδυτέρης προσθέσιος δέονται,[20] στερεὴ τροφή.

51. Μύες στερεώτεροι [δυσεύτηκτοι][21] τῶν ἄλλων, παρὲξ ὀστέου καὶ νεύρου· δυσμετάβλητα τὰ γεγυμνασμένα, κατὰ γένος αὐτὰ ἑωυτῶν ἰσχυρότερα ἐόντα,[22] διὰ τοῦτο αὐτὰ ἑωυτῶν δυστηκτότερα.

52. Πύον τὸ ἐκ σαρκός· πυῶδες τὸ ἐξ αἵματος καὶ ἐξ ἄλλης ὑγρασίης. πύον τροφὴ ἕλκεος· πυῶδες[23] τροφὴ φλεβός, ἀρτηρίης. |

120 53. Μυελὸς τροφὴ ὀστέου, διὰ τοῦτο ἐπιπωροῦται.

54. Δύναμις πάντα αὔξει καὶ τρέφει καὶ βλαστάνει.

55. Ὑγρασίη τροφῆς ὄχημα.

---

[18] Σημήϊα—ὑγιείης M: om. A
[19] ἡ εὐπρόσθετος—ὁκόσοι M: om. A
[20] ὑγρὸν ἴημα—δέονται M: δι᾽ ὀσφρήσιος ταχυτέρης ὑγρὸν ἴημα τρέφει ξενεχέως A
[21] δυσεύτηκτοι M: δυσεκτικοὶ A: del. Heidel Joly Deichgräber

48. Pulsations of veins and respiration of the lung, according to age, both in agreement and in disagreement: signs of disease and of health; and of health more than of disease, and of disease more than of health. For breath too is a nutriment.

49. Liquid nutriment (sc. is) easier to change than dry; dry nutriment (sc. is) easier to change than liquid. Nutriment difficult to change is difficult to digest, while that which it is easy to add to is easy to digest.

50. Persons who are in need of a rapid restorative: a liquid remedy is most effective for the restoration of strength. Those (sc. who need) something even faster (sc. are helped) by smelling. For those who need a slower restorative: solid nutriment.

51. Muscles are more solid than the other parts, except for bone and cord. Parts that are trained are difficult to alter, being more strongly developed in their essence; on this account, they are more difficult to melt than others.

52. Pus is a product of flesh; purulence is a product of blood and other moisture. Pus is the nutriment of a lesion; purulence is the nutriment of a vein and of an artery.

53. Marrow is the nutriment of bone; for this reason it forms a callus on the surface.

54. Potency augments all things, nourishes them, and makes them sprout.

55. Moisture is the vehicle of nutriment.

---

22 ἐόντα Littré: τοῦ ὄντος MA
23 πυῶδες Littré: πύον MA

INDEX

INDEX